# ICME-13 Monographs

**Series editor**

Gabriele Kaiser, Faculty of Education, Didactics of Mathematics, Universität Hamburg, Hamburg, Germany

Each volume in the series presents state-of-the art research on a particular topic in mathematics education and reflects the international debate as broadly as possible, while also incorporating insights into lesser-known areas of the discussion. Each volume is based on the discussions and presentations during the ICME-13 conference and includes the best papers from one of the ICME-13 Topical Study Groups or Discussion Groups.

More information about this series at http://www.springer.com/series/15585

Sepideh Stewart · Christine Andrews-Larson
Avi Berman · Michelle Zandieh
Editors

# Challenges and Strategies in Teaching Linear Algebra

Springer

*Editors*
Sepideh Stewart
Department of Mathematics
University of Oklahoma
Norman, OK
USA

Avi Berman
Department of Mathematics
Technion—Israel Institute of Technology
Haifa
Israel

Christine Andrews-Larson
College of Education
Florida State University
Tallahassee, FL
USA

Michelle Zandieh
Arizona State University
Mesa, AZ
USA

ISSN 2520-8322 ISSN 2520-8330 (electronic)
ICME-13 Monographs
ISBN 978-3-319-66810-9 ISBN 978-3-319-66811-6 (eBook)
https://doi.org/10.1007/978-3-319-66811-6

Library of Congress Control Number: 2017955649



Printed on acid-free paper

This Springer imprint is published by Springer Nature
The registered company is Springer International Publishing AG
The registered company address is: Gewerbestrasse 11, 6330 Cham, Switzerland

# Foreword

Linear algebra is arguably one of the more interesting (and complex) domains that students encounter in their first 2 years at university. The reasons for this include both the nature of a typical first course in linear algebra and the vitality of linear algebra beyond the first course. At the course level, linear algebra is one of the first opportunities for students to wrestle with definitions and proofs. In comparison to much of their prior mathematical experiences that emphasize procedural competency, linear algebra includes a rich array of new ideas, including linear independence, span, linear transformations, eigen theory, vector spaces, invertibility, rank, kernel, etc. These new concepts require students to carefully and consciously use definitions and to prove fundamental statements related to these ideas, all of which is something new for first-year university students. The extensive use of definitions and reliance on theorems often gives the first course an abstract and theoretical flavor, something that experts relish but which many students find distasteful. Linear algebra is also one of the richest domains for making connections between course concepts. For example, what many refer to as the invertible matric theorem relates over a dozen equivalent concepts. Thus students must not only understand the ideas themselves, but they must also develop reasons for how and why ideas are related. It is no wonder then that the literature is replete with studies that examine the challenges and difficulties that students encounter in linear algebra.

The importance of connections extends well beyond course-specific concepts. Indeed, linear algebra is also a vital area of mathematics, both within the discipline and across disciplines. For example, in differential equations, eigen theory plays an essential role in understanding linear homogeneous systems, which then provide useful tools for analyzing nonlinear systems. The concepts in linear algebra also play important roles in more advanced mathematics, including functional analysis and abstract algebra. Linear algebra also plays a vital role in other disciplines such as physics, engineering, and economics.

Despite the growth of research focused on the learning and teaching of linear algebra, there is still tremendous need for work that further examines students' difficulties, the underlying reasons for these difficulties, and instructional sequences and pedagogical approaches that have promise to promote student progress and deep understanding of the ideas in linear algebra and its widespread applicability. This book makes a significant contribution in addressing these needs that span research and practice. In terms of research, several of the chapters in this volume illuminate particular theoretical developments about learning as they relate to linear algebra, while other chapters offer a wide range of interesting and challenging problems that promise to engage students and promote deep understanding of core ideas. Just as these problems will be interesting for students, so will this volume be for readers.

San Diego, USA

Chris Rasmussen
San Diego State University

# Contents

# Introduction

This book stems from the work of a Discussion Group (*Teaching Linear Algebra*) that was held at the 13th International Conference on Mathematics Education (ICME-13). The organizers of this Discussion Group (who are also the co-editors of this volume) aimed to orchestrate a conversation that would highlight current efforts regarding research and practice on teaching and learning of linear algebra from around the world. Their ultimate goal was to initiate a multinational research project on how to foster conceptual understanding of Linear Algebra concepts. This conversation was organized around a theme of problems and issues, with a particular focus on mathematical problems that are productive for learning. Key questions and issues discussed were as follows:

a. How can applications of Linear Algebra be used as motivation for studying the topic?
b. What are the advantages of proving results in Linear Algebra in different ways?
c. In what ways can a linear algebra course be adapted to meet the needs of students from other disciplines, such as engineering, physics, and computer science?
d. How can challenging problems be used in teaching Linear Algebra?
e. In what way should technology be used in teaching Linear Algebra?
f. What is the role of visualization in learning Linear Algebra?
g. In what order (pictures, symbols, definitions, and theorems) should we teach Linear Algebra concepts?
h. How can we educate students to appreciate the importance of deep understanding of Linear Algebra concepts?

While this rich list of questions was motivating, at the time of the 2-day meeting at ICME, the conversations gave rise to a common theme focusing on problems and issues in Linear Algebra instruction and ultimately the making of this book.

This volume offers insights into recent work related to the teaching and learning of linear algebra across a range of countries and contexts, drawing on expertise of mathematics educational researchers and research mathematicians with experience teaching linear algebra. The 18 chapters of this book represent work from nine countries: Austria, Germany, Israel, Ireland, Mexico, Slovenia, Turkey, USA, and Zimbabwe. Chapters share a thread of commonality in their focus on the use of challenging problems or tasks that are supportive of student learning. The chapters are organized in four sections: Chapters highlighting a theoretical perspective on the teaching and learning of Linear Algebra, chapters based on empirical analyses related to learning of particular content in linear algebra, chapters focusing on the use of technology and dynamic geometry software, and chapters featuring examples of challenging problems that experienced practitioners have found to be pedagogically useful.

## Theoretical Perspectives Elaborated Through Tasks

The first three chapters in this volume focus on pedagogical aspects of Linear Algebra theoretically. In his chapter, Guershon Harel builds on his instructional framework, which is organized around notions of Duality, Necessity, and Repeated reasoning (DNR). Specifically, he considers the role of cognitive and pedagogical aspects of Linear Algebra through the lenses of two main DNR concepts, namely, *intellectual need and epistemological justification*, and exemplifies them through a variety of Linear Algebra tasks. Harel invites the mathematics community to reflect on whether instruction that is organized around this theoretical viewpoint will have an effect in advancing Linear Algebra students' performance.

Continuing the theoretical conversation, Maria Trigueros's chapter proposes a teaching approach that builds on theory about Actions, Processes, Objects, and Schema (APOS) through the use of several challenging modeling situations and tasks designed to introduce some main linear algebra concepts. The results reveal crucial moments as students develop new strategies, resulting in further understanding of the concepts.

Based on her work with research mathematicians, Sepideh Stewart believes that creating opportunities to move between Tall's (2013) Worlds of mathematical thinking will encourage students to think in multiple modes of thinking and increases their abilities in dealing with problems from different angles. In her chapter, she proposes a set of Linear Algebra tasks designed to move learners among Tall's Worlds.

## Analyses of Learners' Approaches and Resources

The empirical analyses section of this book offers an exciting variety of findings across populations and topic areas in linear algebra. Data is taken from populations ranging from middle and high school students in Mexico to undergraduates in North and Central America, to current teachers updating their certification through undergraduate coursework in Zimbabwe. Topics include systems of linear equations, matrix multiplication, determinants, vector spaces, eigenvectors, and eigenvalues.

Asuman Oktac provides a synthesis of three previously unpublished thesis studies examining student reasoning about systems of linear equations across middle school, high school, and university contexts (Mora Rodríguez 2001; Cutz Kantún 2005; Ochoviet Filguieras 2009). All three studies were conducted in Mexico and written in Spanish. This chapter identifies points of commonality across these studies and leverages a common theoretical framework, making these findings available to an English-speaking audience.

John Paul Cook, Dov Zazkis, and Adam Estrup point to conceptual underpinnings entailed in matrix multiplication as motivation for analyzing how matrix multiplication is introduced and motivated in 24 introductory linear algebra textbooks. This work provides a timely update to Harel's (1987) textbook analysis and expands the corresponding framework to include computational efficiency. Additionally, this piece offers insight into the variety of ways current texts address the issue of matrix multiplication, considers aspects of reasoning emphasized and valued in each approach, and draws connections between textbook approaches and current research on student reasoning.

The chapter by Cathrine Kazunga and Sarah Bansilal, as well as the chapter by Lillias Mutambara and Sarah Bansilal, draws on data from a population of current mathematics teachers who were part-time students at a Zimbabwean university to meet new teacher certification requirements in the country. Their chapters provide analyses of participants' understanding of determinants and vector spaces, respectively.

The chapter by David Plaxco, Michelle Zandieh, and Megan Wawro, and the chapter by Khalid Bouhjar, Christine Andrews-Larson, Muhammad Haider, and Michelle Zandieh both offer insights into student reasoning about eigenvectors and eigenvalues in the context of inquiry-oriented instruction. The Plaxco et al. chapter offers insights into student reasoning in a guided reinvention approach drawn from classroom data, whereas the Bouhjar et al. chapter documents the effectiveness of this approach by comparing written assessment data of students who learned through this approach with students who learned the material in more standard ways.

## Dynamic Geometry Approaches

Three of the chapters in this book discuss ways that technology can influence the learning of Linear Algebra. Hamide Dogan's chapter compares learners who were exposed to dynamic visual representations to those who were exposed to the traditional instructional tools. She found notable differences in the nature of the mental schemes displayed by learners in the two groups. In addition, those students exposed to dynamic visual representations were able to use this geometry-based knowledge to make sense of more abstract algebraic ideas. Melih Turgut's chapter uses the theory of semiotic mediation to describe how the tools and functions of a dynamic geometric system affect student learning. In particular, he focuses on how these tools mediated the evolution of student reasoning about linear transformations from personal meanings based on work in $R^2$ to new mathematical meanings in $R^3$ and $R^n$. Ana Donevska-Todorova's chapter takes a broader perspective in considering which technology-enhanced environments may best affect student learning of different competencies. She suggests a nested model that illustrates how three modes of thinking in linear algebra can be related to the design of tasks or teaching environments.

## Challenging Tasks with Pedagogy in Mind

The last six chapters of the book involve challenging tasks that illustrate the beauty and usefulness of linear algebra and feature many applications. Barak Pearlmutter and Helena Smigoc show how nonnegative factorization of data matrices can motivate the study of basic Linear Algebra. They give a simple example stopping at discussion points. Their chapter includes an example of factoring a data matrix of module descriptors for 62 mathematics modules that were taught in their school. The chapter by Avi Berman uses formulas on Fibonacci numbers, a proof of the uniqueness of Lagrange polynomials, a periodicity two property of neural networks, and a computer game on lights as examples of challenging problems that can be used as motivation in teaching Linear Algebra. Franz Pauer describes a computational approach to teaching systems of linear equations. He gives an example of electric circuits and concludes his chapter with geometric interpretation. Frank Uhlig demonstrates his successful experience of holistic teaching and holistic learning with a Linear Algebra example of plane rotation. David Strong suggests how to motivate a course, how to motivate a chapter, and how to motivate an idea. He describes many motivational applications including systems of equations, discrete dynamical systems, QR factorization, traffic flow, and investments. Damjan Kobal describes how basic linear algebra concepts can be used for a smooth transformation from intuitive to abstract cognition and to deepen students' understanding. The applications in his chapter include Brower fixed point theorem, projective spaces, and barycentric and trilinear coordinates.

In its breadth of perspectives, this book offers a tremendous number of resources on teaching linear algebra, while also bringing together a community of those interested in pedagogical issues in linear algebra from around the world. It is our intention to continue the work started with the ICME-13 Discussion Group on *Teaching Linear Algebra* as we meet at other international conferences to further these discussions.

## References

Cutz Kantún, B. M. (2005). *Un estudio acerca de las concepciones de estudiantes de licenciatura sobre los sistemas de ecuaciones y su solución*. Unpublished masters' thesis. Cinvestav-IPN, Mexico.

Harel, G. (1987). Variations in Linear Algebra content presentations. *For the learning of mathematics, 7*(3), 29–32.

Mora Rodríguez, B. (2001). *Modos de pensamiento en la interpretación de la solución de sistemas de ecuaciones lineales*. Unpublished masters' thesis. Cinvestav-IPN, Mexico.

Ochoviet Filgueiras, T. C. (2009). *Sobre el concepto de solución de un sistema de ecuaciones lineales con dos incógnitas*. Unpublished doctoral thesis. Cicata-IPN, Mexico.

Tall, D. O. (2013). *How humans learn to think mathematically: Exploring the three worlds of mathematics,* Cambridge University Press.

# Part I
# Theoretical Perspectives Elaborated Through Tasks

# The Learning and Teaching of Linear Algebra Through the Lenses of Intellectual Need and Epistemological Justification and Their Constituents

**Guershon Harel**

**Abstract** *Intellectual need* and *epistemological justification* are two central constructs in a conceptual framework called *DNR-based instruction in mathematics*. This is a theoretical paper aiming at analyzing the implications of these constructs and their constituent elements to the learning and teaching of linear algebra. At the center of these analyses are classifications of intellectual need and epistemological justification in mathematical practice along with their implications to linear algebra curriculum development and instruction. Two systems of classifications for intellectual need are discussed. The first system consists of two subcategories, *global* need and *local need*; and the second system consists of five categories of needs: *need for certainty, need for causality, need for computation, need for communication,* and *formalization,* and *need for structure.* Epistemological justification is classified into three categories: *sentential epistemological justification* (*SEJ*), *apodictic epistemological justification* (*ASJ*), and *meta epistemological justification* (*MEJ*).

**Keywords** Intellectual need · Epistemological justification

*DNR-based instruction in mathematics* (*DNR*, for short; Harel, 1998, 2000, 2008a, b, c, 2013a, b) is a theoretical framework for the learning and teaching of mathematics—a framework that provides a language and tools to formulate and address critical curricular and instructional concerns. *DNR* can be thought of as a system consisting of three categories of constructs: *premises*—explicit assumptions underlying the *DNR* concepts and claims; *concepts*—constructs defined and oriented within these premises; and *claims*—statements formulated in terms of the *DNR* concepts, entailed from the *DNR* premises, and supported by empirical studies.

G. Harel (✉)
University of California, San Diego, USA
e-mail: harel@math.ucsd.edu

 
S. Stewart et al. (eds.), *Challenges and Strategies in Teaching Linear Algebra*,
ICME-13 Monographs, https://doi.org/10.1007/978-3-319-66811-6_1

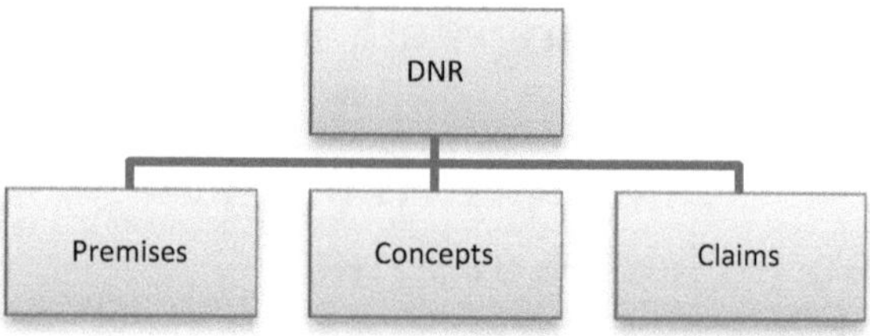


**Fig. 1** *DNR's* three categories of constructs

The main goal of this paper is to discuss cognitive and pedagogical aspects of linear algebra through the lenses of two central *DNR* concepts: *intellectual need* and *epistemological justification*. As the above list of references indicates, *DNR* has been discussed extensively elsewhere, and so in this paper we only reiterate briefly the definitions of these concepts along with their essential constituent elements: the concepts of *ways of understanding* and *ways of thinking* and four out of the eight premises of *DNR*.

We begin in Sect. 1 with the concepts of *ways of understanding* and *ways of thinking*. Following this, in Sect. 2, we discuss the four *DNR* premises. With these concepts and premises in hand, we turn, in Sect. 3, to the definition of *intellectual need* and *epistemological justification*. The fourth and fifth sections present, respectively, more refined analyses into various categories of the latter two concepts. The sixth, and last, section concludes with reflections and research questions. In each section, the discussion is accompanied with observations made in teaching experiments in linear algebra we have conducted during the years. In this respect, this is a theoretical, not empirical, paper. That is, the purpose of the paper is to theorize and illustrate the role and function of *intellectual need* and *epistemological justification* and their constituent elements in the learning and teaching of linear algebra.

To help the reader navigate through the various *DNR* terms introduced in this paper, we end each section with a figure depicting the network of terms accrued up to that section. Figure 1, for example, depicts the three categories of constructs comprising *DNR* outlined in this introduction. The rest of the figures in the paper will be expansions of this figure.

## 1 Ways of Understanding and Ways of Thinking

The notions of *way of understanding* and *way of thinking* have technical definitions (see Harel, 2008c). However, for the purpose of this paper it is sufficient to think of them as two different categories of knowledge, the first refers to one's conceptualization of "subject matter," such as the way one interprets particular definitions, theorems, proofs, problems and their solutions; and the second refers to "conceptual tools," such as deductive reasoning, empirical reasoning, attention to structure and precision, and problem-solving approaches (e.g., heuristics). One of the central claims of *DNR*, called the *duality principle*, asserts that (a) one's ways of thinking impacts her or his ways of understanding; and, (b) it is the acquisition of appropriate ways of understanding that brings about a change and development in one's ways of thinking.

To illustrate, consider the following example. A mathematically mature student who possesses *definitional reasoning*—the way of thinking by which one examines concepts and proves assertions in terms of *well-defined* statements—is likely to understand the concept of *dimension of a subspace* as intended—the number of vectors in a basis of the subspace—but he or she would also realize that such a definition is meaningless without answering the question whether all bases of a subspace have the same number of vectors. Another student, for whom definitional reasoning has not yet reached full maturity, may have the same understanding without realizing the need to settle this question. Yet another student whose conceptualization of mathematics is principally action-based (in the sense of APOS theory),[1] is likely to understand the concept of dimension in terms of a rule applied to n-tuples. For such a student, the dimension of a span of a set of vectors in $R^n$ amounts to carrying out a procedure of, for example, setting up these vectors as the columns of a matrix, row reducing the matrix, and determining, accordingly, the number of pivot columns the matrix has. We observed each of these three conceptualizations among students on various occasions, even in upper division linear algebra courses. And scenarios corresponding to these three conceptualizations have occurred throughout our teaching experiments when attention to a well-defined concept was called for. For example, when the instructor concluded that the *projection matrix onto a subspace* $V$ *of* $R^n$ is the matrix $P = W(W^TW)^{-1}W^T$, where $W$ is a basis matrix[2] of $V$, there were a few students who fully understood, and some even independently raised, the concern that $P$

[1]APOS theory (Arnon et al., 2014; Dubinsky, 1991) will be used to provide conceptual bases for some of these observations. Given how widely this theory has been studied during the last three decades, there is no need to allocate more than a brief illustration to the four levels of conceptualizations, *action, process, object, and schema* offered by the theory and used in this paper. Briefly, consider the phrase "the coordinates of a vector of $x$ with respect to a basis-matrix $A$ in $R^n$," denoted by $[x]_A$. At the level of action conception, the learner might be able to deal with $[x]_A$ only in the context of a specific vector and a specific suitable basis-matrix, by following step-by-step instruction to compute the respective coordinate vector. At the level of process conception one is capable of imagining taking *any* vector $x$ in $R^n$, representing it as a linear combination of the columns of $A$, and forming a column vector whose entries are the coefficient of, and are sequenced in the order they appear in, the combination. With this conceptualization, the learner is able to carry out this process in thought and with no restriction on the vector $x$ considered. At the level of object conception, one is aware of the process of relating the two coordinate vectors as a totality, for example, in finding the relation between two coordinate vectors of $x$, one with respect to a basis-matrix $A_1$, $[x]_{A_1}$, and one with respect to a basis-matrix $A_2$, $[x]_{A_2}$, whereby being able to express the relation in terms of a transition matrix $S = A_2^{-1}A_1$ between the two vectors. Among the ways of thinking that are essential to cope with linear algebra, in particular, and mathematics, in general, are the abilities to construct concepts at the levels of process conception and object conception, as it is demonstrate throughout the paper. (See also Trigueros, this volume.)

[2]A matrix whose columns form a basis for a subspace.

might be dependent on the choice of $W$. For most of the students, however, the conclusion engendered no concern.

The implication of the second part of the duality principle is that students acquire a particular way of thinking only by repeatedly dealing with specific ways of understanding associated with that way of thinking. For example, students develop definitional reasoning not by preaching but by repeatedly using definitions in the process of mathematical argumentations and by dealing in a multitude of contexts with the question whether a concept is well defined.

The examples of ways of thinking we have listed above are general—they pertain to mathematics as a discipline. Different areas or sub-areas of mathematics, however, can be branded by ways of thinking specific to them. The conceptualizations of matrix theory and the theory of general vector spaces share ways of thinking (e.g., axiomatic proof schemes (Harel & Sowder, 1998) and structural reasoning (Harel & Soto, 2016), and yet each is branded by a set of ways of thinking unique to it. For example, while thinking in terms of row reduction and block matrices is part of elementary matrix theory, it is often not applicable to coordinate-free, vector spaces.

Problem-solving approaches are instances of ways of thinking (Harel, 2008c). Therefore, "reasoning in terms of __ in solving problems" is an instance of a way of thinking. For example, reasoning in terms functions, reasoning in terms of row reduction, reasoning in terms of block matrices, reasoning in terms of linear combinations are all problem-solving approaches, and hence are ways of thinking. In our experience, the acquisition and application of such ways of thinking is difficult for students. Consider, for example, reasoning in terms of block matrices (known also as partitioned matrices) and linear combination. It is one thing applying block-matrices rules to multiply matrices; it is another using block matrices to represent and solve problems. Typically, students are able to perform at the level of action conception (ala APOS theory) algebraic operations using block matrices, but they experience major difficulties when block matrices are constructed to represent relations and prove theorems. Consider the simple case of the product $A_{m\times n}x_{n\times 1}$ as a linear combination of the columns of $A$. We repeatedly observed students having difficulties representing a common statement such as "$v_1, v_2, \ldots, v_k \in span(u_1, u_2, \ldots, u_m) \subseteq R^n$" in a matrix form: $[v_1 \quad v_2 \quad \ldots \quad v_k] = [u_1 \quad u_2 \quad \ldots \quad u_m]Q$ for some $Q_{m\times k}$.

This difficulty manifested itself on numerous occasions, for example in comprehending the following proof of the theorem, "Any set of $m$ linearly independent vectors in an $m$-dimensional subspace $H$ of $R^n$ spans $H$". The proof presented in class was an elaboration of the following lines:

> Let $u_1, u_2, \ldots, u_m$ be linearly independent vectors in $H$, and set $U = [u_1 \quad u_2 \quad \ldots \quad u_m]$. Let $V = [v_1 \quad v_2 \quad \ldots \quad v_m]$ be a basis matrix of $H$. There exists a matrix $Q_{m\times m}$ such that $U = VQ$. Since the columns of $U$ are linearly independent, $Q$ is invertible, and so $UQ^{-1} = V$. Hence $u_1, u_2, \ldots, u_m$ span $H$.

We point to two obstacles students typically encounter in comprehending this proof. The first revolves around the equation $U = VQ$; students indicate that they do

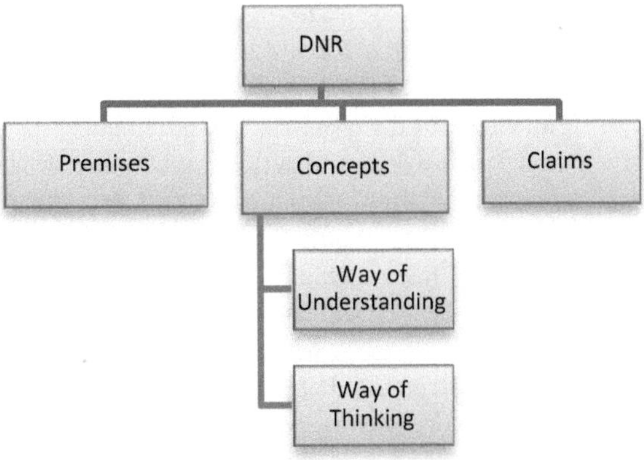


**Fig. 2** Two of the *DNR* concepts: Ways of understanding and ways of thinking—

not understand where the matrix $Q$ came from, even after they are presented with an explanation by their group mates or the instructor. A possible conceptual basis for this difficulty is that the construction of $Q$ requires performance at the level of process conception (ala APOS theory), a form of abstraction known to be cognitively demanding (Dubinsky, 1991). Specifically, one needs to construct, successively and in thought, each column of $Q$ out of the coefficients of the expression representing its corresponding column of $U$ as a linear combination of the columns of $V$ (i.e., $Q_i = [q_{i1} \quad q_{i2} \quad \ldots \quad q_{im}]^T$, where $U_i = \sum_{j=1}^{m} q_{ij} V_i$, $i = 1, 2, \ldots, m$). Even students who overcome this difficulty express discomfort with the claim that the result $UQ^{-1} = V$ completes the proof. At the heart of this claim, and the difficulty, is the fact that linear combination of linear combinations is a linear combination—that since each vector in $H$ is a linear combination of the columns of $V$ and each column of $V$ is a linear combination of the columns of $U$, by $UQ^{-1} = V$, $ColU = H$. Here too performance at the level of process conception seems essential, in that one has to carry out this chain of relations in thought in order to fully bring oneself to a firm conviction about the validity of the claim.

Figure 2 expands Fig. 1 to include the *DNR* constructs discussed in this section.

## 2 *DNR* Premises

*DNR* has eight premises; they are philosophical stands appropriated from existing theories, such as the Piagetian theory of equilibration (Piaget, 1985), Brousseau's (1997) theory of didactical situation, and Aristotle. Relevant to this paper are four of these premises; they are: the *knowledge of mathematics premise,* the *knowing premise,* the *knowledge-knowing linkage premise,* and the *subjectivity premise.*

The ***knowledge of mathematics premise*** states: *knowledge of mathematics consists of two related but different categories of knowledge: the ways of understanding and ways of thinking that have been institutionalized throughout history.*[3] The significance of this premise to mathematics instruction is that while knowledge of and focus on ways of understanding is indispensable for quality teaching, it is not sufficient. Mathematics instruction should also attend to ways of thinking. With this instructional view one would teach, for example, row reduction not only as a tool to solve systems of linear equations but be cognizant of and explicit about the value of this tool in analyzing and answering theoretical questions. In accordance to the duality principle stated earlier, the development of such a way of thinking is facilitated by instruction that persistently models it in proving theorems and solving problems, as the following episode illustrates.

The episode occurred in an elementary linear algebra class. The instructor defined column rank and row rank. It turned out that the class as a whole dealt with these concepts in an add-on Matlab component to the course (entirely not coordinated with the instructional pace of the course), where the students have used the fact that $\dim ColA = \dim RowA$ without proof. Before the instructor turned to prove this statement, one of the students in the class exclaimed publically that she found this fact fascinating—that for any array of numbers, "no matter what" (her words), the maximum number of linearly independent columns equals the maximum number of linearly independent rows. Then she added: "I kept thinking about it for some time until I found why". In response to the instructor's question, "What was the explanation you have found?" she said: "… by reducing the matrix into *rref* … I always bring up *rref* … it helps me solve the homework problems". Then she proceeded by explaining how in *rrefA* the number of columns with a leading 1 is necessarily equal to the number of rows with a leading 1, from which she concluded that $\dim ColA = \dim RowA$, using the previously proved facts that row reduction preserves dependence/independence of the columns of $A$ as well as *RowA*.

We posit that the instructor's explicit and persistent effort to present row reduction as a conceptual tool in proving theorems and solving problems contributed to conceptualizations as the one articulated by this student.

The next two premises are inextricably linked; one is about *knowing* and the other about the linkage between *knowing* and *knowledge*. The ***knowing premise*** states: *The means of knowing is the process of assimilation and accommodation.* According to Piaget (1985), disequilibrium, or perturbation, is a mental state when one fails to assimilate. Equilibrium, on the other hand, is a state in which one perceives success in assimilating. In Piaget's terms, equilibrium occurs when one has successfully modified her or his viewpoint (accommodation) and is able, as a result, to integrate new ideas toward obtaining a solution of a problem (assimilation). The ***knowing-knowledge linkage*** premise states: *Any piece of knowledge humans know is an outcome of their resolution of a problematic situation* (Brousseau, 1997; Piaget, 1985). This premise is an extension of the knowing

---

[3]For the philosophical foundations of this premise, see Harel (2008c).

premise. While the knowing premise is about the mechanism of learning, the knowing-knowledge linkage premise guarantees that row material for the operation of that mechanism (i.e., a problematic situation from the engagement of which knowledge is constructed) exist. Collectively, the last two premises constitute a theoretical foundation for, respectively, the *essentiality* and *viability* of problem-solving based curricula. Namely, these curricula are *essential* because the only way to construct knowledge is by resolution to problematic situations (by the *knowing-knowledge premise*); and they are *viable* because such situations exist (by the *knowing premise*).

The implication for instruction of the view articulated by the last two premises is the *necessity principle,* which states: *For students to learn what we intend to teach them, they must have a need for it, where 'need' refers to intellectual need.* Relevant to curriculum design, the necessity principle entails that new concepts and skills should emerge from problems understood and appreciated as such by the students, and these problems should demonstrate to the student the intellectual benefit of the concept *at the time of its introduction.*

The problematic situations referred to in this premise may or may not be historic. For example, matrices did not grow out of the need to solve systems of linear equations, as typically is done in elementary linear algebra textbooks, but out of the need to develop determinants (in 1848 by J.J. Sylvester). According to Tucker (1993), "array of coefficients led mathematicians to develop determinants, not matrices. Leibniz … used determinants in 1693 about hundred and fifty years before the study of matrices …" (p. 5). Also, most problems studied in linear algebra are not introduced in the context of the field in which they originated initially. For example, Gauss elimination is typically introduced in textbooks in an application-free context, but it initially emerged in the field of geodesy and for years was considered part of the development of this field (Tucker, 1993).

The ***subjectivity premise*** states: *Any observations humans claim to have made are due to what their mental structure attributes to their environment.* This premise orients our interpretations of the actions and views of the learner. It cautions us—teachers—that what might be problematic for one individual or a community may not be so for others. A situation might trigger a mental perturbation with one person and be accepted by another. To illustrate, we continue the discussion about the concept of *dimension* we started in Sect. 1. In one of our teaching experiments, the instructor deliberately defined "dimension" as the number of vectors in a basis without first stating the theorem that all bases of a subspace have the same number of vectors. Following this, he asked the students to discuss in their working groups whether the definition is sound. A while later, when no productive response came from the students, he made the task more explicit by asking whether there is a need to establish a particular property of bases in a subspace for the definition to be meaningful. None of the students found any fault with the definition as stated. Following this, the instructor asked the class to comment on the following hypothetical scenario:

> Two students, John and Mary, are asked to determine the dimension of a particular subspace of a vector space. John identifies a basis of the subspace, counts the number of its elements, and reports that the dimension is 5. Mary identifies a different basis of the subspace, counts the number of its elements, and reports that the dimension is 7.

While some of the students responded as expected—that there is a need to establish that all bases of the subspace have the same number of vectors—astonishingly, there were students who responded by saying something to the effect that for John the dimension of the subspace is 5, and for Mary the dimension is 7. Presumably, these students did not possess the definitional way of thinking, and so the scenario described by their instructor did not cause them the desirable perturbation—their response was an outcome of their current schemes.

The subjectivity premise also cautions us, teachers, that learners' current ways of thinking may lead them to independently generalize faulty knowledge from a correct one. For example, student may, and typically do, erroneously conclude that row reduction preserves the column space, as it does with row space. Furthermore, often due to the level of robustness of certain ways of thinking students possess counterexamples to such faulty generalizations may not be effective (Harel & Sowder, 2007). For example, in our experience, students continue to hold this generalization true even after they are shown counterexamples to the contrary [e.g., for any matrix whose entries are all *1*s, $ColA \neq Col(rrefA)$]. Many scholars (e.g., Confrey, 1991; Dubinsky, 1991; Steffe, Cobb, & Glasersfeld, 1988; Steffe & Thompson, 2000) have articulated essential implications of the *subjectivity premise* to mathematics curriculum and instruction, even if they have not given it an axiomatic status as we do.

Figure 3 expands Fig. 2 to include the *DNR* constructs discussed in this section.

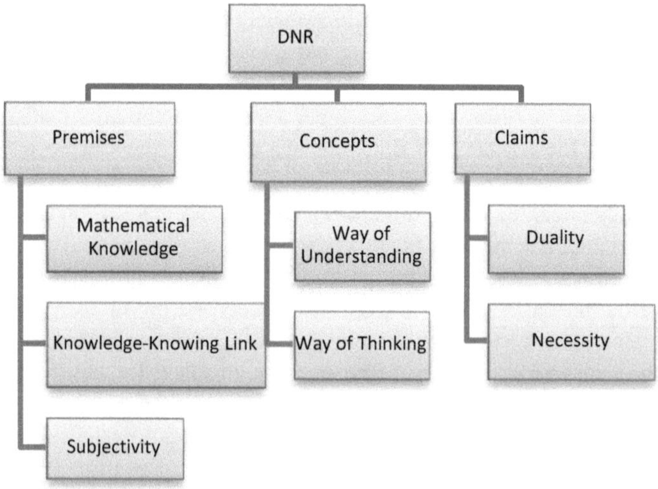


**Fig. 3** Three of the eight *DNR* premises and two of *DNR* foundational principles

## 3 Intellectual Need and Epistemological Justification

With these premises at hand, we now present the definitions of *intellectual need* and its associated concept, *epistemological justification*, as formerly introduced in Harel (2013a), with minor modifications.

[Let] $K$ be a piece of knowledge possessed by an individual or community, then, by the *knowing-knowledge linkage premise*, there exists a problematic situation $S$ out of which $K$ arose. $S$ (as well as $K$) is subjective, by the *subjectivity premise*, in the sense that it is a perturbational state resulting from an individual's encounter with a situation that is incompatible with, or presents a problem that is unsolvable by, her or his current knowledge. Such a problematic situation $S$, prior to the construction of $K$, is referred to as an individual's *intellectual need:* $S$ is the need to reach equilibrium by learning a new piece of knowledge. Thus, intellectual need has to do with disciplinary knowledge being created out of people's current knowledge through engagement in problematic situations conceived as such by them. One may experience $S$ without succeeding to construct $K$. That is, intellectual need is only a necessary condition for constructing an intended piece of knowledge. Methodologically, intellectual need is observed when we see that (a) one's engagement in the problematic situation $S$ has led her or him to construct the intended piece of knowledge $K$ and (b) one sees how $K$ resolves $S$. The latter relation between $S$ and $K$ is crucial, in that it constitutes the genesis of mathematical knowledge—the perceived reasons for its birth in the eyes of the learner. We call this relation *epistemological justification.*

Intellectual need and epistemological justification are two sides of the same coin —they are different but inextricably related constructs. Their occurrence is entirely dependent on one's background knowledge. Consider the question: What is a generator for the *ideal of polynomials annihilating a given operator T over an n-dimensional vector space*? Clearly, such a question wouldn't occur unless one possesses a cluster of ways of understanding for the concepts: *ideal, generator of an ideal, operator annihilating ideal* etc. Less trivial is the question, what ways of thinking facilitate the emergence of such a question with an individual? Or put in another way, how can we educate students to develop the habit of mind of asking such questions? A critical claim of this paper is that attention to epistemological justifications in generating definitions and proving theorems may pave the road to such habit of minds, as we will see in the next sections.

Even if such a question is raised, its answer hinges upon one's understanding and appreciation of the most foundational concept of linear algebra: *linear combination.* Thinking in terms of this concept and its derivative concepts of linear independence and linear dependence, one may recognize that the ideal of polynomials annihilating an operator $T$ over an n-dimensional vector space is not empty, since it contains an annihilator polynomial of degree $n^2$. This may not end here if this individual continues to ask: What is a generator for this ideal? And since the degree of such a polynomial is not greater than $n^2$, can it be $n$? Is there a polynomial of degree $n$ that annihilates $T$? If the search for an answer to this question leads the individual, independently or with the help of an expert, to Cayley-Hamilton Theorem ("Any linear operator on a finite-dimensional vector space is annihilated by its characteristic polynomial"), then by definition, the individual has constructed an epistemological justification for the theorem. An epistemological justification for the proof of the theorem—how the proof might be elicited—is a different matter. Such a proof may require additional or different networks of ways of understanding and ways of thinking.

It is important to highlight two points concerning intellectual need and epistemological justification. First, we iterate a point we made earlier, these constructs are not historical; rather, they are pedagogical (and research) tools. Namely, the need which has originally necessitated a particular concept may not—and is usually not—the one used in a curriculum. For example, in one of our teaching experiment, the concept of *linear independence* was necessitated through the question, When does Gaussian Elimination lead to "loss" of equations (i.e., zero equations in a system obtained through the application of elementary operations)?; and in another experiment through the question, When does a consistent system of linear equation have a unique solution? Historically, this concept emerged from generalizations of spatial relationships by Grassmann (Li, 2008).

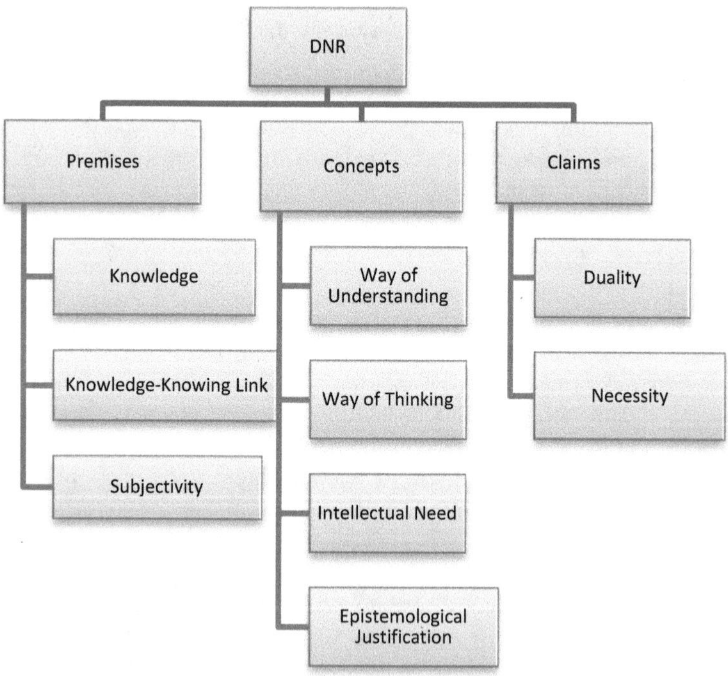


**Fig. 4** Two additional foundational *DNR* concepts: Intellectual need and epistemological justification

Second, while problems outside the fields of mathematics can serve as intellectual need for particular mathematical concepts and ideas, as we know from history, intellectual need is not synonymous with application. Cognitively, the term "application" refers to problematic situations aiming at helping students solidify mathematical knowledge they have already constructed or are in the process of constructing. Intellectual need, on the other hand, aims at eliciting knowledge students are yet to learn.

Figure 4 expands Fig. 3 to include the *DNR* constructs discussed in this section.

# 4 Categories of Intellectual Need

We offer two systems of classifications of intellectual need, each with a particular role in curriculum development and instruction; in this paper, they are instantiated in the context of the learning and teaching of linear algebra. The first system of classification rests on the distinction between *local need* and *global need*; it pertains to the structure of a mathematics curriculum. The second system of classification is more refined, in that it identifies specific types of intellectual needs that emerge in mathematical practice; they are: *need for certainty, need for causality, need for computation, need for communication,* and *need for structure.* These two systems of classifications will be discussed in turn in the next two sections. (For a discussion on the cognitive origins of these needs, see Harel, 2013a.)

## *4.1 Local Need Versus Global Need*

Consider an elementary course in linear algebra structured around a series of investigations, each aimed at answering a particular central question. The course begins with the question: (1) What is linear algebra? And it immediately discusses one of its branches: *systems of linear equations*, both systems in which the unknowns are scalars in a particular field (linear systems of scalar equations) and systems in which the unknowns are functions (linear systems of differential equations). Attending first to linear systems of scalar equations, the course then progressively proceeds by investigating, in this order, the questions: (2) Why is the focus on linear systems? (3) What exactly is the elimination process (which typically students are familiar with its basic form from their high-school mathematics)? (4) Why does the process of elimination work? (5) Why are equations "lost" in the elimination process? (6) Is there an algorithm to solve linear (scalar) systems? (7) What does the reduced echelon form (rref) tell us about the solution set of a system? This is a partial sequence of central questions aimed at helping the students build a coherent global image of the purposes of the study of systems of linear equations. Collectively, not individually, such questions represent a *global intellectual need* for the study of a particular area of mathematics.

An investigation into each of such questions generates specific problems manifesting *local intellectual need*—the need for the construction of particular concepts and ideas. A probe into some of the above questions, generate, for example, the concepts of *linear combination*, *equivalent systems*, *linear independence*, and *basis*, for the purpose of advancing the overarching investigation. To illustrate, consider, for example, Question 4—Why does the process of elimination work? In linear-algebraic terms, this question can be formulated as: Why *elementary operations* preserve the solution set of a system? A probe into the nature of these operations elicits the need for the creation of concepts and ideas. It begins with the following central idea:

Let $S$ be an $m \times n$ system, with equations $\varepsilon_1, \varepsilon_2, \ldots, \varepsilon_m$. For any $m$ scalars $c_1, c_2, \ldots, c_m$, any solution of system $S$ is a solution of the equation $\varepsilon_\Sigma = c_1\varepsilon_1 + c_1\varepsilon_1 + \ldots + c_m\varepsilon_m$.

In turn, this idea elicits the foundational concept of *linear combination* (i.e., the equation $\varepsilon_\Sigma$ is a linear combination of the equations, $\varepsilon_1, \varepsilon_2, \ldots, \varepsilon_m$), and with it, the following conclusion, which gives rise to the concept of *equivalent systems*:

Given two systems $S_1$ and $S_2$ of the same size, if each equation of $S_1$ is a linear combination of equations of $S_2$ and each equation of $S_2$ is a linear combination of $S_1$, then the two systems have same solution set.

Thus,

Two systems of equal size are equivalent if each equation in one system is a linear combination of the equations in the second system, and vice versa.

And so:

If two systems are equivalent, then they have the same solution set.

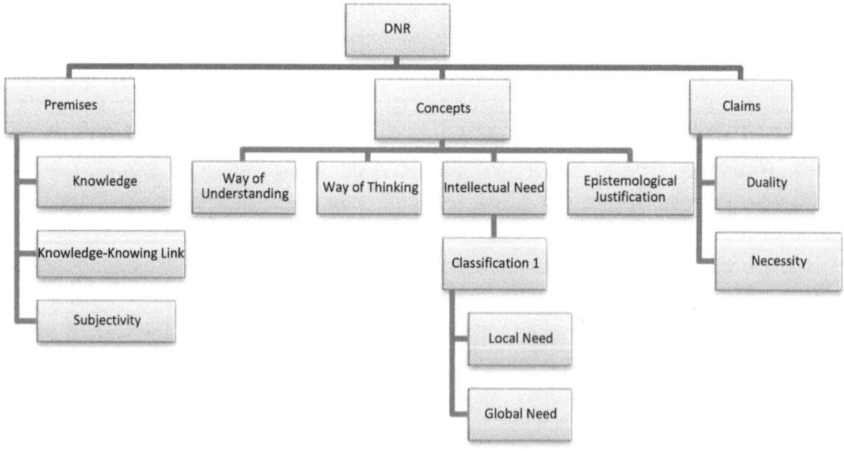


**Fig. 5** First classification of intellectual need

These results, then, lay the foundation for the question under consideration (Question 4), which now can be stated as: Do elementary operations preserve equivalency?

The second half of the course turns to linear systems of differential equations (i.e., $Y'(t) = AY(t)$, $Y(0) = C$) where eigen theory is then introduced through the global need to investigate the question, How to solve such systems? This question leads to local needs, as will be discussed in the next section.

Figure 5 expands Fig. 4 to include the *DNR* constructs discussed in this section.

## *4.2 Intellectual Need in Mathematical Practice*

Based on cognitive and historical analyses, we offered in Harel (2013a) five categories of intellectual needs: (1) *need for certainty,* (2) *need for causality,* (3) *need for computation,* (4) *need for communication,* and (5) *need for structure.*

The first two needs are complementary to each other: understanding cause brings about certainty, and certainty might trigger the need to determine cause. The need for certainty is the need to prove—to remove doubts. One's certainty is achieved when one determines, by whatever means he or she deems appropriate, that an assertion is true. The *need for causality,* on the other hand, is the need to explain—to determine a cause of a phenomenon, to understand what makes a phenomenon the way it is. A student might be certain that a particular assertion is true because a teacher or textbook said so or because he or she verified the assertion empirically. The student might even reach certainty on the basis of a proof, and yet lack an insight as to what makes the assertion true—the proof may not be explanatory for her or him. In the next section, we will discuss explanatory proofs in the context of epistemological justification.

The third need is the *need for computation.* It is the need to quantify or calculate values of quantities and relations among them by means of symbolic algebra. For example, the need to quantify the "size" of a solution set of a linear system $Ax = b$ may be addressed by the concept of *rank:* the smaller the rank of a matrix $A$ is the "larger" the solution set of a consistent system $Ax = b$ becomes. Likewise, the need to reduce the data storage of a digitized image without compromising significantly the quality of the image through its electronic transmission may be responded to by decomposing the matrix representing the gray values of the image into a particular sum of rank-1 matrices, what is known as *singular value decomposition* (*svd*; see below for more discussion on this decomposition).

The fourth need is the *need for communication.* This need consists in two reflexive needs: *the need for formulation*—the need to transform strings of spoken language into algebraic expressions—and the *need for formalization*—the need to externalize the exact meaning of ideas and concepts and the logical justification for arguments. It is common that students experience difficulties formalizing a mathematical statement into a symbolic form. For example, students may understand that to find a least square solution to an inconsistent system $Ax = b$, one needs to replace $b$ by $\hat{b}$, such that $\hat{b}$ is the "closest" to $ColA$. The challenge for students is two-fold: first, they have to reformulate this goal into mathematical statements, verbally or symbolically, such as $\hat{b} \in ColA$ and $b - \hat{b} \perp ColA$; and second they have to express these statements in terms of equation-based expressions, $\hat{b} = Ac$ for some vector $c$ and $A^T(b - \hat{b}) = 0$. This latter step is typically challenging for students. Likewise, students may have an intuitive idea of what dimension is—usually in the context of 2- and 3-dimensional Euclidean spaces, but experience difficulty understanding the formalization of their intuition into a well-defined mathematical concept.

The fifth, and final, need is the *need for structure*. The common meaning of the term *structure* is something made up of a number of parts that are held or put together in a particular way. In mathematics the way these "parts" are held together are relations one conceives among different objects. For example, the expression $Ab = 0$ constitutes a structure for a person when he or she is conceives it as a string of symbols put together in a particular way to convey a particular meaning, such as 0 is a linear combination of the columns of $A$ with the entries of $b$ being the weights of the combination; or $b$ is orthogonal to the row space of $A$.

In mathematics, in general, the need for structure manifests itself as a need to encapsulate (in the sense of APOS theory) occurrences of phenomena. For example, one might encapsulate a series of empirical observations concerning products of square matrices into the patterns, $\det(AB) = \det(A)\det(B)$ or $tr(AB) = tr(BA)$; another may derive such patterns through deduction or may observe them empirically but see a need to establish them deductively. In linear algebra, there is the critical need to encapsulate different structures into a single representation: a vector

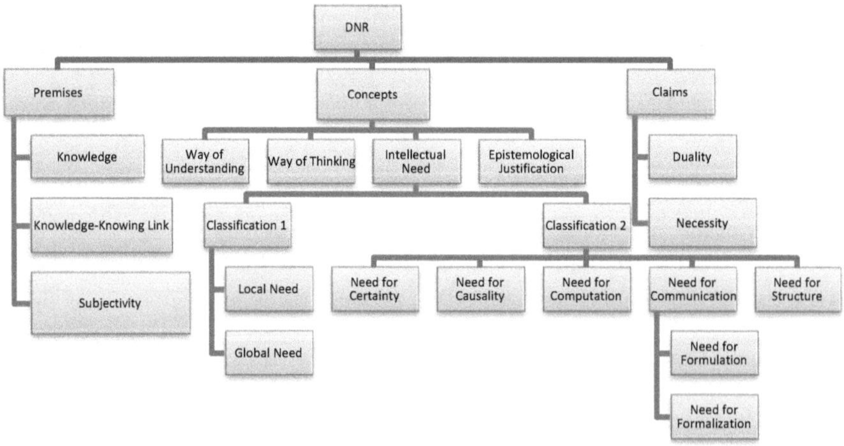


**Fig. 6** Second classification of intellectual need

space over the reals as a single representation of all $n$-tuples of real number, of all polynomials of degree less or equal to $n$ with real coefficients, of all $m \times n$ matrices with real entries, etc. This process of encapsulation assumes, of course, that members of each of these spaces are conceived as conceptual entities (in the sense of APOS theory and Greeno, 1992)) in, respectively, an $n$-dimensional, $n+1$-dimensional, and $mn$-dimensional vector space.

Figure 6 expands Fig. 5 to include the *DNR* constructs discussed in this section.

# 5 Categories of Epistemological Justification

We distinguish among three categories of epistemological justifications: *sentential, apodictic,* and *meta.* While the distinction among these types of epistemological justification is sufficiently clear, as we will now see, it should be noted that they are not mutually exclusive.

## *5.1 Sentential Epistemological Justification*

*Sentential epistemological justification* (*SEJ*) refers to a situation when one is aware of how a definition, axiom, or proposition was born out of a need to resolve a problematic situation. It is called so because it pertains to *sentences* with objective and logical meaning. To illustrate, consider how linear algebra textbooks typically introduce the pivotal concepts of "eigenvalue," "eigenvector," and "matrix diagonalization". A widely used linear algebra textbook motivates these concepts by saying that the concepts of "eigenvalue" and "eigenvector" are needed to deal with the problem of factoring an $n \times n$ matrix $A$ into a product of the form $XDX^{-1}$, where $D$ is diagonal, and that this factorization would provide important information about $A$, such as its rank and determinant. Such an introductory statement aims at pointing out to the student an important problem. While the problem is intellectually intrinsic to its poser (a university instructor), it is most likely to be alien to a student in an elementary linear algebra course, who is unlikely to realize from such a statement the true nature of the problem, its mathematical importance, and the role the concepts to be taught ("eigenvalue," "eigenvector," and "diagonalization") play in solving it.

One of the alternative approaches to this presentation, based particularly on students' intellectual need for computation, is through linear systems of differential equations, which has been experimented successfully several times. In this approach, one begins with an initial-value problem (e.g., a mixture problem) involving a sequence of rate of change functions, $f_1'(t), f_2'(t), \ldots, f_n'(t)$, each is a linear combination of the original position functions, $f_1(t), f_2(t), \ldots, f_n(t)$. This leads to a linear system of differential equations of the form:

$$\begin{cases} AY(t) = Y'(t) \\ Y(0) = C \end{cases} \tag{1}$$

where $A$ is a real square matrix. Students are asked to analogize system (*) to the scalar case:

$$\begin{cases} ay(t) = y'(t) \\ y(0) = c \end{cases} \tag{2}$$

At first, students' typically propose a solution to system (1) that is symbolically analogous to the solution of system (2), which they are familiar with from calculus; that is, corresponding to the symbolic structure of $y(t) = ce^{at}$, students offer $Y(t) = Ce^{At}$ (sic). A discussion of the meaning of the latter expression leads the students to (a) revise the expression $At$ into $tA$ and (b) probe into the definition of the concept of "*e* to the power of a square matrix." This question is resolved by, again, analogizing $e^B$, where $B$ is a square matrix, to $e^b$ where $b$ is a scalar, resulting in the definition, $e^B = \sum_{i=0}^{\infty} (1/i!)B^i$.[4] By considering the sizes of the matrices involved in the product $Ce^{tA}$ in their proposed solution, students come to realize that there is a need to perform a third revision, from $Y(t) = Ce^{tA}$ to $Y(t) = e^{tA}C$. Once the students have verified that their revised proposed solution works, the instructor returns to the solution in its expansion form, $Y(t) = e^{tA}C = \sum_{i=0}^{\infty} (t^i/i!)A^iC$, and points out the following critical observation: If it so happens that there is a relationship between the condition vector $C$ and the coefficient matrix $A$ in the form of $AC = \lambda C$ for some scalar $\lambda$, then the solution to system (1) would be easily computable: $Y(t) = e^{\lambda t}C$. This observation necessitates attention to the relation $AC = \lambda C$, and due to its perceived significance it deserves a name: *C is called an eigenvector of A and λ its corresponding eigenvalue*. Thus, students learn a *sentential* epistemological justification for the emergence of these central linear algebraic concepts; the concepts do not emerge out of the blue, as is typically the case in textbooks.

Following a few examples of solving system (1), the instructor (and in many cases the students themselves) raises the question about the computability of the solution in cases where the condition vector is *not* an eigenvector of the coefficient matrix. This leads, in turn, to the observation that whenever the condition vector $C$ is a linear combination of eigenvectors of $A$, the solution is still easily computable: $Y(t) = e^{tA}C = \sum_{i=1}^{k} a_i e^{\lambda_i t} v_i$, where $C = \sum_{i=1}^{k} a_i v_i$ and $Av_i = \lambda_i v_i$.

---

[4]Questions concerning convergence are not discussed, though on rare occasions were raised by students.

With this background, the instructor turns to the special case where $A$ has a basis of eigenvectors, in which case the solution is easily computable for any choice of $C$. For a more advanced linear algebra course, the proceeding discussion continues the investigation of the computability of the solution to system (1), leading up in a long journey to Jordan Theorem (and its related Canonical Form); namely, that remarkably system (1) is always easily computable since each vector is a linear combination of generalized eigen vectors.

We see here an example of how content presentation in linear algebra can be structured in a way that students develop *sentential* epistemological justifications for the birth of concepts through intellectual need, whereby students become partners in knowledge development, not passive receivers of ready-made knowledge.

## 5.2 Apodictic Epistemological Justification (AEJ)

The second category is *apodictic epistemological justification* (*AEJ*). This pertains to the process of proving; hence, the term *apodictic*. It is when one views a particular logical implication, $\alpha \Rightarrow \beta$, in causality, or explanatory, terms—how $\alpha$ causes $\beta$ to happen; that is, how $\alpha$ explains the presence of $\beta$. Consider, for example, the Spectral Theorem: Any $n \times n$ real symmetric matrix $A$ is orthogonally diagonalizable (i.e., $A = V\Lambda V^T$, where $V$ is orthogonal and $\Lambda$ diagonal). An apodictic epistemological justification of this assertion is present with a student when he or she exhibits an understanding of how the combined features of being real and symmetric are "responsible" for the matrix to be orthogonally diagonalizable—how specifically the absence of one of these features would derail the proof of the assertion. The central characteristic of *AEJ* is that the student is aware of the role that the various conditions in the hypothesis of an assertion play in its proof. The student, however, does not necessarily cognizant of the way the proof was conceived—that is a characteristic of the meta epistemological justification which we will discuss in the next section.

*AEJ* is a way of thinking not addressed sufficiently in mathematics instruction. It is acquired through repeated experience of probing into the specific role the conditions comprising a hypothesis of an assertion play in the proof. We conjecture that through the acquisition of this way of thinking students' ability to produce proofs is advanced. Consider the following episode.

In a matrix-based linear algebra course, a particular attention was paid to epistemological justifications (as well as to other ways of thinking—thinking in terms of block matrices is one of them). At one point during the first half of the course, the instructor presented what is known as the Basis Theorem.

Let H be a $p$-dimensional subspace of $R^n$, and let $v_1, v_2, \ldots, v_p$ be vectors in H.

1. If $v_1, v_2, \ldots, v_p$ span H, then they are linearly independent.
2. If $v_1, v_2, \ldots, v_p$ are linearly independent, then they span H.

As was the standard practice in this course, the students were asked to work in groups on comprehending the theorem, not necessarily proving it. After about 12 min, the instructor initiated a classroom discussion about the theorem. One of the students said something to the effect that she and her working mate thought that for Claim (1) there is a need to express two facts: that H is a $p$-dimensional subspace of $R^n$ and that $v_1, v_2, \ldots, v_p$ span H, and so, she continued, they let $U = [u_1 \quad u_2 \quad \ldots \quad u_p]$ be a matrix basis for H and $U = VA$ for some matrix $A$, where $V = [v_1 \quad v_2 \quad \ldots \quad v_p]$. By this time, we should mention, the class as a whole got accustom to the approach of representing relationship in terms of matrix equations. After some further discussion, the instructor asked what other sufficient conditions in Claim 1 has not been expressed. Another student in the class responded that the fact that the columns of $U$ are linearly independent hasn't been used. With help from the instructor to consider the sizes of the matrices involved in the equation $U = VA$, one of students declared that $A$ must be a square matrix ($p \times p$) and that since the columns of $U$ are linearly independent $A$ must be invertible (a fact which was previously proved and used on several occasions during the course). Following additional time for the students to collaborate on completing the proof, one of the working groups came to the board and completed the proof, saying something to the effect that since $U = VA$ and $A$ is invertible, $V = UA^{-1}$. And since the columns of $A^{-1}$ are linearly independent, the columns of $V$ are linearly independent, as was required.

It is interesting and important to add that this student also indicated at the end of his presentation that he used ideas he learned from the proof of the Dimension Theorem (All bases of a subspace H of $R^n$ have the same number of vectors), which the instructor presented a week earlier. Indeed the proof just presented includes considerations similar to those made in the proof of the Dimension Theorem. The latter proof began by setting two basis matrices $U = [u_1 \quad u_2 \quad \ldots \quad u_k]$ and $V = [v_1 \quad v_2 \quad \ldots \quad v_r]$, aiming at showing that $k = r$. The similar considerations are that since $U$ and $V$ are basis matrices, their columns span H, and therefore there exist two matrices $M$ and $N$ such that $U = VM$ and $V = UN$. By considering the sizes of the matrices involved, it was concluded that $M$ is an $r \times k$ matrix and $N$ is a $k \times r$ matrix. But since the columns of each of these matrices are linearly independent, $r \geq k$ and $k \geq r$, respectively, and hence $k = r$.

## 5.3 *Meta Epistemological Justification (MEJ)*

The third, and final, category is *meta epistemological justification* (*MEJ*). This refers to a situation when one not only views a proof in explanatory terms but also one is aware of how the proof came into being. To illustrate, consider the Singular Value Decomposition (SVD) Theorem. We reviewed the proof of this theorem in five commonly used linear textbooks. In each case the proof is presented with virtually no epistemological justification. Generally, the proof commences with the observation that for any $m \times n$ real matrix $A$ (without loss of generality, $m \geq n$), $A^TA$ is symmetric, and then abruptly the textbook offer three ready-made matrices $V$, $\Sigma$, and $U$ for the decomposition $A = U\Sigma V^T$. In this presentation, even students who fully understand the proof are unlikely to gain an insight as to how the proof might have come into light—students are not partners in figuring out possible sources of the proof ideas.

The following is an alternative presentation used in our classes. Its ultimate goal was not just to prove the SVD theorem, but to help students acquire an *MEJ* for the proof. Of course, the theorem itself was first necessitated through a suitable *SEJ* for its statement. We introduced the theorem in the context of the need to reduce the amount of data in transmitting a digitized image without affecting significantly the quality of the image, by expressing the matrix representing the array of the gray levels of the image's pixels as a sum of rank-1 matrices, i.e., $A = \sigma_1 u_1 {v_1}^T + \sigma_2 u_2 {v_2}^T + \cdots + \sigma_n u_n {v_n}^T$, where $\sigma_1 \geq \sigma_2 \geq \cdots \geq \sigma_n$, and then curtailing a certain number of addends in the tail of the sum.[5] Following this, the proof evolved through the *MEJ* outlined below:

1. At this stage of the course, the students have witnessed the utility of matrix representations in solving problems and proving theorems (e.g., representing a set of differential equations emerging from application problems, such as mixture problems, in terms of matrix equations (see Sect. 5.1) or the Dimension Theorem and Basis Theorem (see Sect. 5.2), and so the students seemed receptive to the idea to represent the desired equation, $A = \sigma_1 u_1 {v_1}^T + \sigma_2 u_2 {v_2}^T + \cdots + \sigma_n u_n {v_n}^T$, in the form of the matrix equation, $A = U\Sigma V^T$, where $U = [u_1 \quad u_2 \quad \ldots \quad u_m]$, $(\Sigma)_{ij} = \sigma_i$ if $i = j$ and $(\Sigma)_{ij} = 0$ if $i \neq j$ for $1 \leq i, j \leq m$, and $V = [v_1 \quad v_2 \quad \ldots \quad v_n]$.
2. Students were then told that we have here an equation with three unknowns, $U$, $\Sigma$, and $V$, and that the goal is to try to eliminate one of the unknowns. We note that students are well familiar with eliminating unknowns as a strategy to solve equations, so this proposed approach by the instructor is unlikely to have been foreign to them.
3. The instructor then wrote $A^T = V\Sigma U^T$ on the board and asked if they can offer an idea as to how to eliminate one of the unknowns. After about 8 min of consultation among students in their working groups, one of the students indicated something to the effect that if $U$ were orthogonal, it would be eliminated in the

[5]Of course other contexts can be used as *SEJ* for the SVD Theorem. The problem of transmitting a digitized image is typically used in textbooks as an application of SVD; we, on the hand, used it as an intellectual motivation (see the distinction between "application" and "intellectual need" in Sect. 3).

product $A^TA$. But, he then added, "we do not know that $U$ is orthogonal". The observation that the product of an orthogonal matrix and its transpose disappears in computing various expressions was well familiar to students at this point of the course; for example, in calculating the power of a diagonalizable matrix, and in computing a *UR*-factorization using the Gram-Schmidt process.

4. The instructor then responded: "Let see what happens if we assume that the unknown matrix $U$ is orthogonal". It is important to point out that this last dialogue between one of the students and the instructor represents a significant way of thinking in mathematical practice, that a desired mathematical result is conditioned a priori by a particular hypothesis. The instructor then pursued the student's proposal to obtain, $A^TA = V(\Sigma^T\Sigma)V^T = V\Sigma_1 V^T$, where $\Sigma_1$ is a diagonal matrix with ${\sigma_1}^2 \cdot {\sigma_2}^2, \ldots, {\sigma_n}^2$ on its diagonal.
5. With no further responses from the students as to how to proceed, the instructor drew students' attention to the fact that $A^TA$ is symmetric, urging them to recall a major result obtained previously in class about symmetric matrices. This, pleasingly, prompted another student to offer taking the missing matrix $V$ as an orthogonal diagonalizing matrix of $A^TA$, whose existence is guaranteed by the Spectral Theorem.
6. None of the students raised the concern that the proposed $V$ entails that the eigenvalues of $A^TA$ must be non-negative. This concern was raised by the instructor, which he then resolved by showing that indeed—remarkably—this was the case. Thus, in collaboration with the students two of the three unknown matrices $V$ and $\Sigma$ were successfully constructed.
7. These results then paved the path for the construction of $U$: that the first $r$ columns of $U$ are necessarily $u_i = \frac{1}{\sigma_i}Av_i$, members of *ColA* and corresponding to the $r$ positive values $\sigma_1, \sigma_2, \ldots, \sigma_r$ (the singular values), and the rest are to be any orthonormal vectors in $(ColA)^{\perp} = NulA^T$. The fact that the vector $u_i = \frac{1}{\sigma_i}Av_i$ turned out to be orthonormal, as needed, fascinated some students.

Surely the reader is familiar with the proof of the SVD Theorem—and all the other proofs, concepts, and ideas discussed in this paper, for that matter. Their appearance in the paper aimed at demonstrating how they can be introduced from the perspective of intellectual need and epistemological justification.

We conclude that students' success in acquiring and applying desirable ways of understanding and ways of thinking in our courses has been correlated with, and therefore, attributed to, the persistent application of *DNR*-based instruction, with particular attention to intellectual need and epistemological justification in their various manifestations. A critical principle of this instruction, beyond the application of the duality principle and the necessity principle we discussed earlier, is the *repeated-reasoning principle,* which states: *Students must practice reasoning in order to internalize, organize, and retain ways of understanding and ways of thinking*. Research has shown that repeated experience is a critical factor in these cognitive processes (Cooper, 1991). Repeated reasoning, not mere drill and practice

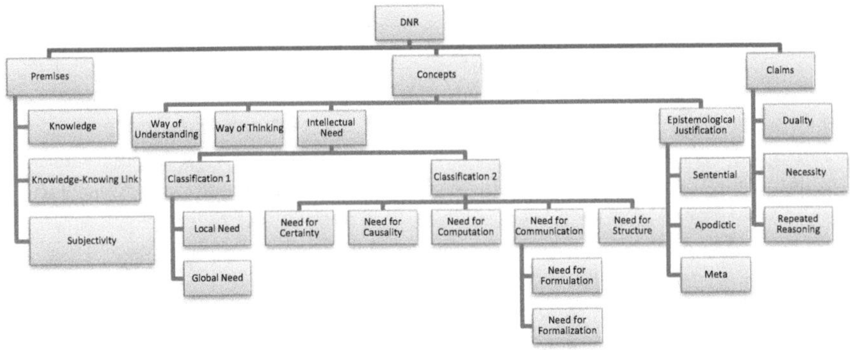


**Fig. 7** Classification of epistemological justification and the third foundational principle of *DNR*

of routine problems, is essential to the process of internalization—a conceptual state where one is able to apply knowledge autonomously and spontaneously—and reorganization of knowledge. The sequence of problems must continually call for reasoning through the situations and solutions, and they must respond to the students' changing intellectual needs.

Consonant with the repeated-reasoning principle, we typically keep the number of theorems presented in elementary linear algebra courses to a minimum, letting students reason and re-reason about various relations and claims (e.g., in terms of row reduction when relevant), rather than stating them as theorems ready to be used. Once a claim is stated as a theorem and proved, there is little incentive for the students to reason about the underlying ideas of its proof (Harel & Sowder, 1998). It should be clear, however, that we are not advocating eliminating theorems from the course; rather, we are advocating preserving the title "theorem" to truly "non-trivial" assertions. For example, the claim "A system $Ax = b$ is consistent iff an echelon form of the augmented matrix $[A|b]$ does not have a row of the form $[0 \;\; 0 \;\; \ldots \;\; 0|c]$, where $c \neq 0$ does not deserve the title theorem, for a brief inspection of the meaning of such a row should be sufficient to conclude that the system is inconsistent. On the other hand, the claim "Eigenvectors corresponding to distinct eigenvalues are linearly independent" is relatively not trivial and so it entitled to the label "theorem". This pedagogical approach is generally antithetical to the approach taken in many current linear algebra textbooks, where even simple claims are stated as propositions or theorems (see for example, the "Invertible Matrix Theorem" with its 25 logically equivalent statements in the widely used linear algebra textbook by Lay, Lay, and McDonald (2016).

Figure 7 expands Fig. 6 to include the *DNR* constructs discussed in this section.

## 6 Summary

In this paper we theorized and illustrated the role and function of *intellectual need* and *epistemological justification* and their constituents in the learning and teaching of linear algebra. We presented two systems of classifications for intellectual need. The first system consists of two subcategories, *global* need and *local need*; and the second system consists of five categories of needs: *need for certainty, need for causality, need for computation, need for communication,* and *formalization,* and *need for structure.* We also presented a classification of epistemological justification into three categories: *sentential epistemological justification (SEJ), apodictic epistemological justification (ASJ),* and *meta epistemological justification (MEJ).* The main constituent elements for intellectual need and epistemological justification presented in this paper are the concepts of *ways of understanding* and *ways of thinking* and four out of the eight premises of DNR: the *mathematical knowledge premise,* the *knowing premise,* the *knowledge-knowing linkage premise,* and the *subjectivity premise.* Figure 7 depicts this web of the *DNR* concepts discussed in this paper. The three foundational principles that articulate best the essence of *DNR*-based instruction are the *duality principle*, the *necessity principle*, and the *repeated reasoning principle*, also depicted in Fig. 7.

The central focus of the paper, however, is the instantiations and role of this network of *DNR* cognitive and epistemological concepts in the learning and teaching of linear algebra. We illustrated how certain ways of thinking (e.g., definitional reasoning) play a critical role in the ways students understand fundamental linear algebraic concepts (e.g., the concept of dimension and projection matrix), claiming that the acquisition and internalization of desirable linear algebraic ways of understanding and ways of thinking can only take place by positioning the intellectual need of the student in the center of the instructional effort (the *necessity principle*), by instruction being cognizant of and explicit about the role and function of ways of thinking in solving problems (the *duality principle*), and by providing the students with opportunity to reason repeatedly about problematic situations that call for the application of such ways of understanding and ways of thinking (the *repeated-reasoning principle*).

We also illustrated how successful students can be in linear algebra when such an instructional approach is applied. We posited that a persistent instructional effort to, for example, present row reduction and block matrices as conceptual tools to represent and solve problems contribute to the emergence of sophisticated linear algebraic conceptualizations among students (e.g., the proofs provided by students for the theorem $\dim ColA = \dim RowA$ and the Basis Theorem).

The underlying approach of focusing on both ways of understanding and ways of thinking, not only the former as is typically the case in traditional linear algebra curricula, is the *knowledge premise*, which provides equal status to these two categories of knowledge in the mathematics discipline. (For a fuller discussion, see Harel, 2008c.) The focus on intellectual need and epistemological justification is theoretically entailed from the *knowing-knowledge linkage premise*, which collectively assert that knowledge construction is (a) possible only through intellectual perturbation and (b) that resolutions of such perturbations always exist for an individual or community who possess suitable mental structures. Entailed from the *subjectivity premise* is that intellectual need and epistemological justification, as well as ways of understanding and ways of thinking, are not fixed; rather, their origin and acquisition vary across individuals and communities. Furthermore, they typically are not historical.

The various classifications of intellectual need aimed at addressing different roles in curriculum development and instruction. While global need pertains to the structure of a mathematics curriculum, as we have demonstrated through an outline of a part of a *DNR*-based elementary course in matrix theory, local need pertains to elicitation of specific concepts and ideas, as we have shown for the concepts of *linear combination*, *equivalent systems*, *linear independence*, *basis*, *eigen value*, *eigen vector,* and *diagonalization.*

Ways of understanding and ways of thinking emerge in a variety of mathematical practices, when mathematicians encounter a need to be certain, to determine cause, to compute, to communicate, and to structure. Humans seem to have been endowed with *cognitive primitive* (pre-conceptualizations whose function is to orient us to the intellectual needs we experience when we learn mathematics; see Harel, 2013a), but in essence these are learned needs. We have illustrated the difficulties students encounter in acquiring and applying some of these needs (e.g., the need to formulate the notion of "closest" in terms of linear algebraic equations and the need formalize the intuitive concept of dimension).

Lastly, epistemological justification represents a higher level of mathematical knowledge—not only does one possess a desirable way of understanding of a particular concept, but also is being aware of how that concept was born out of a need to resolve a problematic situation (*sentential epistemological justification—SEJ*); not only does one understand the proof of an implication, but also understands the proof in terms of cause (*apodictic epistemological justification—AEJ*); and not only does one understand a proof of logical implications in terms of cause but also is aware of how the construction of the proof might have come about (*meta epistemological justification—MEJ*). We have demonstrated the application of each of these categories of epistemological justification in the context linear algebra.

We hope that the analyses presented in this paper will generate interest on the part of mathematics education researchers whose research focus is the learning and teaching of linear algebra to test empirically the central theoretical claim made in this paper: Does instruction that is organized around intellectual need and epistemological justification and their constituent elements, as were articulated in this paper, result in advanced performance by students in linear algebra courses? Elsewhere we discussed the development, application, and utility of *DNR*-based curricula in linear algebra (Harel, in press a), complex numbers (Harel, 2013b), geometry (Harel, 2014), proof (Harel & Sowder, 1998), and mathematical practice (Harel, in press b; Harel & Soto, in press).

## References


Arnon, I., Cottrill, J., Dubinsky, E., Oktac, A., Roa, S. Trigueros, M., & Weller. K. (2014). *APOS Theory—A Framework For Research And Curriculum Development In Mathematics Education*, Springer New York, Heidelberg, Dordrecht, London, 2013.

Brousseau, G. (1997). *Theory of didactical situations in mathematics*, Dordrecht: Kluwer.

Confrey, J. (1991). Steering a course between Vygotsky and Piaget. *Educational Researcher, 20* (8), 28–32.

Cooper, R. (1991). The role of mathematical transformations and practice in mathematical development. In L. Steffe (Ed.). *Epistemological Foundations of Mathematical Experience*. New York. Springer-Verlag.

Dubinsky, E. (1991). Reflective Abstraction in Advanced Mathematical Thinking, in *Advanced Mathematical Thinking* (D. Tall, ed.), Kluwer, 95–126.

Greeno, G. (1992). Mathematical and scientific thinking in classroom and other situations. In D. Halpern (Ed.), *Enhancing Thinking Sills in Sciences and Mathematics* (pp. 39–61). Hillsdale: Lawrence Erlbaum Associates.

Harel, G. (2008a). DNR Perspective on Mathematics Curriculum and Instruction: Focus on Proving, Part I, *ZDM—The International Journal on Mathematics Education*, *40*, 487–500.

Harel, G. (2008b). DNR Perspective on Mathematics Curriculum and Instruction, Part II, *ZDM—The International Journal on Mathematics Education*, 40, 893–907.

Harel, G. (2008c). What is mathematics? A pedagogical answer to a philosophical question. In B. Gold & R. Simons (Eds.), *Proof and other dilemmas: Mathematics and philosophy* (pp. 265–290). Washington, DC: Mathematical Association of America.

Harel, G. (2013a). Intellectual Need. In *Vital Direction for Mathematics Education Research*, Leatham, K. (Ed.), Springer.

Harel, G. (2013b). DNR-based curricula: The case of complex numbers. *Journal of Humanistic Mathematics*, 3 (2), 2–61.

Harel, G. (2014). Common Core State Standards for Geometry: An Alternative Approach. *Notices of the AMS, 61* (*1*), 24–35.

Harel, G. (1998). Two Dual Assertions: The First on Learning and the Second on Teaching (Or Vice Versa). *The American Mathematical Monthly*, *105*, 497–507.

Harel, G. (2000). Three principles of learning and teaching mathematics: Particular reference to linear algebra—Old and new observations. In Jean-Luc Dorier (Ed.), On the Teaching of Linear Algebra, Kluwer Academic Publishers, 177–190.

Harel. G. (in press a). The learning and teaching of linear algebra: Observations and generalizations. *Journal of Mathematical Behavior.*

Harel. G. (in press b). Field-based hypotheses on advancing standards for mathematical practice. *Journal of Mathematical Behavior.*

Harel, G., & Soto, O. (in press). Structural reasoning. *International Journal of Research in Undergraduate Mathematics Education.*

Harel, G., & Soto, O. (2016). Structural reasoning. *International Journal of Research in Undergraduate Mathematics Education*, *3*(1), 225–242.

Harel, G., & Sowder, L. (1998). Students' proof schemes. In E. Dubinsky, A. Schoenfeld, & J. Kaput (Eds.), *Research on Collegiate Mathematics Education* (Vol. III, pp. 234–283). AMS.

Harel, G., & Sowder, L (2007). Toward a comprehensive perspective on proof, In F. Lester (Ed.), *Second Handbook of Research on Mathematics Teaching and Learning,* National Council of Teachers of Mathematics.

Lay, D., Lay, S., & McDonald, J. (2016). *Linear Algebra and its applications.* Pearson, Boston.

Li, H. (2008). *Invariant algebras and geometric reasoning*, World Scientific Publishing Co. Pte. Ltd.

Piaget, J. (1985). *The equilibration of cognitive structures: the central problem of intellectual development*: Chicago: University of Chicago Press.

Steffe, L. P., Cobb, P., & von Glasersfeld, E. (1988). *Young children's construction of arithmetical meanings and strategies*. New York, NY: Springer-Verlag.
Steffe, L. P. & Thompson, P. W. (2000). Interaction or intersubjectivity? A reply to Lerman. *Journal for Research in Mathematics Education, 31,* 191–209.
Trigueros, M. (this volume). Learning linear algebra using of models and conceptual activities.
Tucker, A. (1993). The growing importance of linear algebra in undergraduate mathematics. *The College Mathematics Journal, 24,* pp. 3–9.

# Learning Linear Algebra Using Models and Conceptual Activities

María Trigueros

**Abstract** In this chapter, an innovative approach, including challenging modeling situations and tasks sequences to introduce linear algebra concepts is presented. The teaching approach is based on Action, Process, Object, Schema (APOS) Theory. The experience includes the use of several modeling situations designed to introduce some of the main linear algebra concepts. Results obtained in several experiences involving different concepts are presented focusing on crucial moments where students develop new strategies, and on success in terms of student's understanding of linear algebra concepts. Conclusions related to the success of the use of the approach in promoting student's understanding are discussed.

**Keywords** APOS theory • Schemas • Systems of equations
Linear independence • Eigenvectors

Research on the teaching and learning of linear algebra (LA) has grown considerably throughout the past two decades. During the last ten years, an innovative approach to teach LA was developed and tested in a small university in Mexico. Challenging modeling situations were designed to introduce students to most of the abstract concepts of this discipline. Students approached these problems by using their knowledge, but to fully respond to the questions posed, they needed something else. At these moments APOS-based activities were introduced to foster the construction of new concepts that may help students move forward when they go back to work on the modeling situation. Through cycles of work on the modelling problem and APOS—based activities, students reflect on their knowledge and strategies and construct LA concepts that can be applied to the solution of new mathematical and extra—mathematical problem situations. This approach has been tested using several modeling situations designed so far. Results obtained have helped to refine both the modeling situations and the accompanying sets of activities. Research on students' learning while using these experiences has shown that

M. Trigueros (✉)
Instituto Tecnológico Autónomo de México, ITAM, México City, Mexico
e-mail: tigue@itam.mx



S. Stewart et al. (eds.), *Challenges and Strategies in Teaching Linear Algebra*,
ICME-13 Monographs, https://doi.org/10.1007/978-3-319-66811-6_2

the approach is effective in fostering the construction of specific concepts. The analysis of students' learning when several modeling activities are used is the topic of this chapter.

Data obtained from the use of three specific modeling activities related to construction of systems of linear equations, linear independence and eigenvectors, in classes taught by three different teachers during six semesters were analyzed. Patterns in students' strategies, the emergence of independent new important ideas, students' needs, and the influence of APOS-based activities and teachers' guidance were identified. The analysis focused on data indicating important transitions in students' work. These transitions were linked to the development of students' LA related Schemas that evidence students' learning.

The research questions for this study are: Do the use of modeling situations together with APOS—based activities promote the development of Schemas associated to concepts involved in a first LA course? Is there a relation between crucial moments, where students or teachers change their strategies, and learning? What is the role of teachers and that of students' engagement in relation to students' learning?

## 1 A Brief Look at Some Antecedents

LA has been recognized as a difficult subject for students due mainly to the abstract nature of the concepts of this discipline (for example, Dorier & Sierpinska, 2001). The mathematics education community has proposed different didactic approaches to improve its teaching and learning and work on different concepts has proved to be effective (Gueudet, 2004; Oktaç & Trigueros, 2010; Wawro, Rasmussen, Zandieh, Larson, & Sweeney, 2012). This chapter focuses mainly on the learning of three important concepts, a summary of research results related to each of them follows.

Systems of linear equations (SLE) play a fundamental role in the possibility to understand most of the concepts introduced in a first approach to LA and their relationship. It is well known that interpreting the solution set of a SLE is difficult for many students, and that this obstacle becomes more difficult to overcome when all the variables in the system are not explicit in every equation. These difficulties have been attributed to students' lack of understanding of the concepts of variable, functions and sets which play an important role in the construction of most of the mathematical structures involved in the learning of LA (Dogan-Dunlap, 2006; Malisani & Spagnolo, 2009; Trigueros & Jacobs, 2008; Trigueros, Oktaç, & Manzanero, 2007; Ursini & Trigueros, 1997). Interpretation of the solution set of a SLE has been related to the need to relate different representation systems for SLE and to the use of different thinking perspectives (Bardini & Stacey, 2006; Oktaç,'s chapter; Sierpinska, 2000). Other studies underline the importance of constructing a coherent SLE Schema which would encompass understanding of solution methods as Processes, solution sets as Objects and the relation of these structures to others that are constructed during a LA course (Possani, Trigueros, Preciado, & Lozano, 2010; Trigueros et al., 2007).

Linear combination (lc), linear independence (li) and linear dependence (ld) are closely related abstract concepts that are very difficult for students to understand. Although research on students learning of these concepts is limited, it has been shown that their introduction by using geometric representations may be difficult for students, particularly when they need to relate geometric and algebraic representations, and when they need to use these concepts in the case of vector spaces that cannot be represented geometrically (Dogan, 2010; Harel, 1999; Maracci, 2008). Modelling approaches, on the other hand, have proved to be more effective (Trigueros & Lozano, 2010; Trigueros & Possani, 2013; Wawro et al., 2012).

Research on the learning of eigenvalues, eigenvectors and eigenspaces has shown that their learning presents multiple obstacles for students, since they tend to concentrate in the procedures to handle them (Dogan, 2010). Using different representations while teaching these concepts has proved to help students to make sense of some of their properties (Gol Tabaghi, 2012; Stewart & Thomas, 2007; Thomas & Stewart, 2011) while the use of models stimulates students' understanding of these concepts (Larson, Rasmussen, Zandieh, Smith, & Nelipovich, 2007, 2008; Salgado & Trigueros, 2014). However, these studies have demonstrated the importance to choose a modeling situation where the need to use these concepts is directly related to the desired goal, and that, when the teaching is based on a theory of learning, it is possible to help students to overcome the difficulties found in other studies.

## 2 Theoretical Framework

APOS Theory (Arnon et al., 2014) provides elements to model and describe the construction of cognitive structures that are useful to infer students' learning and to design didactic materials to teach effectively. The basic structures used in APOS to describe the construction of mathematical knowledge are Action, Process, Object and Schema. It is considered that a student's understanding of a concept starts with performing Actions physically or mentally on previously constructed Objects. When reflecting on those Actions students can interiorize them into Processes which can be encapsulated into new Objects when they are perceived as a totality. A Schema for a mathematical concept or topic is a structure which is constructed by building conscious or unconscious relations and transformations among several Actions, Processes, Objects and previously constructed Schema. In order to describe the constructions involved in learning mathematical concepts, a model of the needed construction is designed. This model is known as Genetic Decomposition (GD) and it constitutes an attempt to predict the construction of concepts in terms of the structures of the theory. This model needs to be validated or refined with experimental data and can also serve as a basis for research and instructional design. In APOS Theory Schemas are evoked by individuals in the solution of problem situations. Schemas evolve continuously by the mechanisms of assimilation and accommodation. The development of a Schema can be described by three levels: Intra-, Inter- and Trans-, known as triad. These levels are characterized by the relations constructed among the structures that constitute the Schema or by the assimilation or accommodation of new structures into it. Intra- level is characterized by superficial relations among components of the Schema, in general being aware that these elements appear together in specific problems. In inter-level, there is awareness of transformation relations between the elements of the Schema, and in Trans—level these relations are developed and the Schema is coherent in the sense that it is possible for the individual to distinguish those problems where the Schema can be applied form those where it cannot be applied. This study will focus mainly on the construction of Schemas, since their development evidences students' learning. Schemas considered in this chapter are the SLE Schema, the Schema for li and ld (LID Schema) and that for eigenvalues and eigenvectors (Eigen Schema). APOS theory includes in its framework methodological and didactical components. The fist includes cycles of research where the GD is tested and refined until it is validated. The second consists of what is called the ACE-cycle which consists of collaborative work of students in teams during class hours on activities (A) designed with the GD, whole group discussion with the teacher on the previous work (C) and homework exercises (E).

Although modelling is not included in APOS theoretical framework it is consistent with APOS structures (Trigueros, 2014). When students face a modelling problem, they use the mathematical Schemas and other Schemas to approach the problem they face. They take elements of those Schemas to choose variables, to formulate hypothesis needed and to work with the situation. Through Actions and Processes on some of the components of the Schema, and through coordination of Processes, a mathematical model emerges. This mathematical model is encapsulated into an Object. New Actions or Processes on it, and relations among the components of the Schema contribute to its development and to the solution of the modelling problem.

## 3 Methodology

This study focuses on data obtained during a five-year project designed to test a specific methodology to introduce LA concepts to third semester university students in a small private university in Mexico. Students' work, together with the theoretical framework, were considered in the design and in the refinement of tasks sequences that guided students' reflection towards the definition and formalization of concepts. Participating students' majors were engineering, applied mathematics and economics. The instructional approach used was based on APOS ACE-Cycle. Different topics were introduced by posing open modeling situations for the students to work on. Students worked collaboratively in teams of three students. Work on the problems was intermingled with work on APOS-based activities designed with a specific GD for each of the concepts. Detailed results about the construction of these concepts along with students' instances of the project have been published elsewhere (Possani et al., 2010; Salgado & Trigueros, 2014; Trigueros & Possani, 2013).

The analysis for this part of the project was performed by looking at the data obtained from the use of at least three modeling situations during six one semester courses taught by three different teachers. The selected data include groups where modeling situations were used to teach SLE, li and ld, and eigenvectors and eigenvalues. These topics were selected because they correspond to different moments in the course. The courses started by the introduction of SLE, li and ld concepts were introduced approximately at the middle of the course and eigenvalues and eigenvectors were introduced towards the end of the course. The analysis of all the data obtained through class observation, students' work and interviews focused mainly on the commonalities associated with crucial moments where students change their strategy and argue it with or without the help of the teacher. We called these moments transitions and analyzed them in terms of the Schemas mentioned above to have a general view of students' learning. The data were analyzed by the author and discussed with the teachers for validation. Initials such as AT, or PB refer to the name and surname of the participating student.

The SLE Schema plays an important role in learning all the concepts included in this study. Its development encompasses students' learning throughout the whole course. Table 1 shows the GD for those Schema used in this study, together with the modeling activity used to introduce each of them.

## 4 Important Moments in Students' Work

In what follows we describe briefly results obtained in the analysis of data. As explained above, the focus of the description is on those transitions that can be related to the specific development of Schemas together with aspects that limited some students' development.

### *4.1 Linear Systems of Equations*

Every time this problem was used, students started by exploring the streets' network and trying to discover how cars can move around. After discussing possibilities, a transition occurs. Several teams (average 5/10) focused their attention on intersections as key in describing if a street can be closed:

AT: *...so we need to name the streets and those will be our variables, I mean, the number of cars that pass by them, and in this intersection, all the cars that come in must go out.*

GP: *Yes.... Cars cannot appear or disappear from the whole network... we can write this for each intersection as an equality.*

**Table 1** Genetic decompositions and modeling problems

| Concept | Schema GD | Modeling situation |
|---|---|---|
| Systems of equations | *Intra-SLE*: Students identify variables, equations, and equality as elements related in problems dealing with solution of equations. Students can manipulate variables in simple square SLE. They have constructed equations through equality relations among algebraic expressions and solutions as Actions conducted to find specific values that satisfy the equations. Relations among equations are superficial, equations are related because they appear in the same system. They are not clear about the fact that solving a system of equations means finding solutions that satisfy all the equations. Even if they use this fact in their procedures, they are not conscious of it<br>*Inter-SLE*: Students become aware that all equations in systems may be satisfied by the same solutions. They can accommodate the notions of set and function in the Schema by considering solution sets, and multiple solutions in terms of functions. In making this accommodation they start to think about the procedures used to solve systems of equations as transformations of the equations that, when used, help them to find solutions. They relate equations with equality by being aware that transformations of equations are restricted to those that satisfy equality properties and can find solution sets of different types of systems. When a restriction is introduced students are able to determine the conditions the restrictions impose on the solution set of the system<br>*Trans-SLE*: Students can consider systems of equations as a totality; they are aware that the solution of the system may be found by transforming the equations using properties of the equality and that through these transformations equations are changed but the solution set of the system is conserved. Coherence of the schema is demonstrated by students' possibility to recognize equivalent systems of equations and of the invariance of their solution set under the appropriate transformations | The following diagram (shown in the results section and Table 2 in Appendix) shows a map of a sector of streets in the downtown area of a city. The traffic control center has installed sensors to detect the number of vehicles that transit by the sector. In the figure the arrows represent the direction of each street. In each intersection, we can consider that there is a roundabout that enables a continuous flux of traffic around the sector. Parking is not permitted. Can a street be closed without causing a traffic jam? What is the minimum number of cars that can be allowed to circulate through a street, to avoid traffic jams? |
| Linear independence and dependence | *Intra-LID*: The set of linear combination, linear independence and linear dependence vectors are related to each other to solve specific problems with vectors in $R^n$. Relation to SLE is considered only in terms of a procedure, as a Process, to determine li of sets of vectors<br>*Inter-LID*: The linear independence, linear dependence, and linear combination Processes are | A group of three industries produce goods to satisfy their own demand (i.e. to satisfy the demands of the industries in the group) and to satisfy external consumers demand. Supposing that the quantity of the good produced by each |

(continued)

**Table 1** (continued)

| Concept | Schema GD | Modeling situation |
|---|---|---|
| | also related to Processes describing properties of li and ld sets as Objects and to the generalization of these properties to other vector sets, for example to matrices of vectors in $Z_p{}^n$. Vectors sets can be transformed by adding or removing vectors into li or ld sets as needed in mathematical activity or in applications. The relation to SLE is considered in terms of properties of its solution set<br>*Trans-LID*: The Processes describing properties of li and ld sets are coordinated and the resulting Processes are encapsulated into Objects in such a way that the Schema can be used to determine when a set of vectors is li or ld, independently of its elements, and what properties of sets are conserved through different operations. Relations to SLE and other LA concepts are understood in terms of linear independence, that is, students can compare sets and use properties of a set of vectors in different spaces by determining which of them correspond to li sets | industry satisfies the needs of all the other industries, its own demand and the consumers' needs, and that you know the production of the industries for nine periods of internal and external demand, how would you find the fractions of the production of each industry to satisfy those demands? How many data would you need to respond the former question and how would you chose them? Can you use what you found to predict the production needed in the 6th and 10th periods? (see Tables 3, 4 and 5 in Appendix) |
| Eigen-values, eigen-vectors and eigen-spaces | *Intra-Eigen:* Eigenvalues and eigenvectors are related to each other by considering that they are always used together in the solution of specific problems related to a matrix. Students can apply Action or Processes to a given matrix to find its eigenvalues. When applying the Actions to find eigenvectors for a specific eigenvalue, the eigenspace is not considered as such, but only as a set of eigenvectors from which one is chosen arbitrarily<br>*Inter-Eigen*: Eigenvalues, eigenvectors and eigenspaces are considered as Processes and as the result of a matrix transformation; the geometric properties of eigenspaces are used in the solution of problems<br>*Trans-Eigen*: Eigenvalues, eigenvectors and eigenspaces are considered as the result of a specific transformation. Properties such as li and ld of sets of eigenvectors can be applied to specific problems, relations of complex eigenvalues to periodic functions are constructed. Coherence of the Schema is demonstrated by students' possibility to determine which problems are related to it and which are not, and the invariance of the Schema under changes in eigenvectors sets corresponding to a matrix | In an economy, there are some employed persons and some unemployed persons at a certain time. The number of employed and unemployed persons is considered as the labor force for the economy and can be considered constant. If the probability that an unemployed person finds a job at any time and the probability that an employed person continues in a job at the same period is known, find a mathematical model that describes the dynamics of employment. According to your model, what would be expected to happen with the number of employed and unemployed persons in the long term? (see Table 6 in Appendix) |

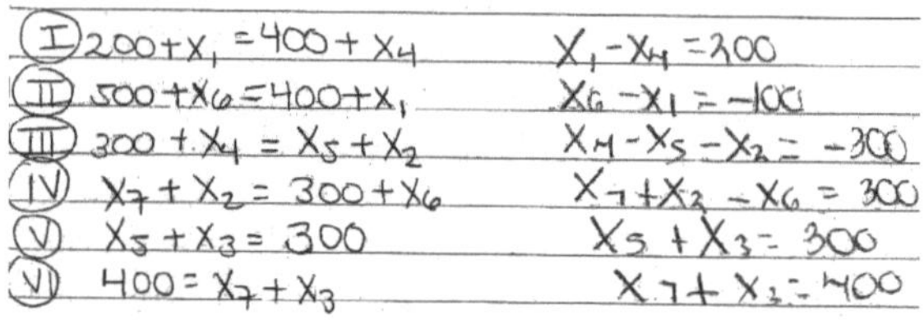


**Fig. 1** A system of equations is proposed and manipulated

The change of focus made it possible for students to select the appropriate variables and to pose an important hypothesis. Not all students in the teams were immediately convinced. This transition took long arguments from some members of these teams to convince the others. Students were forced to think about and refine their mathematical arguments until all team members agreed. Once this happened students wrote one equation for each intersection and considered the resulting model as a SLE. Students evoked the SLE Schema to bear with the problem and could apply Actions or Processes on the SLE Object to solve the problem. Figure 1 shows the system posed by a team and some Actions they performed on it.

After group discussion, most teams understood the mathematical model proposed by other students; however, we found that, in every instance, there were students who did not make this transition. Results show that students who could not engage with the problem evidenced a poor understanding of variables and their SLE Schema consisted mainly of equations as equalities upon which they could perform memorized Actions. They could not explain why procedures used in the solution of systems work and found it very difficult to pose a system with many variables as those presented by other students to the whole group.

Once systems proposed were discussed in the class, all teachers we observed asked students to solve them. This was a difficult task for all students. Students who demonstrated the construction of an Intra-SLE Schema could perform some Actions on the systems' equations. They were confused with so many equations and unknowns. They showed they could use methods they had previously learned. This implied they were able to assimilate more general problems to the SLE Schema and to construct relations between variables, equations and solution set to consider transformations of equations as related to equality and to find solutions that were not considered as solution set. Some students in each group showed they had constructed an Inter-SLE Schema. They could assimilate the problem to their SLE Schema, and showed they had constructed Processes for solving this complex system, but they were confused when tautological results were obtained when applying these procedures. They showed they had constructed transformation relations and could think of some solutions as sets. When these difficulties appeared, teachers introduced APOS-based activities to help students reflect on what SLE are; on solution methods and solution sets; on the interpretation of equations, SLE and solution sets, and to construct Gauss method to solve and to compare it to their original procedures. The analysis of students' work and interviews showed that some students who had evidenced difficulties at the beginning, participated actively in discussions after finishing the activities.

When students went back to work with the model a transition took place. Students expected to find multiple solutions for this system, which has more unknowns than equations. Most students found the solution set for the SLE they had posed. Each team had chosen different names for the variables, so solution sets did not look the same. Recorded discussions of members of several teams evidenced that another transition occurred at this moment:

PB: *This solution is not unique... if all the equations for streets are satisfied, I mean all of them at the same time... it says that you can chose any of those values and you will have a solution, all the unknowns depend on two variables, although some of them depend only on one... that is you can select different ways to go through the streets and now we need to think about what street can be blocked or how many cars should be the minimum possible that may be allowed to go through it...*

RS: *I would say the set of solutions contains many, an infinite number of solutions... but I don't know yet what we need to do to find out if we can block a street.*

PB: *What we should do, I guess... is, ah... is to check if all those variables can or cannot take the zero value...the conditions ... if we use zero cars for these variables, we can find where to close, and the number of cars must be positive...*

We interpreted this as meaning that after work with the activities most students accomodated function and sets as Processes into the SLE Schema. Many of them manipulated functions in the solution set so that each unknown depended at most on two variables and considered those functions as the result of transforming the system of equations to find the simplest form of the solution set. Among these teams, several considered parameters in the solution set as free variables and referred to them, quite naturally, as functions of two different variables. This is an important transition considering that the only functions these students had met before were one variable functions. They extended their Schema for function to include these two types of functions. This evidenced that these students' SLE Schema developed to an Inter-SLE level.

Interpretation of the solution set in terms of the problem constraints was an obstacle for most students. It required students to change their focus. As the example above shows, students needed to consider the joint variation of variables in each function of the solution set. Frequently, students focused only in the solution set as Process and did not take into account the restrictions involved in the context of the modelling problem. When they did, they found it difficult to include the restrictions into the solution set.

This transition is related to the development of the Schema to Trans-SLE level. Students who considered and interpreted the solution set together with the restrictions imposed by the nature of the streets' network, could take restrictions into account in their solution set and to interpret it in terms of the number of cars circulating through the network being conserved. They were thus able to take a decision about the number of cars that could circulate through each street. Although most of the students in the groups participating in the experience could solve and interpret different types of systems of equations and to apply the SLE Schema to different non- related modeling problems, only six or seven students in each group interpreted and clearly explained the restrictions of the problem. These students could use similar arguments in their solution of different problems that required the use of restricted SLE.

It is important to underline that, both, the decision of the teachers to guide students' attention to the role of parameters in the solution of complex SLE through APOS based activities and group discussion played an important role in the development of students' SLE Schema.

In general, all students benefited from the experience. Most of them enriched their SLE Schema by constructing relations between variables, equations, functions, solution procedures and solution sets, with differences among them. We found that persistent difficulties with variables limited students' engagement in the solution of the modelling situation. They could use some methods but could neither explain them nor apply them to complex problems.

## 4.2 Linear Independence and Linear Dependence

Students explored the modeling situation by drawing diagrams to understand the problem, find variables and relations among them. Several diagrams were used throughout the experience, however, at some point there are teams (2/10) that drew a specific diagram (essentially the same as that shown in Fig. 2a) which becomes a crucial tool for students to think productively. We considered this the first transition. Students in these teams focused their attention in internal and external demands for the whole system of industries, they drew them as arrows and used variables to name them (Fig. 2b). They introduced the hypothesis of production being equal to consumption, which is fundamental in finding a simple mathematical model consisting of a SLE These students expressed this relation in terms of a proportion of the total production demanded by each industry plus that of the external sector for each industry. At first these students were not clear about the meaning of the introduced parameters: *This is the production of industry A, that industry j needs for its production so that the external demand is satisfied.* In class discussion, most students were convinced of the usefulness of the diagram and the proposed mathematical model. They all started using it in their own work. As in the SLE experience, students bring the SLE Schema to bear with the situation. Those students who correctly verbalized the problem but did not draw any diagram had more difficulty in converting from their verbal representation of the situation to a mathematical model they could work with.

During group discussion teachers asked students to compare the different proposals and guided discussion towards the use of a simpler symbolization and to the interpretation of the selected SLE model such as:

**(a)** **(b)**

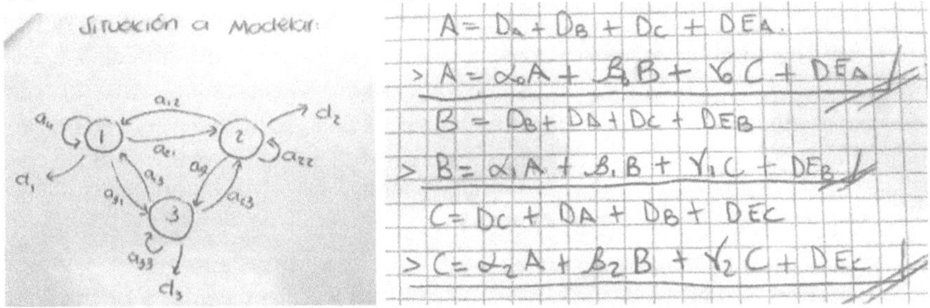


**Fig. 2** **a** Productive students' diagram. **b** An algebraic strategy

$$x_1 = a_{11}x_1 + a_{12}x_2 + \cdots + a_{1n}x_n + e_1$$

$$x_2 = a_{21}x_1 + a_{22}x_2 + \cdots + a_{2n}x_n + e_2$$

$$x_3 = a_{31}x_1 + a_{32}x_2 + \cdots + a_{3n}x_n + e_3$$

where it was clarified that $x_j$ is the production level for each industry in the economy, $a_{ij}$ is the number of units produced by industry $i$ that is necessary to produce one unit by industry $j$, that is, for each industry there is a unitary consumption vector which consists of the necessary inputs per unit of production of the industry, and $e_j$ is the external demand of product $j$. This presentation is similar to that presented in textbooks as Leontief's model, however, in these experiences students developed the model by themselves and the focus was not on the SLE itself.

Once a mathematical model was proposed, students asked for information to work on it. It is interesting that some of the teachers always gave all students the whole list of data organized in a table, but most of them followed the suggestion of one teacher who asked students how many periods of data they needed and accordingly selected from the whole list the number of periods solicited by students. A transition occurred. Students decision was discussed among members of most teams. Their attention was focused on the model, one for each industry, and on the characteristics of the data. Students arguments led them to realize that unknowns of the systems are the parameters $a_{ij}$. Many students showed difficulties to focus on the coefficients of the equations as unknowns: they generally consider coefficients as known constants. Comments as: *What we have in the table are the xs isn't it? ... we know those, then what we need to find are these as, the constants... I guess those are unknowns, of the system for each industry...we don't know them...* appeared in each instance and seem to be convincing to most students, but teams that persisted in using productions as unknowns and data for the coefficients found inconsistencies in the SLE's solution set. Through these discussions many students extended their notions of equations, and their conception of variables became flexible when considering the possibility to decide which variables in the problem are the unknowns of the system. Their construction of the relation of the equation Process or Object and unknowns Objects developed at the same time. Observations show that most teams asked for three periods of data since the model for each industry has three equations and three unknowns. Teachers selected different periods for students that consisted of an ld set of data, considered as vectors, for some teams and an li set for other teams. One or two teams, however, were not able to make the selection and asked for the whole table, or guessed a number, without reflection. These students' Schemas didn't show a development. They were usually lost when trying to solve the systems.

Some teams solved the system by hand, others used the calculator or the computer. Results obtained by students were compared in whole class discussion. Some teams obtained the same unique solution while other teams obtained multiple solutions, or unexpected numbers in the solution set. The teachers asked students to explain what happened and, after some confusion, some students, only two or three in each course, focused on data as the possible cause of the problem, this signalled a new transition: *We had different data and obtained the same solution, but others, with different data obtain different solutions*... all teams checked their work and no mistakes were found. Several students became aware that the difference in solutions was related to the data used. Teachers suggested to argument what caused the differences. Some students focused both on the data used and the properties of SLE demonstrating they were now considering SLE as Objects. They also used vectors as Objects by referring to the data with comments such as ... *these data are all on the same plane*... or, *In this set this vector is multiple of this other.* Another transition occurred when, in every experience, there was a student who explained... *when there is no unique solution, I can see that there is some redundant information in the data that does not give useful information. Some of the information is repeated somehow.* The expression "redundant information" emerged as the cause of differences found. This transition is linked to the students' evoking a Schema that related properties of vectors with the properties of SLE. A new Schema we called the linear independence (LID Schema) appeared. In all cases teachers took advantage of the "redundant information" idea in the introduction of APOS -based activities to promote the construction of the concepts of lc, li and ld, and included work on the sets of data used in the modelling problem to be worked in teams.

Work on activities favoured the development of the LID Schema and the formalization of new relations among sets of vectors. After working with these activities most students showed the construction of an Intra-LID Schema; they were convinced that a unique solution for SLE was related to li sets of vectors. After more work with the mathematical model using the new concepts, teachers introduced APOS-based activities to explore properties of li and ld sets of vectors, their geometric representation, when possible, and their relation to solution of SLE. Students applied these properties to find data-vectors that could be written as lc of others in the whole data table. They concluded that there was "redundant information" in those data and in some of its subsets and unique solution was related to "no redundancy", so they were li. This idea became a powerful tool in understanding these abstract concepts, it gave students a "concrete" meaning for them.

After this discussion, students validated the model with different sets of data and predicted demands for not given periods. Throughout these experiences students were able to relate the LID Schema and the SLE Schema through accommodating the new relations in each of them. All the students in this experience showed to have constructed a LID Schema at least as an Intra- LID level of development and more than a half demonstrated its construction at an Inter-LID level later in the course by using it in other contexts, such as columns and rows of a matrix or matrices, and to explain, for example, why li is needed for the inverse matrix to exist. Some of the students who had shown difficulties with SLE were found to have progressed in their understanding SLE's solution sets. Others continued using memorized algorithms and had difficulties making sense of other students' arguments.

Although the production problem was difficult for students to interpret and work with, every time it was used students developed powerful ideas, as "redundant information" to make sense of the relation between data, vectors, and SLE. When the concepts of lc, li and ld were formalized by using the APOS- based activities, most students had already constructed them by themselves as Processes. Teachers found APOS Theory to be "a guiding light" in understanding students' suggestions and in giving students freedom to use their own ideas which were later formalized according to their needs.

## 4.3 *Eigenvalues, Eigenvectors and Eigenspaces*

Again, students explored the modelling problem presented. Students evoked once more the SLE Schema and used ideas from economics courses and their common sense and wrote a system, but faced new elements they had not met before: the system depended on time. Students were confused, so teachers guided their attention to a population model with one linear difference equation they had worked with before. Students used the population model as example but needed some guidance from teachers who generally decided to discuss with the whole group. It was then that a student, at each instance, asked something as: *Unknowns are functions, if we write a linear system, do the same methods for solving systems work?* We considered this comment as signalling a transition. The attention of the group changed from focusing on algebraic variables to functions that appeared as variables in the model. In some instances, the discussion continued and other students joined the discussion before teachers replied. Comments such as the following were registered:

CC: *Ok, so it* (referring to the system) *should be for the employment, x at t plus one is equal to q times x at t, plus p times y at t* (the teacher writes on the board $x_{t+1} = qx_t + py_t$).

ET: .... *Are p and q the probabilities? They should add 1 and we have that 1 − q is the probability that a person who was employed loses his job, and the same for conserving the job but with the other...* (The teacher writes $y_{t+1} = (1 - q)\ x_t + (1 - p)\ y_t$),

Teacher: *So, the model is this system? Can you interpret it?*

In other instances, a team proposed the system, presented it to the whole group and in both cases, students decided to use the proposed model and could interpret the variables involved in it. However, even when the dependence of time was introduced, students were using a SLE Schema, although they considered unknown as functions.

Another transition occurs when students tried to find a solution to the system. In every instance, students showed difficulties but, each time, some teams (2/10) thought differently: *I am confused... time appears twice... I mean the unknowns are functions of time, but as in population, the equations involve two times... Two time periods t and the next t + 1, how can we solve it, do we use exponential functions as solutions as before?* Students demonstrated that they were now aware of the temporal relation both in unknowns and equations. These teams looked for specific functions of time that could describe the behaviour of the number employed and unemployed persons, and students made an analogy with the population model by proposing an exponential solution. This transition changed again students' attention to focus in verifying this idea by substituting the exponential function in the system. Students evoked two elements of the SLE Schema: the solution set and the matrix form of a system of equations. Substitution was not easy, some teams used an exponential matrix, others a solution of the form $x_n = k^n x_0$ Teachers usually suggested trying the later.

Until this moment students had not considered the need to use vectors in the solution of the system, the SLE equations they were using was clearly at the Intra-SLE level, even though they clearly had two different unknowns: employed and unemployed people. After facing some problems and asking the teacher several teams went back to the dependence on time of the equations, but now considering explicitly that time periods were the same for the employment and unemployment equations since time was de independent variable of a function whose dependent variable was a vector, even though they had not been introduced to this kind of functions: *It might be something like ... (writes: $\overline{x}_t = k^t \overline{v}$)... t is the independent variable, the vector (u, v) is a function of t, with that solution... Is $\overline{v}$ the vector of initial conditions as in the population model?*

Students went back to substitute the solution proposed. Their strategy consisted of Actions needed in verifying that the solution proposed satisfied the SLE since the original difference equations system was transformed into a SLE, the dependence on time was not explicit in this system. In all instances, there were at least two teams that used everything they had learnt about SLE in the solution process. These students showed that their SLE Schema had developed to a Trans-SLE level by the accommodation of different Processes, Objects and Schemas they had constructed along the course and through restructuring the necessary relations among them. They realized the system was homogeneous, and that it should have multiple solutions, they wrote the system as $A\overline{v} - k\overline{v} = \overline{0}$ and easily transformed it into the matrix form $(A - kI)\overline{v} = \overline{0}$ they realized both $k$ and $\overline{v}$ were unknowns: *...there are two unknowns here, the vector $\overline{v}$ and the scalar k.* They used the determinant of matrix $(A - kI)$ being zero to find the solution set of this SLE, together with possible values of the parameter $k$.

Focusing on the SLE was an important transition. Students were looking for a solution to the difference equation. In doing so, they realized they could concentrate only on the SLE obtained when substituting the proposed function into the system of difference equations, since the SLE solution would give them the clues to find the solution set of the difference equations. In all the instances, several teams (3/10) gave evidence of having constructed a Trans-SLE level Schema. They could write the equation defining eigenvalues and eigenvectors, and to find eigenvalues and eigenvectors without being introduced to them. Teachers discussed students' work in group discussion, formalized the definition of the new concepts by referring to students' work, and introduced APOS-based activities for students to interiorize these concepts and analyse their properties. Teachers went back to the modelling situation and asked students to interpret their findings in terms of the modelling situation, in the context of specific probability values, which were provided for guidance.

Students identified two eigenvalues and computed their corresponding eigenvectors. They found out that the system had multiple solutions in each case and that those solutions could be considered as the span of each eigenvector. All this work, mainly done independently by students, contributed both to the development of students SLE Schema and to the construction of an eigenvalues, eigenvectors and eigenspaces Schema (Eigen Schema) at an Intra-Eigen level.

Most students could interpret each solution independently but they did not know which of them to use. Interpretation in terms of the problem was difficult for most students although many teams drew graphs of solutions components and used them to discuss the long- term behaviour of each of them. Teachers guided students in the interpretation of the solution of the difference system of equations by showing to them that a lc of the solutions was also a solution, and directed students' focus to the initial vector values as a basis for the space of solutions of the system. They discussed the fact that, if initial conditions were known, they could find a unique solution to the system. Evidence was found that some students, about 3 in each course, constructed eigenvalues and eigenvectors as Objects, and that they developed at least an Inter-Eigen level Schema.

## 5 Discussion

Situating this study at a different analysis level, that of the results obtained from multiple applications of the didactic approach, made it possible to find new evidence of students' learning. When looking back at the data and the reports written for each of the experiences it was possible to detect specific moments when a clear change in the way students were working or in the decisions of the teacher took place. These transitions signal shifts in students' attention to focus on some aspect of the problem that becomes salient at that moment but was not detected before. All transitions seem to appear when students are thinking mathematically and are thus completely involved in the problem at stake. They have been found to happen both when they are working with the modelling situation or with the activities designed to construct and formalize new concepts. By turning their attention to new possibilities students are able to develop their mathematical knowledge by discovering new properties of the situation, or by being able to construct relations among concepts they had not considered before to be linked to the problem. Most students in this experience either did not show the conceptual difficulties described in the literature, or could overcome them while working on the whole course.

Transitions can thus be related to moments where students accommodate new structures in the evoked Schemas by bringing in new structures and restructuring relations among Schema's components. As a result, Schemas develop and learning occurs, as described in the theoretical framework section. Transitions were related to possibilities to extrapolate students' previous knowledge, to imagine and interiorize different Actions that can be performed on Objects, or to change the approach in a way that opens new possibilities of using mathematics. It was always possible to relate these moments to the emergence or development of students' Schemas. It also permitted us to better understand how students learn when they face challenges while learning abstract concepts.

This study shows that some common phenomena appear to be independent of specific concepts the activities were designed to teach. Commonalities were also independent of the groups and teachers involved in the experience. Students' exploration strategies, emergent ideas and strategies, such as the idea of "redundant or useful information" or the consideration of types of functions that were new for students, appeared each time the modeling problems were used.

When students face new challenging situations, they need some time to explore the problem from different perspectives in terms of their experience and previous knowledge to handle them. It is clear from results obtained that students can develop strikingly creative and imaginative ideas and strategies. Teachers' guidance appeared in the data to be essential to support students' interest throughout the course and to help students develop their own strategies to work on it. as can be observed from the results discussed in this chapter.

It was also found that spacing out work on the modelling activity to give students' time to work on APOS-based activities fosters both students' ideas and the constructions predicted by the GD. On the contrary to distractive nature one may consider, these activities promoted reflection and new perspectives useful to look at the problem.

## 6 Final Reflection

Students' previous knowledge together with their attitude towards full participation in team and individual work play an important role in their possibility to fully take advantage of all the affordances offered by the modelling activities used in projects such as the one described in this chapter.

It stands out from this and other similar innovative didactic projects (see Berman's chapter) that when students are challenged and have opportunities to think by themselves, to argument and discuss freely and openly with others and with their teachers they can come up with creative, interesting and intelligent ideas. These ideas emerge from the need to better understand the problematic situation and to convince others of their ideas (Harel's chapter). Transitions detected demonstrate that students can think mathematically and that this possibility results in development of students' Schemas, that is, in learning.

Teachers' guidance and the development of complementary activities for students play, at the same time, an important role in making deep learning possible. Teachers' intervention in discussing and comparing new ideas, to relate students' work with conceptual activities an to formalize concepts involved in the work with modelling situations are essential. All this work helps to underline the mathematical potential of new ideas and strategies.

In the long term the use of these kind of projects helps to better understand students' needs, to identify transition moments and to find opportunities where they can develop their Schemas. It also provides rich opportunities to explore the nature of mathematics learning.

**Acknowledgements** Work funded by Asociación Mexicana de Cultura A.C. and ITAM.

# Appendix

## A Possible Solution to the Traffic (SLE) Problem

The following diagram shows a map of a sector of streets in the downtown area of a city. The traffic control center has installed sensors to detect the number of vehicles that transit by the sector. In the figure the arrows represent the direction of each street. In each intersection, we can consider that there is a roundabout that enables a continuous flux of traffic around the sector. Parking is not permitted. Can a street be closed without causing a traffic jam? What is the minimum number of cars that can be allowed to circulate through a street, to avoid traffic jams? (Table 2).

## A Possible Solution to Production Problem (li, ld)

A group of three industries produce goods to satisfy their own demand (to satisfy the demands of the industries in the group) and to satisfy external consumers demand. Supposing that the quantity of good produced by each industry satisfies the needs of all the other industries, its own demand and the consumers' needs and that you know the production of the industries for nine periods of internal and external demand, how would you find the fractions of the production of each industry to satisfy those demands? How many data would you need to respond the former question and how would you chose them? Can you use what you found to predict the production needed for the 6th (Tables 3, 4 and 5).

## A Possible Solution to Employment Problem (Eigen)

In an economy, there is a certain number of employed persons and a certain number of unemployed persons at a certain time. The number of employed and unemployed persons is considered as the labor force for the economy and can be considered constant. If the probability that an unemployed person finds a job at any time and the probability that an employed person continues in a job at the same period is known. Find a mathematical model that describes the dynamics of employment. According to your model, what would be expected to happen with the number of employed and unemployed persons in the long term? (Table 6).

**Table 2** Solution traffic problem

| | |
|---|---|
| A possible diagram<br>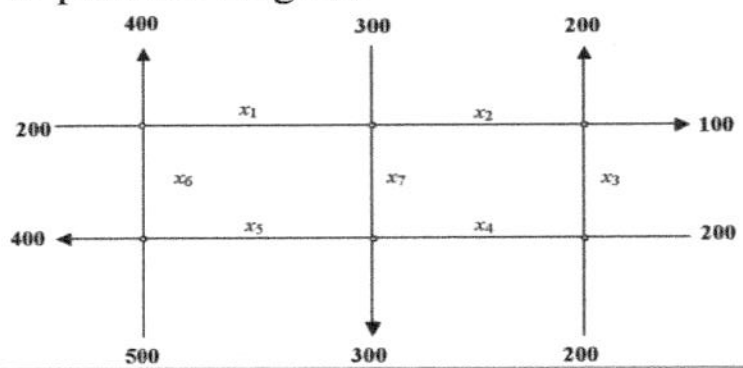 | Equilibrium in the traffic flux leads to<br>Roundabout 1: $200 + x_6 = 400 + x_1$<br>Roundabout 2: $x_1 + 300 = x_2 + x_7$<br>Roundabout 3: $x_2 + x_3 = 200 + 100$<br>Roundabout 4: $200 + 200 = x_3 + x_4$<br>Roundabout 5: $x_7 + x_4 = x_5 + 300$<br>Roundabout 6: $500 + x_5 = 400 + x_6$ |
| Simplifying:<br>$-x_1 + x_6 = 200$<br>$x_1 - x_2 - x_7 = -300$<br>$x_2 + x_3 = 300$<br>$-x_3 - x_4 = -400$<br>$x_4 - x_5 + x_7 = 300$<br>$x_5 - x_6 = -100$ | Final augmented matrix<br>$\begin{pmatrix} 1 & 0 & 0 & 0 & 0 & -1 & 0 \\ 0 & 1 & 0 & 0 & 0 & -1 & 1 \\ 0 & 0 & 1 & 0 & 0 & 1 & -1 \\ 0 & 0 & 0 & 1 & 0 & -1 & 1 \\ 0 & 0 & 0 & 0 & 1 & -1 & 0 \\ 0 & 0 & 0 & 0 & 0 & 0 & 0 \end{pmatrix} \begin{pmatrix} x_1 \\ x_2 \\ x_3 \\ x_4 \\ x_5 \\ x_6 \\ x_7 \end{pmatrix} = \begin{pmatrix} -200 \\ 100 \\ 200 \\ 200 \\ -100 \\ 0 \end{pmatrix}$ |
| Vector solution<br>$(t_1 - 200, t_1 - t_2 + 100, -t_1 + t_2 + 200, t_1 - t_2 + 200, t_1 - 100, t_1, t_2)$<br>Constrains<br>$t_1 \geq 200$<br>$t_2 \leq t_1 + 100$<br>$t_2 \geq t_1 - 200$ | A graph made by students to analyze possible values of parameters:<br>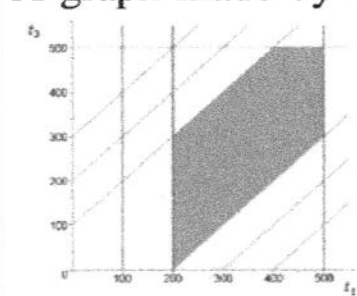 |

**Table 3** External demand for 9 months is millions of pesos

| Industry/period | 1 | 2 | 3 | 4 | 5 | 6 | 7 | 8 | 9 |
|---|---|---|---|---|---|---|---|---|---|
| A | 30 | 20 | 30 | 20 | 15 | 50 | 10 | 10 | 10 |
| B | 20 | 20 | 20 | 30 | 10 | 50 | 0 | 0 | 10 |
| C | 20 | 20 | 30 | 30 | 15 | 60 | 10 | 0 | 0 |

**Table 4** Production for 9 months in millions of pesos

| Industry/period | 1 | 2 | 3 | 4 | 5 | 6 | 7 | 8 | 9 |
|---|---|---|---|---|---|---|---|---|---|
| A | 53.515 | 40.470 | 57.202 | 47.661 | 28.601 | 104.86 | 16.732 | 13.044 | 16.548 |
| B | 36.967 | 34.847 | 39.687 | 50.150 | 19.843 | 89.836 | 4.838 | 2.120 | 14.704 |
| C | 40.470 | 37.150 | 53.422 | 52.408 | 26.711 | 105.83 | 16.271 | 3.319 | 5.623 |

**Table 5** Solution production problem

| | |
|---|---|
| Model where unknowns are $a_{ij}$ 's:<br>$a_{11}x_1 + a_{12}x_2 + a_{13}x_3 + b_1 = x_1$<br>$a_{21}x_2 + a_{22}x_2 + a_{23}x_3 + b_2 = x_2$<br>$a_{33}x_1 + a_{32}x_2 + a_{33}x_3 + b_3 = x_3$<br>Students chose how many data they need The teacher selects the data from the tables, data chosen can be when columns are taken as vectors, li or ld<br>Where $a_{ij}$ represents the necessary input needed by an industry one to produce a unit of the produce of the other industry and $b_j$ represents the external demand for each industry | A possible demand matrix (the first three columns of production Table 4). 3 systems of equations are needed, one for each industry. This matrix has li columns but students can have matrices with ld columns<br>53.515 36.967 40.47<br>40.47 34.847 37.15<br>57.202 39.687 53.422 |
| Vector **b** corresponds to the external demand of Table 4 corresponding to the periods included in the matrix<br>For the first system: $\mathbf{b} = \begin{pmatrix} 30 \\ 20 \\ 20 \end{pmatrix}$ | Solving this system, the coefficients of the first column of the production matrix are obtained. The same procedure is repeated for the other columns of the production matrix:<br>$\begin{pmatrix} 0.150027755 & 0.200020743 & 0.199952999 \\ 0.099945247 & 0.150024132 & 0.150049122 \\ 0.200102992 & 0.099862999 & 0.149984011 \end{pmatrix}$ |
| If data vectors are not li students find either an infinite number of solutions or nonsense solutions as the matrix due to computer calculation errors:<br>$\begin{pmatrix} 0.85 & -0.2 & -0.2 \\ -0.1 & 0.85 & -0.15 \\ -0.2 & -0.1 & 0.85 \end{pmatrix}$ | Students can calculate production using several methods, for example, the inverse matrix:<br>$\begin{pmatrix} 1.304448029 & 0.35031113 & 0.36874856 \\ 0.212030422 & 1.25835446 & 0.271952063 \\ 0.331873704 & 0.23046785 & 1.295229316 \end{pmatrix}$ |

**Table 6** Solution employment problem

| | |
|---|---|
| Model:<br>$x_{t+1} = qx_t + py_t$<br>$y_{t+1} = (1-q)x_t + (1-p)y_t$<br>Or $\overline{x}_{t+1} = A\overline{x}_t$<br>Where $p$ is the probability that an unemployed person finds a job at any time and $q$ the probability that an employed person continues in a job | Proposed solution:<br>$\overline{x}_t = \lambda^t \overline{v}$<br>Verification by substitution leads to<br>$\lambda^t(A\overline{v}) - \lambda^t\overline{v} = \lambda^t(A\overline{v} - \lambda I\,\overline{v}) = \overline{0}$<br>If $\lambda \neq 0$ then $(A - \lambda I)\overline{v} = \overline{0}$ |
| $\lambda$ y $\overline{v}$ should satisfy the equation but $\overline{v} = \overline{0}$ makes no sense in the problem, then $\lVert A - \lambda I \rVert = 0$, must have an infinite number of solutions for $\overline{v}$<br>If $p$ = 1/2 and $q$ = 1/3, $\lambda_1 = 1$ and $\lambda_2 = \frac{1}{6}$. Then<br>$\overline{v} = \left(x_1, \frac{3}{2}x_1\right)$ and $\overline{v} = (x_1, -x_1)$ respectively | General Solution:<br>$\overline{x}_t = c_1\lambda_1^t\overline{v}_1 + c_2\lambda_2^t\overline{v}_2$<br>$= c_1 1^t \begin{pmatrix} 2 \\ 3 \end{pmatrix} + c_2 \left(\frac{1}{6}\right)^t \begin{pmatrix} -1 \\ 1 \end{pmatrix}$ |
| Geometric representation<br>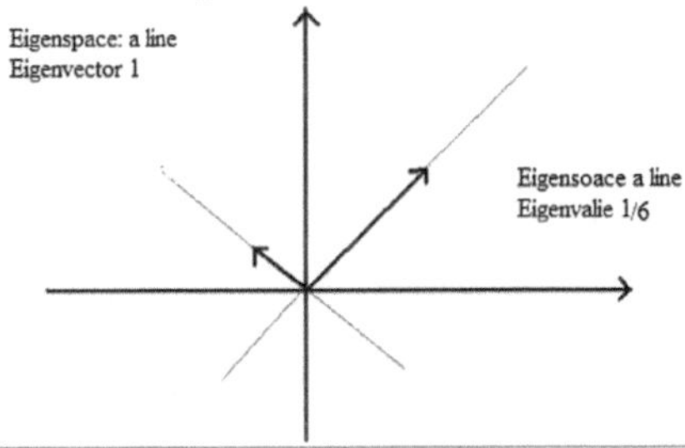 | Long time behavior independently of initial conditions:<br>If $t \to \infty$ then $\left(\frac{1}{6}\right)^t \to \infty$ and<br>$\overline{x}_t \approx c_1 1^t \begin{pmatrix} 2 \\ 3 \end{pmatrix} = c_1 \begin{pmatrix} 2 \\ 3 \end{pmatrix}$<br>This means that the rate is constant, there are 2 employed persons per 3 unemployed persons |

## References


Arnon, I., Cottrill, J., Dubinsky, E., Oktaç, A., Roa Fuentes, S., Trigueros, M., & Weller, K. (2014). *APOS Theory: A framework for research and curriculum development in* mathematics education. New York: Springer Verlag.

Bardini, C., & Staccy, K. (2006). Students' conceptions of m and c: How to tune a linear function. In J. Novotna, H. Moraova, M. Kratka & N. Stehlikova (Eds.), *Procedings of the 30th conference of the international group for the psychology of mathematics education* (Vol. 2, pp. 113–120). Prague, Czech Republic: Charles University.

Dogan, H. (2010). Linear Algebra Students' Modes of Reasoning: Geometric Representations. *Linear Algebra and Its Applications*, 432, 2141–2159.

Dogan-Dunlap, H. (2006). Lack of Set Theory-Relevant Prerequisite Knowledge. *International Journal of Mathematical Education in Science and Technology* (IJMEST). *37*(4), 401–410.

Dorier, J. L., & Sierpinska, A. (2001). *Research into the teaching and learning of linear algebra.* In D. Holton, M. Artigue, U. Krichgraber, J. Hillel, M. Niss, & A. Schoenfeld (Eds.), The Teaching and Learning of Mathematics at University Level: An ICMI Study (pp. 255–273). Dordrecht, Netherlands: Kluwer Academic Publishers.

Gol, S. (2012). Dynamic geometric representation of eigenvector. In S. Brown, S. Larsen, K. Marrongelle, & M. Oehrtman (Eds.), *Proceedings of the 15th annual conference on research in undergraduate mathematics education* (pp. 53–58). Portland, Oregon.

Gueudet, G. (2004). Should we teach linear algebra through geometry? *Linear Algebra and its Applications*, 379, 491–501.

Harel, G. (1999). Students' understanding of proofs: A historical analysis and implications for the teaching of geometry and linear algebra. *Linear Algebra and Its Applications*, 302–303, 601–613.

Larson, C., Rasmussen, C., Zandieh, M., Smith, M., & Nelipovich, J. (2007). Modeling perspectives in linear algebra: a look at eigen-thinking. http://www.rume.org/crume2007/papers/larson-rasmussen-zandieh-smith-nelipovich.pdf.

Larson, C., Zandieh, M., & Rasmussen, C. (2008). A trip through eigen-land: Where most roads lead to the direction associated with the largest eigenvalue. Paper presented at the 11 Research in Undergraduate Mathematics Education Conference, San Diego https://www.researchgate.net/profile/Chris_Rasmussen/publication/253936179.

Malisani, E., & Spagnolo, F. (2009). From arithmetical thought to algebraic thought: The role of the "variable". *Educational Studies in Mathematics*, 71, 19–41.

Maracci, M. (2008). Combining different theoretical perspectives for analyzing students' difficulties in vector spaces theory. *ZDM*, 40, 265–276.

Oktaç, A., & Trigueros, M. (2010). ¿Cómo se aprenden los conceptos de álgebra lineal? *Revista Latinoamericana de Investigación en Matemática Educativa*. 13, 373–385.

Possani, E., Trigueros, M., Preciado, G., & Lozano, M. D. (2010). Use of models in the Teaching of Linear Algebra. *Linear Algebra and its Applications. 432*(8), 2125–2140.

Salgado, H., & Trigueros, M. (2014). Una experiencia de enseñanza de los valores, vectores y espacios propios basada en la Teoría APOE. Educación Matemática 26, 75–107.

Sierpinska, A. (2000). On some aspects of students' thinking in Linear Algebra. In J. Dorier (Ed.), *On the Teaching of Linear Algebra*. 209–246.

Stewart, S., & Thomas, M. (2007). Eigenvalues and eigenvectors: formal, symbolic, and embodied thinking. *The 10th CRUME (RUME)*, 275–296.

Thomas, M., & Stewart, S. (2011). Eigenvalues and eigenvectors: embodied, symbolic and formal thinking. *Mathematics Education Research Group of Australasia*. 23, 275–296.

Trigueros, M. (2014). Vínculo entre la modelación y el uso de representaciones en la comprensión de los conceptos de ecuación diferencial de primer orden y de solución. *Educación Matemática*. 25 años (número especial) 207–226.

Trigueros, M. & Jacobs, S. (2008). On Developing a Rich Conception of Variable. In M. P. Carlson & C. Rasmussen (Eds.) *Making the Connection: Research and Practice in Undergraduate Mathematics*. MAA Notes#73, Mathematical Association of America, pp. 3–14.

Trigueros, M., & Lozano, M. D. (2010). Learning linear independence through modelling. In M. F. Pinto & T. F. Kawasaki (Eds.), *Proceedings of the 34th Conference of the International Group for the Psychology of Mathematics Education* (Vol. 4, pp. 233–240). Belo Horizonte, Brazil: PME.

Trigueros, M., & Possani, E. (2013). Using an economics model for teaching linear algebra. *Linear Algebra and its Applications*. 438, pp. 1779–1792.

Trigueros, M., Oktaç, A., & Manzanero, L. (2007). Understanding of systems of equations in linear algebra. In *Proceedings of the 5th CERME*, pp. 2359–2368.

Ursini, S., & Trigueros, M. (1997). Understanding of Different Uses of Variable: A Study with Starting College Students. Proceedings of the XXI PME International Conference, vol. 4, pp. 254–261.

Wawro, M., Rasmussen, C., Zandieh, M., Larson, C., & Sweeney, G. (2012). An inquiry-oriented approach to span and linear independence: The case of the magic carpet ride sequence. PRIMUS: Problems, Resources, and Issues in Mathematics Undergraduate Studies *22*(8), 577–599.

# Moving Between the Embodied, Symbolic and Formal Worlds of Mathematical Thinking with Specific Linear Algebra Tasks

**Sepideh Stewart**

**Abstract** Linear algebra is made out of many languages and representations. Instructors and text books often move between these languages and modes fluently, not allowing students time to discuss and interpret their validities as they assume that students will pick up their understandings along the way. In reality, most students do not have the cognitive framework to perform the move that is available to the expert. In this chapter, employing Tall's three-world model, we present specific linear algebra tasks that are designed to encourage students to move between the embodied, symbolic and formal worlds of mathematical thinking. Our working hypothesize is that by creating opportunities to move between the worlds we will encourage students to think in multiple modes of thinking which result in richer conceptual understanding.

**Keywords** Tall's Worlds · Linear Algebra · Tasks
Moving between Tall's Worlds

## 1 Introduction

Over the last decade, we have employed Tall's (2004, 2008, 2010, 2013) framework of embodied, symbolic and formal mathematical thinking along with Dubinsky's (Dubinsky & McDonald, 2001) Action, Process, Object and Schema (APOS) theory to build a framework (Stewart & Thomas, 2009), namely the Framework of Advanced Mathematical Thinking (FAMT). This framework (see Fig. 1) has enabled us to investigate students' conceptual understanding of major linear algebra concepts (Hannah, Stewart & Thomas, 2013, 2014, 2015, 2016; Stewart & Thomas, 2009, 2010; Thomas & Stewart, 2011). The natural blend of these two learning theories provides an ideal platform to analyse students' thinking in the context of primary concepts in linear algebra (e.g., vectors,

S. Stewart (✉)
University of Oklahoma, Norman, USA
e-mail: sepidehstewart@ou.edu

 
S. Stewart et al. (eds.), *Challenges and Strategies in Teaching Linear Algebra*,
ICME-13 Monographs, https://doi.org/10.1007/978-3-319-66811-6_3

linear combinations, linear independence, basis, span and eigenvalues and eigenvalues). Tall (2010) defines the worlds as follows: The *embodied world* is based on "*our operation as biological creatures, with gestures that convey meaning, perception of objects that recognise properties and patterns ... and other forms of figures and diagrams*" (p. 22). Embodiment can also be perceived as giving body to an abstract idea. The *symbolic world* is based on practicing sequences of actions which can be achieved effortlessly and accurately as operations that can be expressed as manipulable symbols. The *formal world* is based on "*lists of axioms expressed formally through sequences of theorems proved deductively with the intention of building a coherent formal knowledge structure*" (p. 22). Dubinsky and McDonald (2001) define action, which is somewhat external and requires either explicit or from memory, step-by-step instructions and rules on how to perform a certain task. Once an action is repeated and it is reflected upon by the individual, it may be interiorized into a process. The individual can successfully think of a process as an object, when he or she is able to "*reflect on operations applied to a particular process, becomes aware of the process as a totality, realizes that transformations can act on it, and is able to actually construct such transformations. In this case, the process has been encapsulated to an object*" (Asiala et al., 1996, p. 11).

In this chapter I will present the results of some of the research that I have done in regards to movements between the worlds as well as proposing a set of tasks that may facilitate some potential movements between the worlds of mathematical thinking in linear algebra.

| Worlds APOS | Embodied World | Symbolic World Algebraic Rep. | Symbolic World Matrix Rep. | Formal World |
|---|---|---|---|---|
| Action | Can draw two specific linearly independent vector 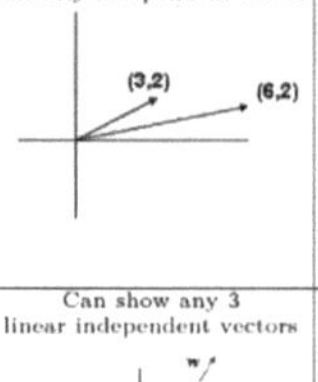  | Can arrange $c_1\mathbf{v}_1 + c_2\mathbf{v}_2 + c_3\mathbf{v}_3 = 0$ to get a linear combination $v_1 = -\frac{c_2}{c_1}v_2 - \frac{c_3}{c_1}v_3, c_1 \neq 0$ (To show dependent) | $c_1\begin{bmatrix}1\\3\\1\end{bmatrix} + c_2\begin{bmatrix}1\\-1\\1\end{bmatrix} + c_3\begin{bmatrix}3\\0\\1\end{bmatrix} = \begin{bmatrix}0\\0\\0\end{bmatrix}$ This equation has trivial solution, where $c_1 = c_2 = c_3 = 0$. Also $\begin{pmatrix}1 & 1 & 3\\3 & -1 & 0\\1 & 1 & 1\end{pmatrix} \sim \begin{pmatrix}1 & 0 & 0\\0 & 1 & 0\\0 & 0 & 1\end{pmatrix}$ Vectors are not multiples, or linear combination of each other. | |
| Process | Can show any 3 linear independent vectors 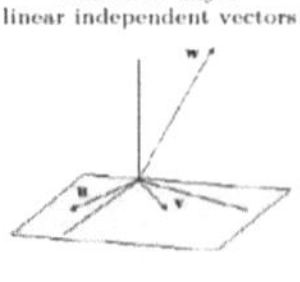  | Can see that for linearly dependent vectors one can always be written as a linear combination of the others. | Can relate linear independence and dependence to row-reduced echelon form of a relevant matrix. | Can see processes that relate linear independence to other linear algebra concepts such as: linear combination, span, rank and basis. |
| Object | Any two vectors define a plane, and if the third vector does not lie on the same plane means vectors are independent. | Can think of a set of linearly independent vectors, $\mathbf{v}_i$ as an entity and can use it eg as a basis. | Can think of a matrix as a set of linearly independent vectors $(a_{1i}, a_{2i}, a_{3i}, \ldots, a_{ni})$ and as an entity, and can use it eg as a basis. | Understands the formal definition, where the equation $c_1\mathbf{v}_1 + c_2\mathbf{v}_2 + \ldots + c_n\mathbf{v}_n = 0$ has only the trivial solution, $c_1 = c_2 = c_3 = 0$. $\mathbf{v}_i \in \mathbf{V},\ c_i \in F$ |

**Fig. 1** Framework of advanced mathematical thinking

## 2 Moving Between the Worlds

Hillel (1997, p. 232), delineates three type of languages or levels of description in linear algebra: “The language of the general theory (vector spaces, subspaces, dimension, operators, kernels, etc.); The language of the more specific theory of $R^n$ (n-tuples, matrices, rank, solutions of the system of equations, etc.); The geometric language of 2- and 3-space (directed line segments, points, lines, planes and geometric transformations)”. He believes that these languages are interchangeable but are definitely not equivalent, stating that “A vector (linear operator) in a finite-dimensional vector space is represented as an n-tuple. A 2- or 3-tuple can be represented as a geometric vector.” (p. 234). Dorier and Sierpinska (2001) add the ‘graphical’, ‘tabular’ and the ‘symbolic’ modes of languages to the above list. The ‘Cartesian’ and the ‘parametric’ representations of subspaces too, are part of a typical linear algebra course. Of course, teachers and text books move between these languages and modes very naturally and rapidly, not allowing students time to discuss and interpret their validities. They assume that students will pick up their understandings along the way, but the linguistic and epistemological studies show how these assumptions are rather deceiving. Hillel (1997, p. 233) suggests that “knowing when a particular language is used metaphorically, how the different levels of description are related, and when one is more appropriate than the others is a major source of difficulty for students”. As Duval (2000b, pp. 150–155, cited in Duval 2006, p. 114) declares: “…in the classroom we have a very specific practice of simultaneously using two registers. It is spoken in natural language, while it is written in symbolic expressions as if verbal explanations could make any symbolic treatment transparent”. Tall and Mejia-Ramos (2006, p. 3) declare that the word ‘world’ is carefully chosen and has a ‘special meaning’ in order to represent “not a single register or group of registers, but the development of distinct ways of thinking that grow more sophisticated as individuals develop new conceptions and compress them into more subtle thinkable concepts”. As Dreyfus (1991a, p. 32) declares “One needs the possibility to switch from one representation to another one, whenever the other one is more efficient for the next step one wants to take… Teaching and learning this process of switching is not easy because the structure is a very complex one.” I hypothesis that most students do not have the cognitive structure to perform the switch that is available to the expert. Duval (2006) noted that to construct a graph, most students have no difficulties as they follow a certain rule (Fig. 2), “but one has only reverse the direction of the change of register to see this rule ceases to be operational and sufficient” (p. 113).

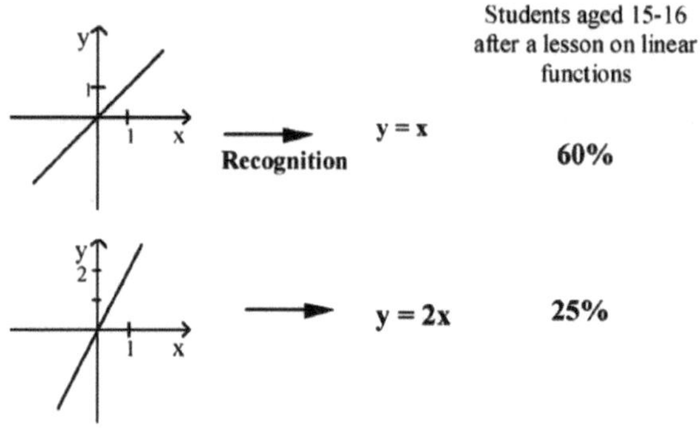


**Fig. 2** Difficulties going from one register to another

## 2.1 Bridging the Embodied and Symbolic Worlds of Mathematical Thinking

In a study by Thompson, Stewart, and Mason (2016), the authors conjectured that, physics must "bridge" the embodied and symbolic worlds (Fig. 3). Their hypothesize was that novice students struggle to embody the symbols and symbolically express the embodiments. They believed that the physics instructor created a bridge for his students to move between the embodied and symbolic worlds. He put several connected support pillars in place, including classroom demonstrations of physical phenomena, a student response system that allowed real-time communication with the instructor, and peer instruction. The experienced instructor acted as a guide for his novice students as they crossed uncharted territory. He often broke more complex problems down into smaller, more manageable pieces. He noted the importance of students creating visualizations on a regular basis. He believed: "Sometimes students' main obstacle to crossing the embodied-symbolic bridge is simply a lack of mathematical knowledge. I wish I could guarantee that my students had vector calculus when we were talking about some of this." Students' self-generated drawings on the final exam revealed gaps in students' embodied understanding even though their overall exam grades showed that they had a firm grasp on how to symbolically solve related problems.

## 2.2 From Intuition to the Formal World of Mathematical Thinking: A Geometric Topologist's Thought Processes

In a study, we examined a geometer's thought processes while teaching Algebraic Topology over a semester (Stewart, Thompson, & Brady, 2017). We spent the following semester coding his teaching journals which he wrote after each class and

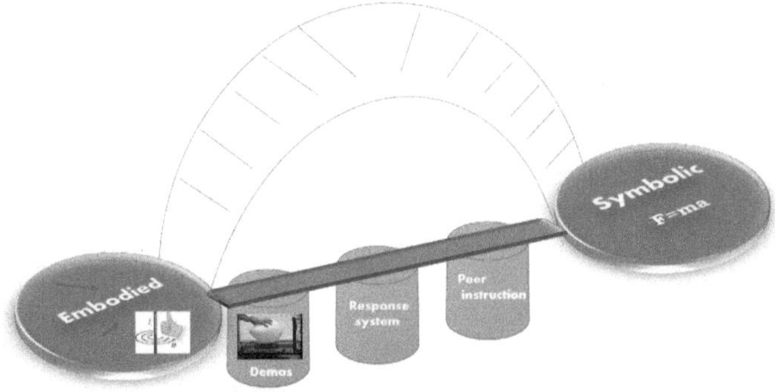


**Fig. 3** The process of embodying the symbolism and symbolizing the embodiment in physics (Thompson, Stewart, & Mason, 2016, p. 1341)

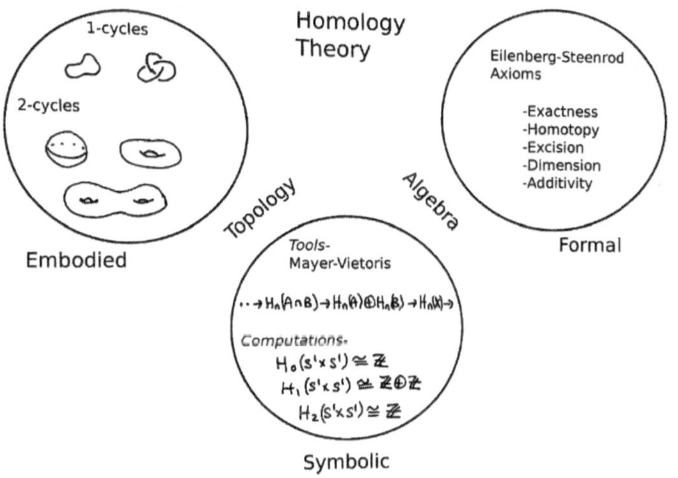


**Fig. 4** The three-lens view of homology theory (Stewart, Thompson, & Brady, 2017, p. XX)

examined them in our weekly research meetings by asking the geometer further questions. We noticed that he was able to confidently move between the worlds of mathematical thinking (see Fig. 4). How do we train students to perform in the same way?

## 2.3 *The Importance of Visualization in Mathematics Education*

Many mathematics educators have long been fascinated by the power of visualization for learning and teaching mathematics. For example, Tall and Vinner's

(1981) *Concept image and Concept definition* paper, has been a useful theoretical standpoint for many researchers (over 1936 citations). Presmeg's (2006) extensive review of papers on research on visualization from the *Psychology of Mathematics Education* (*PME*) *Proceedings* over the last 20 years, shows significant research interest in visualization over many years. For example, Dreyfus (1991b) stated at his plenary paper at PME-15: "It is therefore argued that the status of visualization in mathematics education should and can be upgraded from that of a helpful learning aid to that of a fully recognized tool for learning and proof" (Vol I: p. 33). Presmeg's review concluded with the statement that: "An ongoing and important theme is the hitherto neglected area of how visualization interacts with the didactics of mathematics. Effective pedagogy that can enhance the use and power of visualization in mathematics education is perhaps the most pressing research concern at this period." Almost two decades later, her proposed list of 13 "*Big Research Questions*" on visualization still remains unanswered. Some of her questions include: "How can teachers help learners to make connections between visual and symbolic inscriptions of the same mathematical notions? How may the use of imagery and visual inscriptions facilitate or hinder the reification of processes as mathematics objects? How may visualization be harnessed to promote mathematical abstraction and generalization? What is the structure and what are the components of an overarching theory of visualization for mathematics education?" (Presmeg, 2006, p. 227). Although, there is some research in visualization in mathematics education, specific research on visualization in linear algebra is scarce. Harel (1989, p. 49) investigated the question, "Would an emphasis on the familiar geometric system lead students to a better understanding of the vector space concept than an emphasis on a variety of unfamiliar algebraic systems?" Harel (2000) specifically considers geometry in linear algebra as an "intellectual need". In terms of understanding linear algebra concepts, Harel believes that "In the absence of a concept image that sustains the concept definition, these [linear algebra] students are unable to retain the concept definitions for a long period of time. Once the concept definition is forgotten, they are unable to retrieve or rebuild it on their own." (Harel, 1997, p. 109). In teaching linear independence for example, pictures generated by Geogebra (Fig. 5), which are dynamic, are vital in visualizing the linearly independent vectors (two vectors are on the same plane, whereas the third one is not).

## 2.4 Living in the Formal World of Mathematical Thinking

The overarching aim of this case study was to investigate how mathematicians live and dwell in the formal world of mathematical thinking and, at the same time, communicate their knowledge to their students. We employed Tall's (2013) three-world model to guide this research and to help us understand more about mathematicians as formal thinkers. Although, in theory we have some understanding of his formal world, in reality it is hard to know what actually happens in this world. Our working research questions were:

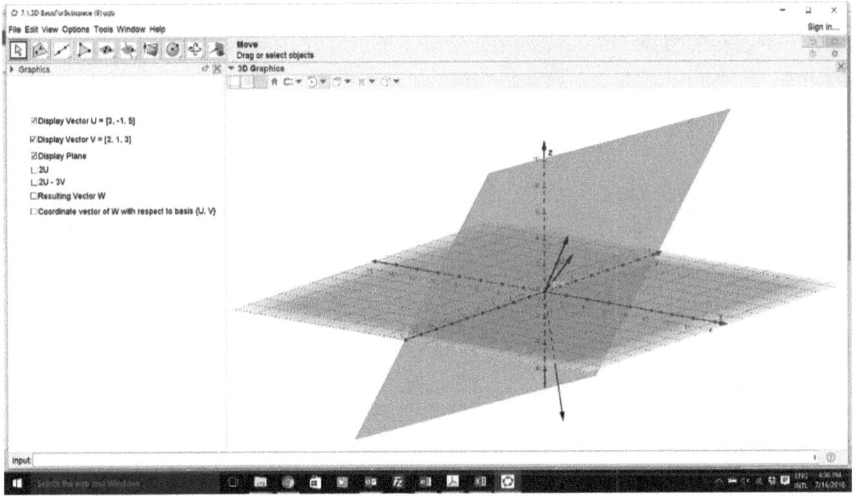


**Fig. 5** Using Geogebra to show 3 linearly independent vectors spanning the entire $R^3$ (http://www.alverno.edu/factorla/)

*What are some of the features of the formal world of mathematical thinking? Given that the mathematician is a formal thinker, how does he invite students to his world and to what extent is he willing to help students to reach the higher level of mathematical thinking? What are the pedagogical challenges faced by mathematician in communicating formal mathematics to students?*

The data for this study came from one of the research mathematician's daily reflections on his teaching of an abstract algebra course, which were made available to the group after each class; the team members' observation of the classes and their comments; weekly discussion meetings of the whole group after reading each of these reflections; the audio recordings of each meeting which were later transcribed and mathematician's reflections while working on a research paper. In addition a student who was doing the abstract algebra course taking mathematician's class, also wrote her daily journals about the classes and made it available to the PI. The main themes emerging from the data were: (a) pedagogical challenges of *communicating* the "greatness" of a concept (e.g. Galois Theory) to a beginner, (b) difficulties of teaching very abstract concepts (e.g. Tensor products) which are hard to explain or break down, (c) having a dynamical class while still being traditional, (d) mediating the disconnect between desire for mathematical elegance and the struggles of a student learning difficult material. Our preliminary qualitative data analyses indicates the disparate thought processes of the mathematician and the student, it was as if they described completely different classes and at times operated in separate worlds of mathematical thought. The preliminary results of investigating mathematician's classes over two semesters and spending another two semesters coding and discussing the data and writing conference papers has provided us with some insight into the formal world (Cook, Pitale, Schmidt, & Stewart,

2013, 2014; Stewart & Schmidt, 2017; Stewart, Schmidt, Cook, & Pitale, 2015). In mathematician's view: *The air in the formal world is much thinner, but also much clearer.*

## 2.5 Linear Algebra in the Embodied, Symbolic and Formal Worlds of Mathematical Thinking: Is There a Preferred Order?

A number of recent studies in linear algebra have considered the relationship between formal and other approaches and have demonstrated that developing teaching approaches that promote formal ideas is valuable. For example, Wawro, Sweeney, and Rabin, (2011) considered students' concept images of the notion of subspace and found that students made use of geometric, algebraic and metaphoric ideas to make sense of the formal definition. In other work, Wawro, Zandieh, Sweeney, Larson, and Rasmussen (2011) found that students' intuitive ideas about span and linear independence could be employed to assist them in developing the formal definitions. However, it may not always be clear in which order—embodied, symbolic or formal—the concepts could be introduced. In Tall's view, "although embodiment starts earlier than operational symbolism, and formalism occurs much later still, when all three possibilities are available at university level, the framework says nothing about the sequence in which teaching should occur" (Tall, 2010, p. 22). Tall explicitly states that the order of the worlds for teaching purposes is not specified. The worlds of mathematical thinking that teachers access to describe the concepts in their courses is completely up to them and the goals of their course. According to Tall, each world offers unique advantages for instruction. Mason's (2002) view is similar because he suggests that some instructors may prefer presenting examples before definitions, whereas others prefer the reverse ordering. In the context of linear algebra, Harel (1999, p. 612) believes a specific ordering of content is necessary: "The sequence in which we present material to students and the way we introduce new concepts are critical learning factors. When geometry is introduced before the concept has been formed, the students view the geometry as the raw material to be studied, they remain, as a result, in the restricted world of geometric vectors, and do not move up to the general case". Although, the existing literature shows various studies on the role of definitions, theorems, proofs, diagrams, and examples in pedagogy, carefully designed studies that investigate the order in which these should be presented are still scarce.

In a study by Hannah, Stewart, and Thomas (2014, 2015), the authors examined two sections of a first year linear algebra course. The research questions (a sample) for this study were: What is the best order (Embodied, Symbolic and Formal) to teach linear algebra concepts? Does exposure to embodied understanding of linear algebra concepts have an effect on students' willingness to embrace the formal world? The study examined the effect of teaching linear algebra concepts in the following eight different orders: ESF, EFS, SEF, SFE FES, FSE, FS, and SF. Figure 6 shows the order that was implemented in one of the classes. The analysis of the data showed (Hannah, Stewart, & Thomas, 2014) that students wanted to see examples (symbolic) of the concepts. We found that student affect was much more positive when concepts were first met in the embodied or symbolic worlds, but that once students have met all three aspects of a concept (ESF) there seemed to be little difference in the level of understanding gained. One of the aims of the class was for students to appreciate for themselves the power of formal world thinking, and that examples alone are often insufficient. By the end of the course, student perspectives on formal aspects of mathematics, definitions, theorems and proofs, were much more positive than at the beginning of the semester. The challenges that faced during this pilot study included being able to find the most effective order (using the ESF model) in which to teach linear algebra concepts and, at the same time, teaching the course well. This study needs to be followed up with less permutations of the worlds and less concepts.

## 3 Moving in and Between the Worlds of Mathematical Thinking in Linear Algebra via Tasks

Based on the work by Duval (2006) and some of the research described in this Chapter, theoretically we believe that movements between different worlds (embodied, symbolic and formal) of mathematical thinking are beneficial. We hypothesize that having the ability to move will be valuable to linear algebra students. In this section we propose nine possible types of movements in embodied and symbolic and formal worlds. In each case we propose a task to help learners to travel from one world of mathematical thinking to another.

| Worlds / Concepts | ESF | EFS | SEF | SFE | FSE | FES | FS | SF |
|---|---|---|---|---|---|---|---|---|
| Subspace | | | | | * | | | |
| Linear Combination | * | | | | | | | |
| Span | | | | | | * | | |
| Linearly Independent/dependent | | * | | | | | | |
| Basis | | | | * | | | | |
| Column space | | | | | | | * | |
| Null space | | | | | | | | * |
| Transformation | | | | | | | | |
| Kernel and range | | | | | * | | | |
| Eigenvalues & eigenvectors | * | | | | | | | |

**Fig. 6** The main concepts and the order in which they were taught

## 3.1 Linear Algebra Tasks

We propose a set of nine possible linear algebra tasks that are designed strategically to enable the learners to move between the worlds (see Table 1). The nine types of movements are: Embodied to embodied, embodied to symbolic, embodied to formal, symbolic to embodied, symbolic to symbolic, symbolic to formal, formal to embodied, formal to symbolic, formal to formal. In Table 2 we will give a conceptual analysis of each task. We anticipate that some of these movements would be more difficult for students (e.g. embodied to formal).

## 3.2 Metaphorical View of the Possible Movements

Metaphorically, for novice students the worlds of mathematical thinking can be thought of as isolated islands. In order to help students to live in all three worlds of mathematics, our proposed tasks (see Table 1) can be thought of as boats carrying students between the worlds. Other related metaphors in play could be the direction of the wind and the intensity of the waves, and the speed of the boats (see Fig. 7). The learners need guidance and navigation skills to travel to the other island, and once they arrive to the next island they will need accommodation and pedagogical support.

## 3.3 Concluding Remarks

Duval (2006) believes that students' difficulties with comprehending mathematical ideas is due to their lack of flexibility between moving between registers, as most students do not have the cognitive structure to perform the switch that is available to the expert. In his view, "changing representation register is the threshold of mathematical comprehension for learners at each stage of the curriculum." (p. 128). Duval asks the question: "does such a register coordination come naturally to pupils and students in the context of mathematical thinking?" (p. 115).

According to Ausubel, Novak, and Hanesian (1978, p. 117) "the learner is simply required to comprehend the material and to incorporate it into his cognitive structure."

Vinner (1988) claims that "very often (and specially in mathematics) the cognitive structure of the learner is not suitable for incorporating the new material. In this case the cognitive structure has to undergo some changes, that is, to accommodate". (p. 594)

He believes that acquisition of new mathematical concepts in more advanced settings requires accommodation, since "a concept which seems quite simple to the mathematician can be difficult for the student to accommodate" (p. 606).

**Table 1** Linear Algebra tasks: nine possible types of movements between the worlds of embodied, symbolic and formal

| | Embodied | Symbolic | Formal |
|---|---|---|---|
| Embodied | **Task 1**. Is vector space $R^2$ a subspace of $R^3$? Justify your answer. | **Task 2**[a]. Which of the following subsets of $R^2$ are subspaces?<br>(a) $W_1$ is the set of all vectors of the form $\begin{bmatrix} x \\ y \end{bmatrix}$, where $x \geq 0$.<br>(b) $W_2$ is the set of all vectors of the form $\begin{bmatrix} x \\ y \end{bmatrix}$, where $x = 0$. | **Task 3**. A Transformation T is said to be one-to-one if each element in the range is the image of just one element in the domain. Show this definition geometrically. |
| Symbolic | **Task 4**. Compute the area of the triangle, with vertices (-1,4), (3,1), and (2,6). | **Task 5**[a]. Determine all values of a for which the resulting linear system has:<br>(a) No solutions, (b) a unique solution, (c) infinitely many solutions.<br>$x + y = 3$<br>$x + (a^2 - 8)y = a$ | **Task 6**[a]. The inverse of a matrix, if it exists, is unique. Write a proof for this theorem. |
| Formal | **Task 7.** The matrix $\begin{bmatrix} 1 & 2 \\ 2 & 1 \end{bmatrix}$ has transformed the vector $\begin{bmatrix} 1 \\ 1 \end{bmatrix}$ into the vector $\begin{bmatrix} 3 \\ 3 \end{bmatrix}$, s.t. $\begin{bmatrix} 1 & 2 \\ 2 & 1 \end{bmatrix}\begin{bmatrix} 1 \\ 1 \end{bmatrix} = \begin{bmatrix} 3 \\ 3 \end{bmatrix}$. | **Task 8**. Consider the following matrices A, B and C and their eigenvalues and eigenvectors.<br>$A = \begin{bmatrix} 1 & 0 \\ 1 & 1 \end{bmatrix}, \lambda_1 = 1 \quad \lambda_2 = 1, V_1 = \begin{bmatrix} 1 \\ 0 \end{bmatrix}$<br>$B = \begin{bmatrix} 1 & 1 \\ -2 & 4 \end{bmatrix}, \lambda_1 = 2 \quad \lambda_2 = 3, V_1 = \begin{bmatrix} 1 \\ 1 \end{bmatrix} V_2 = \begin{bmatrix} 1 \\ 2 \end{bmatrix}$<br>$C = \begin{bmatrix} 1 & 4 & 7 \\ 0 & 8 & 5 \\ 3 & 13 & 3 \end{bmatrix}, \quad \lambda_1 = 1 \quad \lambda_2 = 2, \quad \lambda_1 = 3$<br>$V_1 = \begin{bmatrix} 1 \\ 3 \\ 2 \end{bmatrix} V_2 = \begin{bmatrix} 1 \\ 1 \\ 2 \end{bmatrix} V_3 = \begin{bmatrix} 1 \\ 1 \\ 1 \end{bmatrix}$<br>Write a theorem on diagonalization to generalize the above examples. | **Task 9**[a]. If L: V → W is a linear transformation of a vector space V into a vector space W and dim V = dim W, then the following statements are true:<br>(a) a)If L is one-to-one, then it is onto.<br>(b) b)If L is onto, then it is one-to-one. |

(continued)

**Table 1** (continued)

| | Embodied | Symbolic | Formal |
|---|---|---|---|
| | 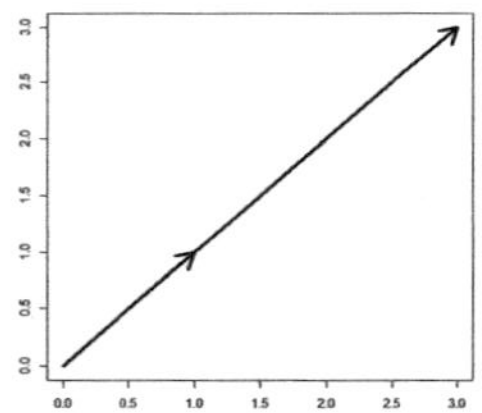 The matrix $\begin{bmatrix} 1 & 2 \\ 2 & 1 \end{bmatrix}$ has transformed the vector $\begin{bmatrix} 2 \\ 1 \end{bmatrix}$ into the vector $\begin{bmatrix} 4 \\ 5 \end{bmatrix}$, s.t. $\begin{bmatrix} 1 & 2 \\ 2 & 1 \end{bmatrix}\begin{bmatrix} 2 \\ 1 \end{bmatrix} = \begin{bmatrix} 4 \\ 5 \end{bmatrix}$ 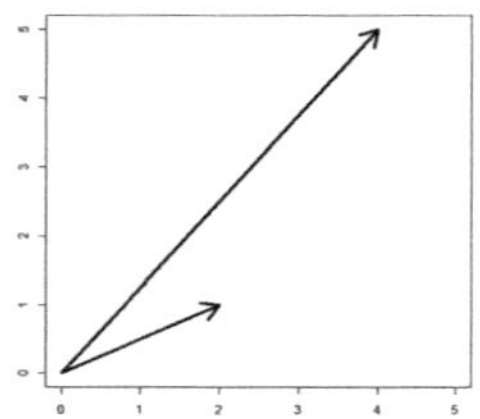 Generalize these examples to establish the definition for eigenvalues and eigenvectors. | | |

[a]Some tasks are taken from the textbook by Kolman and Hill (2007).

**Table 2** Conceptual analysis of the proposed tasks

| | Embodied | Symbolic | Formal |
|---|---|---|---|
| Embodied | 1. One justification is that, the vector space $R^2$ is not even a subset of space. $R^3$, which is one of the requirements in the definition of a subspace.<br>Although, in Duval's (2006) view this movement is a treatment (moving within the same register), thinking within the embodied world is not a trivial matter.<br>This task requires some introduction and understanding of the concept of subspace. | 2. Drawing the required region will position the learners at the right place, however, to answer the questions, knowing the properties of a vector space (in this case $R^2$) being a subspace is essential.<br>This task is beneficial in connecting several ideas. It will also help the learner to visualize the concept of subspace, which is often introduced formally by way of a definition and theorems. | 3. This task is designed to help the learner to think visually about a formal definition. Having visual images of 1-1 functions may help to tie the concept of function to the concept of transformation.<br>We hypothesis that this type of movements will generally be difficult. |
| Symbolic | 4. This task requires drawing a triangle (embodied), using the given information and calculating the area of the triangle (symbolically). | 5. This task requires the knowledge of unique solution, infinitely many solutions and no solution for systems of equations and is done symbolically all the way through. | 6. The proof of this theorem starts with the assumption that B and C are inverses of A. Then:<br>$AB = BA = I_n$ and $AC = CA = I_n$.<br>The following 1-line symbolic proof can then be performed:<br>$B = BI_n = B(AC) = (BA)C = I_nC = C$. |
| Formal | 7. In this task, using a 2 by 2 matrix one can show graphically (and symbolically) the matrix can transform a vector into another vector.<br>This can be followed symbolically as:<br>$\begin{bmatrix} 1 & 2 \\ 2 & 1 \end{bmatrix}\begin{bmatrix} 1 \\ 1 \end{bmatrix} = 3\begin{bmatrix} 1 \\ 1 \end{bmatrix}$.<br>Next, students can be guided to generalize the above scenario and reach a definition for eigenvalues and eigenvectors. | 8. We must assume students know something about diagonalization and are able to do simple calculation to find out whether a matrix is diagonalizable.<br>Distinct eigenvalues produce adequate number of linearly independent eigenvectors. Repeated eigenvalues, may or may not produce enough linearly independent eigenvectors (as shown in the | 9. In this task, students are required to remain in the formal world. The prove requires knowing the definitions of one-to-one and onto, as well as calling on another theorem, namely,<br>dim Ker L + dim range L = dim V. |

(continued)

**Table 2** (continued)

| | Embodied | Symbolic | Formal |
|---|---|---|---|
| | Definition: If A is a square matrix, say n by n, and v is a non-zero vector in $R^n$, and λ is any scalar, such that<br>$Av = \lambda v$, we say that λ is the eigenvalue of the matrix A and v is its associate eigenvector. | examples). These comparisons will lead to the following theorem:<br>Theorem. If the eigenvalues are distinct, the matrix is diagonalizable.<br>No calculation is required for this task. | |

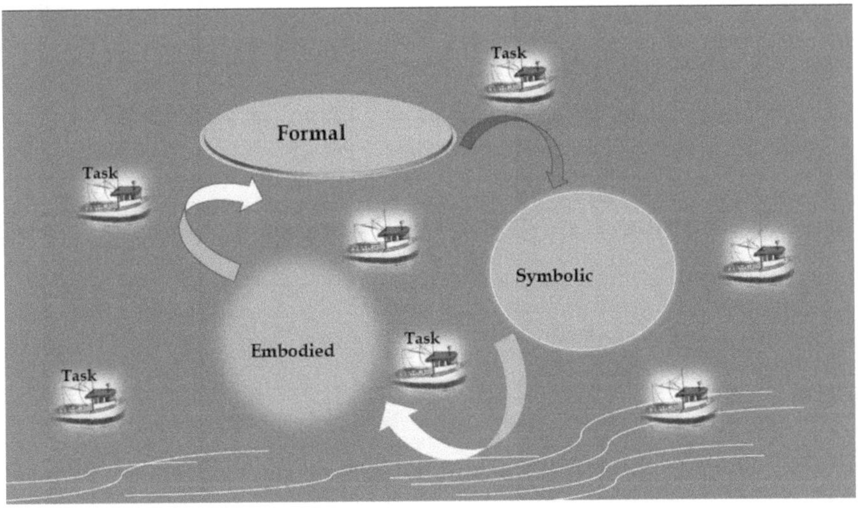


**Fig. 7** Moving between the worlds of mathematical thinking with tasks

Lack of attention to accommodation will lead into situations where "certain concepts are not conceived by the students the way we expected." (p. 593).

We anticipate that by engaging in linear algebra tasks proposed in this chapter students will be exposed to different ways of thinking about the concepts. Research is needed to determine for a given concept what kind of accommodation is ideal and ultimately the optimal pedagogy is. For example, how to accommodate for linear algebra students when they arrive in a different world of mathematical thinking would be of interest.

## References

Asiala, M., Brown, A., DeVries, D., Dubinsky, E., Mathews, D., & Thomas, K. (1996). A framework for research and curriculum development in undergraduate mathematics education. *Research in Collegiate Mathematics Education II, CBMS Issues in Mathematics Education, 6*, 1–32.

Ausubel, D. P., Novak, J. D., and Hanesian, H. (1978). Educational Psychology-A Cognitive view (second edition) (New York: Holt Rinehardt and Winston).

Cook, J. P., Pitale, A., Schmidt, R., & Stewart, S. (2014). Living it up in the formal world: An abstract algebraist's teaching journey. In T. Fukawa-Connolly, G. Karakok, K. Keene, M. Zandieh (Eds.), Proceedings of the 17th Annual Conference on Research in Undergraduate Mathematics Education (pp. 511–516). Denver, CO.

Cook, J. P., Petali, A., Schmidt, R., & Stewart, S. (2013). Talking Mathematics: An Abstract Algebra Professor's Teaching Diaries. In S. Brown, G. Karakok, K. Hah RoH, & M. Oehrtman, *Proceedings of the 16th Annual Conference on Research in Undergraduate Mathematics Education* (pp. 633–536). Denver, CO.

Dorier, J. L., & Sierpinska, A. (2001). Research into the teaching and learning of linear algebra. In D. Holton, M. Artigue, U. Krichgraber, J. Hillel, M. Niss & A. Schoenfeld (Eds.), *The Teaching and Learning of Mathematics at University Level: An ICMI Study* (pp. 255–273). Dordrecht, Netherlands: Kluwer Academic Publishers.

Dreyfus, T. (1991a). Advanced Mathematical thinking processes. In D. O. Tall (ed.) *Advanced Mathematical Thinking*, (pp. 25–41). Dordrecht: Kluwer.

Dreyfus, T. (1991b). On the status of visual reasoning in mathematics and mathematics education. In F. Furinghetti (Ed.), *Proceedings of the 15th PME International Conference, 1*, 33–48.

Dubinsky, E. & McDonald, M. (2001). APOS: A constructivist theory of learning. In D. Holton et al. (Eds.) *The Teaching and Learning of Mathematics at University Level: An ICMI Study* (pp. 273–280). Dordrecht: Kluwer.

Duval, R. (2006). A cognitive analysis of problems of comprehension in a learning of mathematics. *Educational Studies in Mathematics*, 61, 103–131.

Hannah, J., Stewart, S., & Thomas, M. O. J. (2016). Developing conceptual understanding and definitional clarity in linear algebra through the three worlds of mathematical thinking, *Teaching Mathematics and its Applications: An International Journal of the IMA*. 35 (4), 216–235.

Hannah, J., Stewart, S., & Thomas, M. O. J. (2013). Conflicting goals and decision making: the deliberations of a new lecturer, In Lindmeier, A. M. Heinze, A. (Eds.). *Proceedings of the 37th Conference of the International Group for the Psychology of Mathematics Education*, Vol. 2, pp. 425–432. Kiel, Germany: PME.

Hannah, J., Stewart, S., & Thomas, M. O. J. (2014). Teaching linear algebra in the embodied, symbolic and formal worlds of mathematical thinking: Is there a preferred order? In Oesterle, S., Liljedahl, P., Nicol, C., & Allan, D. (Eds.) *Proceedings of the Joint Meeting of PME 38 and PME-NA* 36, Vol. 3, pp. 241–248. Vancouver, Canada: PME.

Hannah, J., Stewart, S., & Thomas, M. (2015). Linear algebra in the three worlds of mathematical thinking: The effect of permuting worlds on students' performance, *Proceedings of the 18th Annual Conference on Research in Undergraduate Mathematics Education*. (pp. 581–586). Pittsburgh, Pennsylvania.

Harel, G. (1989). Applying the principle of multiple embodiments in teaching linear algebra: Aspects of familiarity and mode of representation, *School Science and Mathematics, 89(1)*.

Harel, G. (1997). The linear algebra curriculum study group recommendations: Moving beyond concept definition. In D. Carlson, C. R. Johnson, D. C. Lay, A. D. Porter, A. Watkins & W. Watkins (Eds.), Resources for Teaching Linear Algebra, (Vol. 42, pp. 107–126). MAA Notes, Washington: Mathematical Association of America.

Harrel, G. (1999). Students' understanding of proofs: A historical analysis and implications for the teaching of geometry and linear algebra, *Linear Algebra and Its Applications*, 302–303, 601–613.

Harel, G. (2000). The Linear Algebra Curriculum Study Group Recommendations: Moving Beyond Concept Definition. In J. L. Dorier (Ed.), *The Teaching of Linear Algebra in Question* (pp. 107–126). Dordrecht, Netherlands: Kluwer Academic Publishers.

Hillel, J. (1997). Levels of description and the problem of representation in linear algebra. In J. L. Dorier (Ed.), *L'enseignement de l'algebra en question* (pp. 231–237). Edition Pensee Sauvage.

Kolman, B. & Hill, D. (2007). *Elementary linear algebra with applications*, 9th edition, Pearson.

Mason, J. (2002). *Mathematics Teaching Practice: a guidebook for university and college lecturers*. Chichester: Horwood Publishing.

Presmeg, N. C. (2006). Research on visualization in learning and teaching mathematics: emergence from psychology. In A. Gutierrez & P. Boero (Eds.), *Handbook of Research on the psychology of mathematics education* (pp. 205–235). Rotterdam: Sense Publishers.

Stewart, S., & Thomas, M.O.J. (2009). A framework for mathematical thinking: the case of linear algebra. *International Journal of Mathematical Education in Science and Technology*, 40(7), 951–961.

Stewart, S., & Thomas, M.O.J. (2010). Student learning of basis, span and linear independence in linear algebra. *International Journal of Mathematical Education in Science and Technology* 41 (2), 173–188.

Stewart, S., Schmidt, R., Cook, J. P., & Pitale, A. (2015). Pedagogical challenges of communicating mathematics with students: Living in the formal world of mathematical thinking, *Proceedings of the 18th Annual Conference on Research in Undergraduate Mathematics Education* (pp. 964–969). Pittsburgh, Pennsylvania.

Stewart, S., Thompson, C. & Brady, N. (2017). Navigating through the mathematical world: Uncovering a geometer's thought processes through his handouts and teaching journals, 10th Congress of European Research in Mathematics Education (CERME 10). Retrieved August 10, 2017, from https://keynote.conference-services.net/resources/444/5118/pdf/CERME10_0581.pdf

Stewart, S., & Schmidt, R. (2017). Accommodation in the formal world of mathematical thinking, *International Journal of Mathematics Education in Science and Technology*. 48(1): 40–49. https://doi.org//10.1080/0020739X.2017.1360527

Tall, D. O. (2004). Building Theories: The Three Worlds of Mathematics, *For the Learning of Mathematics*. 24(1): 29–32.

Tall, D. O. (2008). The transition to formal thinking in mathematics. *Mathematics Education Research Journal, 20*, 5–24.

Tall, D. O. (2010). Perceptions Operations and Proof in Undergraduate Mathematics, Community for Undergraduate Learning in the Mathematical Sciences (CULMS) Newsletter, 2, 21–28.

Tall, D. O. (2013). *How humans learn to think mathematically: Exploring the three worlds of mathematics*, Cambridge University Press.

Tall, D. O., & Mejia-Ramos, J. P. (2006). The long-term cognitive development of different types of reasoning and proof, presented at the Conference on Explanation and Proof in Mathematics: Philosophical and Educational Perspectives, Essen, Germany.

Tall, D. O., & Vinner, S. (1981). Concept image and concept definition in mathematics with particular reference to limits and continuity. *Educational Studies in Mathematics, 12*(2), 151–169.

Thomas, M.O.J., & Stewart, S. (2011). Eigenvalues and eigenvectors: Embodied, symbolic and formal thinking, *Mathematics Education Research Journal*, 23, 275–296.

Thompson, C. A., Stewart, S., & Mason, B. (2016). Physics: Bridging the embodied and symbolic worlds of mathematical thinking. *Proceedings of the 19th Annual Conference on Research in Undergraduate Mathematics Education*. In T. Fukawa-Connolly, N. Engelke Infante, M. Wawro & S. Brown (Eds.), (pp. 1340–1347). Pittsburgh, Pennsylvania.

Vinner, S. (1988). Subordinate and superordinate accommodations, indissociability and the case of complex numbers, International Journal of mathematics Education, Science and Technology, 19, 4, pp. 593–606.

Wawro, M., Sweeney, G., & Rabin, J. (2011). Subspace in linear algebra: Investigating students' concept images and interactions with the formal definition. *Educational Studies in Mathematics*. https://doi.org/10.1007/s10649-011-9307-4.

Wawro, M., Zandieh, M., Sweeney, G., Larson, C., & Rasmussen, C. (2011). Using the emergent model heuristic to describe the evolution of student reasoning regarding span and linear independence. Paper presented at the 14th Conference on Research in Undergraduate Mathematics Education, Portland, OR.

# Part II
# Analyses of Learners' Approaches and Resources

# Conceptions About System of Linear Equations and Solution

**Asuman Oktaç**

**Abstract** Understanding of systems of linear equations permeates in the study of several topics of importance in linear algebra, such as rank, range, linear independence/dependence, linear transformations, characteristic values and vectors. After giving an overview of the literature on the teaching and learning of systems of linear equations, research results on student difficulties at different school and university levels are presented, establishing relationships with the way this topic is taught. The conceptions that students develop about 'system' and 'solution' are discussed in synthetic-geometric and analytic contexts in two and three dimensional spaces. Based on these observations, some pedagogical suggestions about planning instruction on this topic are offered. Although the findings reported in this chapter correspond to research undertaken in Mexico and Uruguay, they might be reflecting a more general phenomenon related to conceptions that students develop in relation with systems of linear equations and their solutions.

**Keywords** System · Linear equations · Modes of thinking

## 1 Introduction

Systems of linear equations constitute a topic of study at different school levels, usually starting at the secondary level in many countries around the world. Although the depth and methods of their study vary, they are considered important mainly for two reasons. On the one hand the comprehension of systems of linear equations constitutes an important step for further study in mathematics in general, and Linear Algebra in particular; on the other hand many applications in the fields of engineering and social sciences involve models that make use of this subject.

A review of literature reveals that despite the importance of this topic it has not been extensively studied by mathematics educators. However examples of

A. Oktaç (✉)
Cinvestav-IPN, Mexico City, Mexico
e-mail: oktac@cinvestav.mx



S. Stewart et al. (eds.), *Challenges and Strategies in Teaching Linear Algebra*,
ICME-13 Monographs, https://doi.org/10.1007/978-3-319-66811-6_4

difficulties experienced by students can be found within more general studies. For example Sfard and Linchevski (1994a) mention that when in the process of solving a system the variables "disappear" leading to expressions such as $-1=4$ or $0=0$, the students interpret the situation as the system having no solution. The authors explain this phenomenon in terms of a confusion that results from a lack of an abstract entity (equation as object) that can be transformed by applying algebraic manipulations on it. This gives rise to *out-of focus* (p. 289) use of mathematical language and behavior.

The same authors (Sfard & Linchevski, 1994b) contend that only by considering equations within a functional approach can the students attribute meaning to expressions such as $0=0$ and realize that a system of linear equations can have infinitely many solutions. In the case of systems with parametric equations the student has to understand that each such equation represents a family of functions. Another difficulty arising in this context is the confusion between the unknowns of the system and the parameters, thinking that the solution set consists of those values of the parameters which make the system true, as also reported in Stadler (2011).

In the next section we explain the theoretical framework that guided our study of the concepts of system of linear equations and their solution and then we present the aims of our research.

## 2 Theoretical Framework—Synthetic and Analytic Modes of Thinking

Sierpinska (2000) distinguishes between three modes of thinking: *synthetic-geometric*, *analytic-arithmetic* and *analytic-structural*, tension between these three modes marking the historical development of Linear Algebra. According to her the synthetic mode corresponds to geometrical objects as they are perceived by our senses directly, such as thinking about lines in the plane. Analytic mode on the other hand makes use of symbols and the person who is thinking in this mode has to go through an interpretation in order to work with mathematical concepts; so in this mode the interactions with objects is indirect. In the analytic-arithmetic mode the equations that form a system can be solved using different methods and to interpret the symbols as equations some background is necessary, whereas in the structural mode the emphasis is on the properties of objects; in this mode one can ask for example under what conditions a system has a unique solution. In the synthetic mode the properties describe the object, whereas in the analytic mode the objects are defined through specific relationships between mathematical elements.

Apart from characterizing the historical development of linear algebra concepts, these three modes of thinking also appear in students' reasoning while working on different linear algebra problems. Sierpinska (2000) considers that all the three modes of thinking are useful in different contexts and through their interaction a better understanding of Linear Algebra concepts can be achieved. On the other hand, beyond consisting in ways of thinking, these modes also provide tools for solving problems and points of view that help in comprehending different facets of mathematical concepts.

Although they are both analytic, arithmetic and structural modes differ from each other considerably. In the arithmetic mode the focus is on algorithms to solve problems and arrive at answers that can consist in numbers, ordered pairs or matrices for example. In the structural mode on the other hand the space becomes an axiomatic system whose elements "lose their numerical substance" (Sierpinska, 2000, p. 233). The following example illustrates this difference:

> Given two matrices *A* and *B*. The problem is to check if *B* is the inverse of *A*. One student calculates the inverse of *A* using the well-known formula with the determinant and co-factors, and compares the result with *B*. Another student multiplies *A* by *B*. To deal with the inverse of a matrix, the first student uses a technique for computing the inverse. The second student uses the defining property of the inverse. We shall say that the first argument was analytic-arithmetic, while the second was analytic-structural. (p. 234)

As for the difference between the synthetic and analytic modes, Sierpinska (2000) describes the reaction of a student who, when shown a figure with three planes intersecting in a line, said that the respective system has a unique solution that consists in the line that was common to all planes. The same kind of phenomenon was also observed in our research group, as will be commented later in this chapter. According to Sierpinska this kind of answer is due to a generalization from two to three dimensions in a synthetic-geometric manner, instead of an analytic one. In her study when the student was able to interpret the question in an analytic context, imagining the line as made up of infinitely many points, he gave the correct answer.

However, according to Sierpinska (2000), synthetic-geometric and analytic-structural modes are quite closely related, in their "*independence from a coordinate system*" (p. 236) and being "*based on properties, not calculations*" (p. 236). It must be added that these two modes of thinking are the most absent in current educational practices, analytic-arithmetic thinking reigning at all levels. One way to promote reasoning by properties in students might be to place emphasis on structural argumentation in synthetic contexts at lower educational levels. The research reported in this chapter brings to the front this aspect.

## 3 Aims of the Study

This chapter discusses the synthetic-geometric and analytic conceptions that students develop at different school levels about the notions *system of linear equations* and *solution of a system of linear equations*. Different positions that lines in a two-dimensional plane or planes in a three-dimensional space can have with respect to each other and the kinds of solution sets the respective systems have, provide a context in which the mathematical activities are carried out.

In our research group we have been studying the system of linear equations and solution concepts from different perspectives, including how they are constructed, associated difficulties in geometric and algebraic contexts as well as conceptions that students might develop. The intention of this chapter is to bring to the attention of an international audience insights from this data that has been collected in Spanish about the learning of these notions. The data analyzed for this chapter was collected over a ten year period with a fluctuating team of 3–7 researchers led by the author of this chapter, which presents a reanalysis of portions of this data that allow for insights into student understanding of systems of equations and their solutions across three different settings. Each of these settings was analyzed previously by a different student and relevant results from their work are cited. In addition, if a piece of data was reported previously in another study, the data will be identified by citing the thesis in which they appear first. Considering them together will also help offer an interpretation from the viewpoint of the research project, as to the causes of the obstacles experienced by students. It should be noted however that the interpretations reported in this chapter belong to its author and do not necessarily reflect the opinions of the authors of the theses in which the data appear, unless otherwise indicated by a citation. The findings reported here form part of a larger ongoing project about the understanding of Linear Algebra concepts in which different theoretical approaches are being employed, such as mental constructions, intuition, mathematical work spaces and representations. The framework used in this chapter is that of modes of thinking (Sierpinska, 2000) and it is chosen to explain the phenomena related to the learning of the notions in question. Although the reported data comes from studies conducted in Mexico and Uruguay, the observations might shed light on difficulties experienced by students in other parts of the world as well.

## 4 The Notions of *System* and *Solution* in Literature

Among the factors that have an influence on the conceptions that students develop about the notions of *system of equations* and *solution to a system of equations*, is their previously constructed knowledge about what an *equation* is and what a *solution to an equation* is. The understanding of variables and their different uses in turn has an impact on this knowledge (Borja-Tecuatl, Trigueros, & Oktaç, 2013). In this section by considering the results reported by two studies, one conducted at the middle school level and the other with undergraduate students, we get a glimpse into the consequences of the lack of these prerequisite constructions.

Panizza, Sadovsky, and Sessa (1999) ask whether students can conceive an equation with two unknowns on its own right, outside of a system of equations. The six middle school students who participated in their study had worked previously with equations with one unknown, linear functions and $2 \times 2$ systems, but not with a single equation in two unknowns. Five of them thought that an equation in two unknowns has a unique solution, but adopted a strategy that they tried to generalize from a method of solution for $2 \times 2$ systems. First they solved for one of the unknowns, then they substituted this expression for that unknown in the same equation, arriving at a tautology; at that point they could not give meaning to what they obtained. Only one student who formed a linear function from the equation was able to answer the question correctly. On the other hand when they were asked to solve a system of two linear equations in two unknowns with a unique solution, none of the students presented difficulties. They also said that a pair of numbers is a solution of the system if it satisfies each equation. However, when after a pair of students working together had found the unique solution to the system, they were asked if it was a solution for one of the equations in particular, their answer was negative; they contended that it was a solution "for both [equations] together, because it is a system" (Panizza et al., 1999, p. 458). The students were also asked specifically what would happen if they substituted the solution that they found in the first equation of the system, to which they answered that the solution to the system would not satisfy the equation. The authors conclude that "'the equation with two unknowns in a system' is a different object than 'the equation with two unknowns'" for these students and that "the equation with two unknowns is not recognized by the students as an object that defines a set of infinitely many pairs of numbers" (p. 459). If students have difficulty interpreting the concept of equation, it is more than likely that this will have repercussions on their conceptions about systems that are made up of equations.

DeVries and Arnon (2004) observe that for some university students the solution to an equation is the constant that appears on the right hand side of that equation, since it gives the 'result'. This is in line with what Kieran (1981) reports about a meaning attributed to the equation sign. According to these authors other students equate the notion of solution with the process of solving. They suggest from an APOS Theory perspective that the construction of the concept of solution would improve if it started with actions of substitution, in the sense that a number or a tuple is a solution to an equation or to a system of equations, if when substituted satisfies it. However this is usually not the approach taken in linear algebra courses, focusing instead on solution methods such as Gauss-Jordan algorithms which turn out to be difficult to interiorize, since it is hard to imagine their result without actually going through the steps to arrive at the solution (DeVries & Arnon, 2004).

Knowledge constructed about the prerequisite notions of system and solution to a system of equations permeates the understanding of students in subsequent years. We now turn to explain the general method employed in our research and then to examine how at different educational levels the topics under study are perceived by students and how through time some conceptions persist, despite many years of formal education.

# 5 Method

As it was mentioned before, the three studies on systems of linear equations and their solutions that were conducted in a sequence form part of a project that aims at understanding student difficulties with different topics of Linear Algebra, viewed from different perspectives. These studies in particular represent different phases that correspond to our understanding of a phenomenon at different educational levels. The modes of thinking approach in Linear Algebra (Sierpinska, 2000) provided a rich tool for investigating and explaining student interpretations, tendencies and ways of relating or avoiding particular ways of solving problems.

Our research on these topics started with a concern about university students' comprehension of them in varying representational settings, in particular when these settings interact with each other. The initial study was performed with seven students who worked collaboratively on problems while we observed them and registered their activity. The problems were designed with the purpose of identifying difficulties in synthetic and analytic modes as well as in connecting them. As a result of this first inquiry, we gained a general understanding about the matter and designed a more comprehensive questionnaire to be applied to a larger group of students.

Application of the new questionnaire and subsequent interviews with selected students revealed in a detailed manner the characteristics of synthetic thinking about systems of linear equations and their solutions when the equations represented lines in the plane or planes in the space; it also allowed us to observe more closely how the students interacted with synthetic and analytic modes when the question called for both.

Subsequently we wanted to know what happens at lower educational levels, namely at middle and high school, searching possible causes for the conceptions observed at the university level. This study led us to discover pedagogical strategies employed starting at the middle school level, which might explain partially where the roots of the issue lie.

# 6 Systems of Linear Equations at Different Educational Levels

In what follows we present the results of our study concerning the notions of system of linear equations and solution of a system of linear equations at middle school, high school and university levels. By doing that, we want to evidence the lasting effect that the initial context in which these notions are introduced has, on the conceptions that the students generate. We will see how, the way students first get acquainted with these notions at middle school permeates in their understanding throughout their studies up to university. It can be noted that although the populations are different, and they even come from different countries, the difficulties that they experience related to these notions are very similar. The particular studies are presented in a chronological order; after we observed difficulties at the university level we wanted to inquire into their origins and set up investigation at lower educational levels. Although the questionnaire as a whole differed from one educational level to another because of the mathematical maturity of the students involved and our level of understanding of the issues involved, some of the questions that are examined in this chapter were essentially the same at all levels and together they help explain a phenomenon related to the understanding of the concepts involved as well as allow a comparison in terms of the characteristics of their reasoning. In particular, the fact that the three modes of thinking presented in the theoretical framework section appear in students' reasoning at different levels makes this approach especially suitable for a study undertaken at different educational levels and with students of different mathematical maturity.

## *6.1 University*

In this section we present results concerning two groups of students coming from different universities in Mexico. All of the students' responses that appear in quotes are translations from Spanish.

### 6.1.1 First Group

In this preliminary study we worked with seven students from a public university in Mexico in a collaborative setting, asking them to solve some problems prepared previously. All the students were in their third semester of the nutrition engineering program; they had seen the topics of systems of equations and matrix algebra in their first year of studies in the Mathematics I course, but they had not taken Linear Algebra. They were chosen based on their interest in participating in the study.

The complete questionnaire applied to the students can be consulted in Appendix 1; here we will present the results from the first, second and fourth questions. First part of the first question asked them to draw in the Cartesian plane two lines whose equations were given and whose graphs coincided; in the second part they were supposed to solve a system corresponding to the graph, but the question did not specify the relationship between the graph and the system given algebraically. One of the students who used a row reduction method and arrived at a row of zeroes wrote: "The system of equations has no solution; because when solving we look for the points in which they intersect, and since it doesn't have a solution, it has infinitely many solutions" (Mora Rodríguez, 2001, p. 76). In this seemingly contradictory interpretation we can see two conceptions of solution: one as *the point of intersection geometrically* and the other as *the points that satisfy both equations*. It is as if the student is saying that since the system does not have the first kind of solution (a point of intersection), than it should have the second kind. Another student, after arriving at $0=0$ by substitution method said that the system had no solution, since "when trying to solve it we don't find the value for $x$ and $y$" (p. 78). Here we can observe the preconceived solution as *a unique ordered pair* and difficulty in interpreting the tautology. Two other students working together used the elimination method and arrived at $0=0$, after which one of them explained that "the system has an infinite solution, since when solving it you look for the point where the equations intersect and in this case the graphs intersect at all of their points" (p. 76). The other student on the other hand interpreted this result in the following way: "the system has no solution which means that they are the same or they are parallel" (p. 76). Two other students used determinants to solve the system and came to the conclusion that since the determinant is zero and the graphs coincide, there are infinitely many solutions. The remaining student made a mistake and arrived at a unique solution and did not comment on it. The conception of solution as an intersection point of two lines geometrically and as *the* ordered pair that results from applying some solution method works out in the case of a system with a unique solution, but fails otherwise. Furthermore lack of structural thinking makes it difficult to interpret the arithmetic results that do not consist in an ordered pair.

After the students worked on the problem, there was a discussion period in which the answers were compared. The comments made by the students during this discussion gave an idea of the conceptions that students have about solution as an intersection point or an ordered pair that results from the solving process: "When you solve a system of equations you always look for the intersection"; "When you solve a system of equations you always look for the $x$ and $y$ values"; "It is the same line and there is no solution because there are no $x$ and $y$ values"; "If it is the same equation and you have only one line then it does not have a solution, because there is no other line that intersects it"; "But when they ask you to solve it, how are you going to solve it? We need to have two equations, right?"; "We cannot work with the two equations because they are the same, that's why we cannot find values for $x$ and $y$, so we cannot have a system" (Mora Rodríguez, 2001, pp. 78–79). Some of these comments were made by students who had given the correct answer, which shows that they have doubts about the process.

The second question of the questionnaire was similar to the first one, with the exception that in this case the system corresponded to two parallel lines. After graphing the lines and solving the associated system, the students arrived at $0=1$; trying to interpret this result, one student wondered: "Could this 1 be the distance between the two lines?" (Mora Rodríguez, 2001, p. 83). Another student, who was trying to reconcile the visual information with the expression he obtained, asked: "Supposing that I don't know how to graph; how can I conclude the same thing with the analytic method?" (p. 84). Another student commented: "When there is no point of intersection you arrive at an inconsistency" (p. 84). We can see in these answers the attempts that students make in order to relate the geometric and analytic aspects of the solution set, however the lack of an immediate interpretation of the arithmetic result makes it difficult; actually the interpretation would require structural thinking based on properties of the objects involved.

The fourth question of the questionnaire showed the case of three lines intersecting two at a time in the Cartesian plane, forming a triangle. All but one student thought that there were three solutions to the system; the remaining student did not give an answer. One student wrote down the equations of the lines. Afterwards, using all three equations, he eliminated one of the unknowns and found the value of the other one. Next he substituted that value in each of the three equations to solve for the other unknown and thought that those three ordered pairs were the three solutions that he was looking for. This was an incorrect generalization of a solution method from a system with two equations to one with three equations. If he had graphed those points perhaps he would have realized that they all lied on a straight line and this did not coincide with the information given in the graph. However when they started to discuss among themselves about the meaning of a solution, three students forming a group came to the conclusion that the system had no solution. They explained that at the beginning they thought about three systems with a unique solution each, but they realized that the solution to the system should be at the intersection of all three lines.

This preliminary study gave us an idea about the kinds of difficulties that students might experience when learning the concepts of system and solution, as well as the conceptions that they might develop. We were especially intrigued by the answer of "three solutions" given by university students in the context of a graph containing three lines intersecting two at a time and wondered about the causes; the characteristics of the synthetic mode and its interpretation by the students seemed to be involved. Considering that reasoning in terms of properties is an important skill for university students to develop and that the synthetic mode provides opportunities in this direction, based on these observations we decided to apply a more extensive questionnaire to another group of university students in order to inquire more deeply into their conceptions about the notions of system and solution.

### 6.1.2 Second Group

We applied a questionnaire to all the 27 students enrolled in an introductory Linear Algebra course at a public university in Mexico, who were majoring in agricultural engineering and were in their first year of studies.

The questionnaire (Appendix 2) consisted in four parts. In the first part (questions 1 and 2) we included figures that showed different configurations of lines in two dimensions or planes in three dimensions and asked them to tell how many solutions a corresponding system of equations would have. This time the figures were presented without coordinate axes in order to focus on the properties of the figures rather than providing elements to students that would allow them to find the respective equations, as it had happened with the previous group. In this part the questions were presented in a synthetic mode, asking for an answer in an analytic mode. In the second part (question 3) we asked the students to come up with a system that might represent the information given in each figure. This way we wanted to see how the synthetic and analytic-arithmetic modes would be connected, employing properties related to the structural mode. The questions in the third part (questions 4 through 7) gave specific conditions in terms of the number of unknowns and solutions and asked if a system satisfying those conditions might exist. Here the student could employ the synthetic and structural modes. In the fourth part of the questionnaire (question 8) the students were asked to solve a system of three equations in two unknowns, representative of the triangular formation case mentioned earlier; the intention was to see what kinds of solution methods would be used. In general we were interested in the students' synthetic arguments, elements of structural thinking involved in the synthetic mode and in what ways the students could relate the synthetic mode to the analytic modes, both arithmetic and structural.

According to the responses given to the questionnaire we divided the students into three categories. In the first category there are 9 students whose answers provided clues about their conceptions regarding systems of linear equations and their solutions; only five of these students answered the questions concerning lines as well as planes, the remaining four answered only the questions about lines. In the second category there are 10 students whose answers to the first part were guided by geometric elements of the figures such as angle, perimeter and distance that did not have a direct relationship with what was being asked for. The third group consisted of 8 students who did not answer or gave responses that had nothing to do with the questions. The findings that we present here concern the first group of students for the first part of the questionnaire; for the second and third parts the second group is included as well.

In the first part of the questionnaire all 9 students said that there were three solutions in the case where three lines intersecting two at a time were shown; in the case of two parallel lines and one intersecting them, they thought there would be two solutions. The following was a typical explanation in the case of "three solutions": "Since there are three points of intersection then each point represents a solution for the system" (Cutz Kantún, 2005, p. 55). One student argued that the three solutions are found by solving the equations corresponding to two lines at a time. This reasoning resonates with the conception of a system as a collection of equations that are made up of subsystems of two equations, similar to the one that we will observe at the middle school level, result of relying exclusively on $2 \times 2$ systems in the initial teaching of the subject.

Four students thought that in the following case (Fig. 1) there is one solution, arguing in the following way: "It is the region in which the two figures intersect with each other" (Cutz Kantún, p. 43). One student thought that there are two solutions, considering the end points where the planes touch each other.

Three students said there are three solutions in the case shown in Fig. 2. One student said that there was no solution "because the three figures do not intersect at any point" (p. 43). Given that this student had answered "three solutions" for the case of three lines, his answer reflects a conception of solution as a single intersection point. The student who had said there were two solutions for Fig. 1, said here there are 6 solutions, using the same argument. Apart from the conception of

**Fig. 1** Cutz Kantún (2005, p. 244)

**Fig. 2** Cutz Kantún (2005, p. 245)

solution as the point of intersection of two lines, this reasoning also shows that the planes are being perceived as finite.

Another problem showed a line and the question specified that there were three lines; one student answered that there were no solutions and wrote: "There is only one line and there is no way to solve one equation without another, or there are several lines that are united but even then they cannot be solved" (p. 46). Similar to this response is the one where a student said there was no solution because there is only one equation. Yet another student who thought that there is no solution responded: "The line is only one, there is no other one intersecting it" (p. 62). This kind of response is related to the conception of solution as a point or region of intersection; if visually there is no intersection perceived, the case is associated to one with no solution. Also in play is the impression that there must be more than one equation in order for a solution to exist; this has to do with the solution methods that students practice. In case of three coincident lines, what is perceived synthetically might be very different from an analytic interpretation, where three unidentical (but equivalent) equations might be associated to the same line.

For the same question, another student said that "since it is only one line, then the solution would be unique" (p. 55). This student was perceiving the line synthetically as one entity, instead of analytically as an object consisting of infinitely many points. In the same line of reasoning another student commented the following: "the lines don't intersect, in this case they only unite in a parallel way" (p. 43) and considered that there is no solution.

The answers to the first part of the questionnaire indicate that figures given in a synthetic mode provoke answers in a synthetic mode even if the question demands an analytic answer, unless the student had the opportunity to practice different situations involving the two modes and reflect on the connections between them.

In the second part of the questionnaire students had lots of difficulties in providing the systems; usually the type of solution set did not coincide with the visual information given in the figures. There was confusion about parallel objects (no intersection) and objects that were coinciding. Some students used two unknowns to represent the planes. The answers to this part show the difficulty in transiting from a synthetic thinking to an analytic thinking. Although the answer is being asked for in an arithmetic language, structural thinking has to be employed to arrive at the result because of the properties of the mathematical objects involved.

In the third part of the questionnaire the correct response rate was very low; the systems that were provided by students did not comply with the given conditions in their majority. Some students even rejected the questions as plausible, arguing that they needed as many unknowns as the number of equations. The highest rate occurred in the case of some systems with unique solutions; part 4(a) had 7 correct responses, part 5(a) had 3 correct responses and part 6(a) had one correct response. Parts (b) combined of all four questions received 7 correct answers and part (c) had only one correct answer. These findings indicate that students have difficulties in making the visual information of the graphs correspond to the equivalent properties of equations, which implies once more the lack of connection between the three modes of thinking.

In the last question the solution methods varied greatly. One student solved the equations two by two, arriving at the three points of intersection. Another student added the three equations together, then formed a system with the resulting equation and the third equation, coming up with an ordered pair. Two students solved the first two equations for $x$ and substituted it in the third equation to find $y$. Another student used matrices and row reduction, but could not interpret the result. Yet another student solved each equation for $x$ and then substituted it in the same equation arriving at three tautologies. Three students solved for two equations and reported the result that they obtained. One student solved two of the equations and reported that the third equation is not compatible with the others, however did not mention that the system has no solution. Only one student gave a correct answer, solving two equations simultaneously and substituting the ordered pair in the third equation, observing that it does not satisfy it and concluding that there is no solution. It can be noted that the type of conception a student has of what a system is and what a solution is plays a determining role on the success with this question. A purely arithmetic thinking leads to applying some solution methods inappropriately, if it is not combined with structural reasoning.

Before the course started we had interviewed the instructor and in particular we had mentioned that students in general thought that there are three solutions in the case of the triangular configuration mentioned above. He had given us permission to apply the initial questionnaire before the course started and to interview some of the students when the course would be over. We shared the results of the questionnaire with him before he started the course. Afterwards we found out that during the course he put special emphasis on the "three solution" problem to make sure that the students would not commit the usual error.

As planned, after the course was completed five students were selected to be interviewed based on their questionnaire results. The interview questions were taken directly from the questionnaire with the exception of one question that was completely new. The intention was to inquire into the conceptions of students regarding the notions of system and solution. The new question showed two lines in a plane intersecting at one point, and asked if it was possible to draw another line so that the system of equations represented by the figure would have (a) two solutions, (b) three solutions and (c) no solution. Since in the questionnaire the students thought that three lines showing three points of intersection represented three solutions, we wanted them to reflect on the no solution case and we were curious about the drawings that they might produce. It should be noted that the new question had the intention to involve the student in the construction of a situation based on the properties that it should satisfy, hence promoting structural thinking in a synthetic context, rather than just observing a figure and giving an answer. This way we hoped that the connections between the different modes of thinking would be motivated more actively.

From the five students only one evidenced conceptions compatible with mathematical theory throughout the interview, in the sense that a system was viewed as consisting of equations with a common solution set; other students repeated the kinds of responses that were obtained from the questionnaire. Let's first discuss the findings for the new question.

One of the students produced the drawing shown in Fig. 3 for part (a) and a triangular figure for part (b), as we expected. His arguments evidenced that he was thinking that each pair of intersecting lines formed a system.

For part (c) he produced the drawing shown in Fig. 4 and said that he was thinking of a line that should not touch any of the given lines. When the interviewer asked what would happen when the lines are extended, the student said that he would make sure not to extend them that much. The interviewer insisted by asking if he was considering that the lines had finite length, to which the student responded that in the Cartesian plane he could make sure that the two given lines stay in one quadrant and the third line in another quadrant.

Another student read all the parts of the question and started commenting about them in a mixed order. He said that three solutions would be impossible to obtain, because in the case where the lines form a triangular shape there is no solution, since there is no point of intersection to all the lines; he added that this case would be an example of no solution. However something curious happened when he considered part (a); he produced the drawing shown in Fig. 5. Probably because of the instruction he received he had memorized that the case of three lines represented no solution, but when the problem was changed he went back to his initial conception. The interviewer tried to confront the student with the answers he gave to the different parts of the question. After insisting for a while on the conflicting answers, finally the student said that there is no solution in the figure he drew, but it is not clear whether he really understood where the problem lied.

Two other students said that in part (c) since the two lines have a point of intersection, the system already has a solution and it would be impossible to add another line to result in a no-solution case. Only one student responded in a satisfactory manner to all parts of the question.

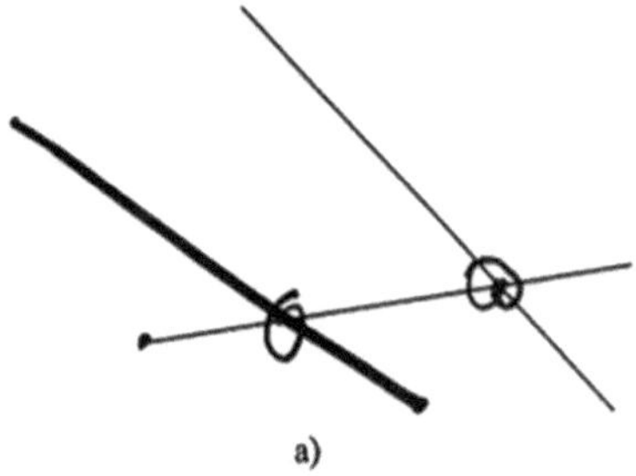


**Fig. 3** Cutz Kantún (2005, p. 103)

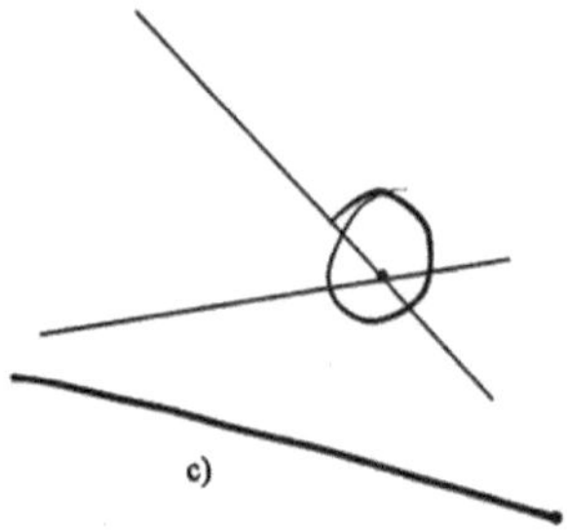


**Fig. 4** Cutz Kantún (2005, p. 105)

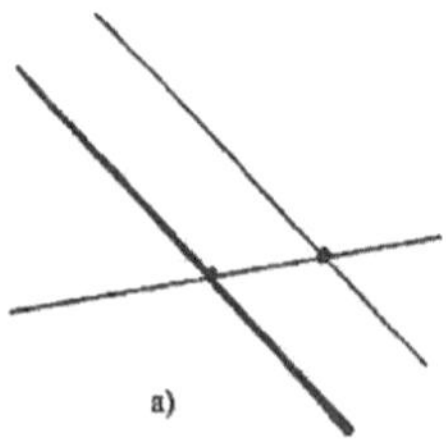


**Fig. 5** Cutz Kantún (2005, p. 109)

Four out of five students, when shown a figure similar to Fig. 1 but with three planes intersecting in a line (question 2–8 in Appendix 2), responded that there is a unique solution. One of those students explained his reasoning in the following manner, showing how he was generalizing from two to three dimensions: "Representing this graphically, since now they are planes, the point will not be seen as a point, rather as a line" (Cutz Kantún, 2005, p. 128). This is an evidence of thinking about the line as a synthetic object, as we also saw discussing Sierpinska's (2000) example. The remaining student said that the answer would be equal to the number of points included in the line segment shown in the figure where the planes intersect, but could not determine how many of those points there are.

When the students were asked to propose a system of equations representative of the figures that showed three lines in different relative positions, different kinds of difficulties emerged. Some students thought that they needed to use three unknowns because there were three equations. Others said that they could come up with a system if there were two lines instead of three. Some said that without the Cartesian coordinates they could not find a system, although the interviewer clarified that they should concentrate on the relative positions of the lines and the type of solution set. This also has to do with the fact that the questions were being presented in a synthetic mode emphasizing the visual properties of objects, with the intention for the students to reflect about the problems in an analytic-structural way, whereas the students thought about them in an analytic-arithmetic way (Sierpinska, 2000) emphasizing formulas and algorithms to find the equations.

From this discussion we can see that in general the underlying conception that the students have of a system is one of a set of two equations in two unknowns that are solved together, that is operating when they work on problems. The conception of solution also fits this scheme, being identified geometrically as the point of intersection of two lines, and algebraically as an ordered pair that satisfies both equations that form a system. This "definition" of a system naturally leads to affirming that three lines intersecting at three points represent a system with three solutions. These conceptions are very persistent even after instruction.

As we had mentioned earlier, after obtaining these results we wanted to see what happens at middle and high school levels with the teaching and learning of these concepts, hoping that conducting a study at those levels might help identifying the causes of the difficulties observed at the university level.

## 6.2 *Middle School*

At this school level the topic is usually introduced in the context of $2 \times 2$ systems, that is two equations in two unknowns. Different solution methods are presented to students: elimination, substitution and graphical, the last one in general receiving less attention than the other two. Students are expected to practice their solution skills with respect to these methods. Most often these systems have a unique solution and students' success is determined by their ability to arrive at that unique ordered pair. Less common but still used in teaching are word problems where students have to translate the given information to a system of equations and then solve it.

Normally students at this level are not given the opportunity to experiment with non-square systems and the conceptions generated in relation with the notion of *solution of system of equations* is strongly influenced by this initial context. Let's consider the following question presented to a group of middle school students in Uruguay who had studied the subject in the context described earlier (Fig. 6):

In the following figure three lines associated to a system of three first-degree equations in two unknowns are graphed. How many solutions does this system have? Justify your answer.

From the 22 students who answered the question, 12 responded that the system has three solutions; 7 said it does not have a solution and 3 gave other kinds of responses. Since the only kind of system with which these students had worked previously was one with two equations in two unknowns, their answers reflect the generalization that they had to apply to the case of a system of three equations in

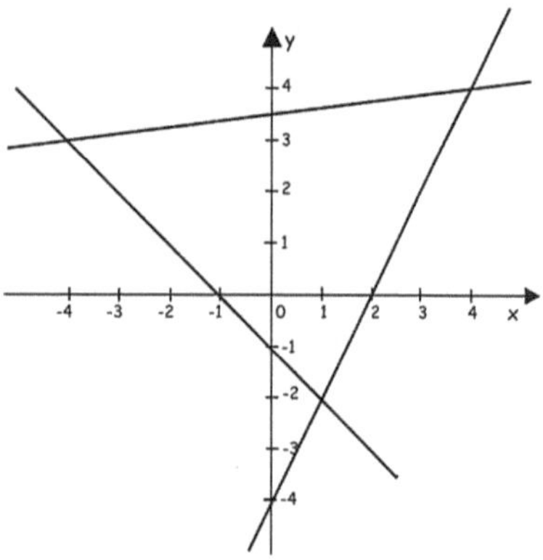


**Fig. 6** Ochoviet Filgueiras (2009, p. 82)

two unknowns. In some cases, as the responses of the 12 students indicate, this generalization was based on the idea that the intersection point of two lines is a solution to a system, no matter how many equations it is comprised of. It should be mentioned that some students, when asked what it means for a pair to be a solution to a system, correctly mention that it should satisfy all the equations in the system; some declare that this system does not have a solution, but that considering the equations two at a time, it does. This later kind of response indicates an attempt to generalize the previous knowledge to the new case, conserving the essential aspect —intersection of two lines as a solution—but combining it with the meaning of the solution of a system of equations, as a general notion.

The above discussion provides us with some elements as to the conceptions that the students generate for the concept of solution of a system of equations, in the context of instruction based on $2 \times 2$ square systems with emphasis on unique solution. The tendency to equate 'an intersection point of two lines' in a geometric context to 'a solution of a system of equations in which those two lines are represented' is very strong. Knowing this tendency can inform instructional design and form the basis of pedagogical ideas for dealing with this topic.

Following this initial exploration a questionnaire (Appendix 3) was designed and applied to two groups of middle school students after the topic of systems of linear equations was covered. The questionnaire involved 18 questions, the first one corresponding to the triangular configuration shown in Fig. 6. The last two questions asked what a system of equations is and what a solution to a system of equations is for the student. Table 1 shows the number of students who answered in one of the three ways the first question.

For students who responded that there are three solutions, the following explanation was typical: "In my opinion it has three solutions, because the lines intersect at three different points, which at least in systems of two equations indicate the solution" (Ochoviet Filgueiras, 2009, p. 147). Another kind of answer reflects the tendency to separate the bigger system into smaller $2 \times 2$ systems: "It has three solutions but each one is for two equations" (p. 164).

**Table 1** Types of response to the triangle configuration problem at the middle school level

| Answer to the first question | Number of students |
|---|---|
| Three solutions | 24 |
| No solution | 12 |
| Other answers | 12 |
| Total | 48 |

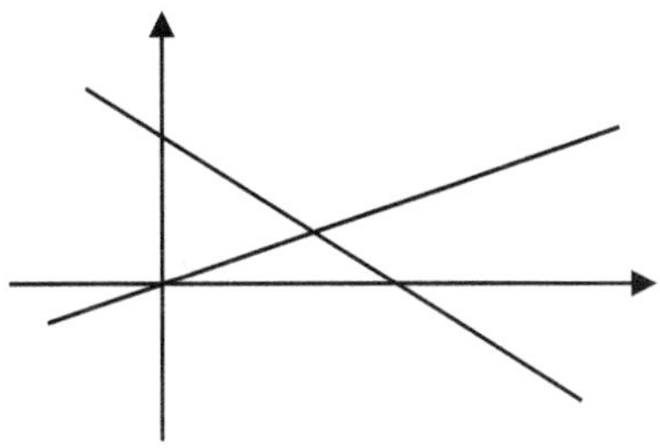

**Fig. 7** Ochoviet Filgueiras (2009, p. 159)

The sixth question in the questionnaire was the following: Can you put another line in Fig. 7 so that the system of equations associated to all the lines has no solution?

Some students, when they came to this question, answered it correctly and they also realized that their response to the first question was wrong. Something in this question made them reflect on the problem and on the meaning of a solution. One student, who had given the response of three solutions, answered this problem correctly; he explained what a system is and what a solution of a system is in the following way, underlying the word 'common': "A set of 2 or more equations to which you should find one (or more) common solutions" (Ochoviet Filgueiras, 2009, p. 159); "a solution of a system of equations is a common result of each one of the equations that there is in that system" (p. 160). We should point out that this question is practically the same as the extra question that was asked to the second group of university students during an interview, who did not show this kind of reflection. The difference is that those students had already constructed a conception about these notions, whereas the middle school students were doing it while solving this questionnaire. This shows the importance of designing situations that motivate transitions between modes of thinking, allowing the students to make the necessary connections.

Another question that had the same effect on some students was the following: Can a system of three linear equations in two unknowns have (a) a unique solution? (b) exactly two solutions? (c) exactly three? (d) infinitely many solutions? (e) no solution? Explain each answer and illustrate it by means of a graphical representation. One student said that when she had to do it herself she realized that she had given a wrong answer to the previous questions. This student also said that at the beginning she was thinking about two lines whose intersection point is the solution of the system as they had seen in class, which provoked her to give the answer of three solutions to the first question.

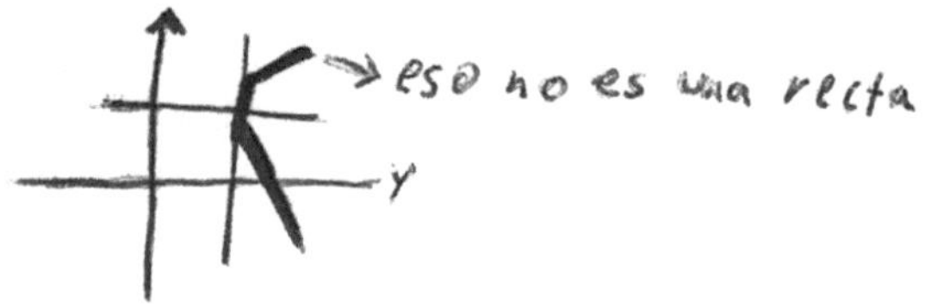


**Fig. 8** Ochoviet Filgueiras (2009, p. 141)

Reflecting about this phenomenon, we note that the triangle configuration system is presented in a synthetic-geometric mode, which provokes an intuitive response in students; the visual clue of three intersection points is quite strong in terms of associating them with intersection points as solution. Ochoviet Filgueiras (2009) sustains that the fact that in the sixth problem (Fig. 7) the third line is not given but it is the student who has to produce it provokes a more analytic thinking.

Some students used a synthetic reasoning to explain why a system of three linear equations in two unknowns cannot have exactly two solutions, as one of the questions demanded. One student drew the following graph and wrote "this is not a line", explaining it as follows: "A line either crosses at one point or all, infinitely many, if it crossed at two, it would not be a line" (Fig. 8).

Students were asked to state what a system of equations is for them and what a solution of a system of equations is. Some students who responded correctly to the triangle figure stating that the corresponding system has no solution, gave definitions that are in line with mathematical theory. For example one student said that a system is "a set of two or more equations in which you cannot look for an individual solution, but a common one for all" (Ochoviet Filgueiras, 2009, p. 143). The same student said: "A solution for me is a result or pair that satisfies all the equations that constitute the system at the same time" (p. 143). But this was not always the case; in general what the students stated as a definition was not helpful in guessing how the student would actually answer the remaining questions. There were students who, although gave correct definitions, did not apply them when answering the questions. And in the other direction some students who identified the notion of solution with points of intersection of lines when they were asked for a definition, responded the triangle configuration problem correctly, probably because they had it clear that they were referring to the points that were common to all the lines. This shows that teachers cannot rely on students citing definitions as an indication of learning.

Some students, although they answered 'three solutions' to the triangular configuration problem, in the algebraic context verified that there was no solution common to all equations and hence the system had no solution. This shows the influence of the mode in which the question is presented in how students reason about the problem. A synthetic mode provokes an immediate, intuitive answer based on the perceived properties of the figures, whereas an analytic mode can motivate analytic thinking and the two modes would not necessarily be connected by the student.

## 6.3 High School

The situation reported above where three lines intersecting two at a time being associated to a system of equations with three solutions was observed earlier in our research group at the high school level (Eslava & Villegas, 1998). It was also observed that many students interpreted geometrically the solution to a system of equations as the intersection of two lines, sometimes one of the lines being a coordinate axis.

The same questionnaire (Appendix 3) that was applied at middle school was also applied at the high school level in Uruguay. Analytically, the same phenomenon was observed when students, after completing instruction on general systems of linear equations (with varying number of systems and number of unknowns) took two equations at a time from a system with three equations, forming three smaller systems and solved them separately, arriving at three solutions (Ochoviet Filgueiras, 2009). These students had studied matrices, determinants and Cramer's method. We now present a Table 2 similar to the one for the middle school level, showing the number of students who responded in a particular way to the first problem of the questionnaire.

A phenomenon observed at the middle school level was witnessed here as well. Several students who answer graphically that there are three solutions to the triangle configuration problem, algebraically are conscious of the fact that in order to be a solution an ordered pair has to satisfy all three equations, and if there is no such pair, there is no solution to the system. As explained above, this kind of behavior might indicate the separation of the synthetic and analytic modes of thinking (Ochoviet Filgueiras, 2009). One student, in the geometric case reasons the following way: "This system has three solutions, since there are three intersection points. The system gives us the intersection points" (p. 174). For the equivalent problem given analytically she says: "This system does not have a solution, since the $x$ and $y$ values for the first two equations don't coincide with the ones for the third" (p. 174). For this student, a solution is a point of intersection of two lines synthetically and an ordered pair that satisfies all the equations analytically.

We also observed another phenomenon, noticed earlier at the university level, where the lines in the plane are imagined as having a finite length, that is, what is seen in the figure, a line segment, is what the line is. The same student mentioned in the above paragraph, when asked if it is possible to have a system of three equations in two unknowns with two solutions, produced the graph in Fig. 9.

Another student, knowing that the only possibilities for solutions for a system of linear equations are unique solution, no solution and infinitely many solutions, responded that the triangle configuration problem has infinitely many solutions, since it obviously did not have a unique solution because of the three points of intersection (more than one) and it was not the case of no solution, because there were points of intersection (Ochoviet Filgueiras, 2009). The only remaining option was infinitely many solutions.

**Table 2** Types of response to the triangle configuration problem at the high school level

| Answer to the first question | Number of students |
|---|---|
| Three solutions | 7 |
| No solution | 6 |
| Other answers | 8 |
| Total | 21 |

**Fig. 9** Ochoviet Filgueiras (2009, p. 175)

One student explained what a system is in the following way: "A set of equations that have unknowns, the system can have only one solution, none or infinitely many" and what a solution is in the following way: "It is the point in which the lines intersect if it has one solution, if it has infinitely many it is the same line. If it doesn't have any it is because the lines never intersect" (Ochoviet Filgueiras, 2009, p. 147). This student responded correctly the triangle configuration problem. Ochoviet Filgueiras explains that although the idea of a solution as an intersection point prevails in these descriptions, the structural thinking in which the types of solution sets are present and this interpretation is associated to the unique solution case is what causes the student to give correct answers; geometric and structural modes of thinking interact to give rise to a correct interpretation.

## 7 Discussion

Learning of a concept cannot be considered adequate if it only involves one type of representational system (Duval, 2006). Similarly thinking about the objects of Linear Algebra in only one mode restricts the understanding to only one aspect of the structure. The notion of system of linear equations is representable in at least two registers at lower dimensions that can be employed in designing problems starting at the middle school level. Making connections between them is essential to bring forth the characteristics of the mathematical objects involved. The findings reported in this chapter point out to some important educational phenomena in this direction, which have consequences for teaching practices as well as the conceptions that students develop as a result of the nature of the concepts involved and didactical strategies that are employed in designing instruction.

We have seen that starting with middle school where students are introduced for the first time to the systems of linear equations and their solutions, they begin developing a conception about systems consisting of two equations in two unknowns that are solved together to give an ordered pair as a result. As a consequence of this system conception, they associate the solution of a system with the ordered pair obtained by solving two equations simultaneously, and relate it to the intersection of two lines in the plane geometrically. Later they try to generalize these two conceptions acquired in two-dimensional plane to three-dimensional space, arriving at interpretations such as conceiving the line of intersection of planes as a special point (unique solution). Another type of generalization consists in thinking that a system should involve as many unknowns as equations. We posit that the tendency to separate a system with three equations in two unknowns into three square systems to solve them separately also has to do with these conceptions. The interpretation of a solution as a point of intersection of two lines works perfectly well in the $2 \times 2$ context; however it becomes an obstacle when the context changes.

We also observe that at all the educational levels that we studied, students were coming from a heavily emphasized analytic-arithmetic background, which meant algorithmic methods to solve systems of equations. The fact that only one student out of five who were interviewed at the university level showed progress toward understanding the meaning of a system of equations and was able to transit between different modes of reasoning as the problems required it, is evidence that even after a Linear Algebra course the first conceptions acquired in the context of $2 \times 2$ systems are difficult to modify and need specially designed instruction.

Ochoviet Filgueiras (2009) observed a middle school classroom when the topic of systems of linear equations was introduced, and at the end of the first class the following notion was institutionalized through interaction between the teacher and a student: "the solution of the system is the pair of numbers where the lines intersect each other" (p. 103). It is not surprising then, that the same student, when asked how many solutions there were in the figure where the lines formed a triangle in the plane, gave the following response: "three solutions: Because the lines intersect three times forming three points (solutions)" (p. 104). Although the students received instruction where the textbook and the teacher stated that a solution should satisfy all the equations of a system, for these students "all the equations" meant "all the two equations" and their conceptions were formed accordingly.

Synthetic-geometric, analytic-arithmetic and analytic-structural modes differ in nature. In this chapter we focused especially on the synthetic thinking and ways to promote its relationship with the structural mode, since these two modes are the least favored ones in the educational system, and paradoxically, structural mode being necessary for conceptual understanding. We detected the intuitive tendency of students to generalize from two dimensions to three dimensions in a certain way, if not guided in this process of discovery. We also saw that these intuitive generalizations persist through years up to the university level, with consequences for understanding the mathematics involved. We noted that the separation of the synthetic and analytic ways of reasoning in the student's mind can lead to inconsistent interpretations of different aspects of the notions involved, and because of the difference in nature of these modes students are not likely to realize it. We also point out the importance for teachers of noticing these phenomena, in order to develop suitable strategies for dealing with them in the classroom and helping with the learning process.

Further research on the topic might involve the conceptions related to parametric equations in the space; the conceptions related to other notions in Linear Algebra such as Linear Transformations within the framework of modes of thinking; how synthetic thinking can be carried out in higher dimensions; furthering the understanding about the relationships and transiting between different modes in different directions.

## 8 Didactical Suggestions

We suggest the inclusion of systems other than the $2 \times 2$ case when students are first being introduced to this topic, emphasizing the meaning of a system and a solution as well as exemplifying them in contexts such as $3 \times 2$ systems. We also suggest addressing the case of systems with infinitely many solutions, such as the ones composed of equivalent equations. Similarly, systems with different geometric positioning that give rise to "no solution" should be treated in order not to give the idea that the only way that no solution is obtained is when the lines or planes are parallel. All these suggestions are also important in order to help generalizing the two-dimensional situation to the three-dimensional space. $2 \times 2$ systems are too restrictive in that the case where a point appears as an intersection point but is not a solution cannot be exemplified (Ochoviet Filgueiras, 2009).

In order to help converge the synthetic and analytic interpretations of solution, we suggest that activities that require transiting from one mode to another be included in the design of instruction. In particular, activities where associations are made between 'line' and 'equation', between 'point that lies on all the lines' and 'point that satisfies all the equations' (Ochoviet Filgueiras, 2009) would be adequate. In particular we suggest providing students with the opportunity to pass from a geometric representation to an algebraic one; usually students are comfortable with drawing systems given as algebraic equations, but not vice versa. Here a synthetic context might prove useful, placing the emphasis on the relative positions of lines or planes, and the types of solution sets that can be obtained from the respective systems. This might help in developing a structural viewpoint, emphasizing relations and properties.

When students work with systems of equations, we suggest that even when an unknown such as $x$ cancels during the solution process, it is kept as $0.\,x$ instead of 0. So an expression like $0=0$ actually comes from $0x+0y=0$, which is easier to interpret in terms of the set of all ordered pairs that satisfy the equation. Similarly, the expression $0x+0y=8$ instead of $0=8$ shows in a clearer way that there is no ordered pair that satisfies it.

The different modes of thinking provoke different kinds of conceptions in students. For example in a drawing it might be immediate to see if there is a unique intersection point or infinitely many, whereas in an analytic context some steps might have to be taken before coming to a conclusion about the type of solution set involved. Articulating these interpretations and different ways of reasoning can motivate conceptions that take into account different aspects of the mathematical objects involved.

**Acknowledgements** I would like to thank Bonifacio Mora, Blanca Cutz, Cristina Ochoviet, Irving Alcocer, Carina Ramírez and Juan Guadarrama for their collaboration on parts of the project and data collection as well as the insights they brought to the project meetings.

## Appendix 1: Questionnaire Applied to the First Group of University Students

1. (a) Draw the graphs of the following equations on the same coordinate system: $4y = 3x - 5$ and $y = \frac{6x-10}{8}$
   (b) Compare the graphs and write down your comments.
   (c) Use any method that you may know to solve the following system of equations:

$$\begin{cases} 4y = 3x - 5 \\ y = \frac{6x-10}{8} \end{cases}$$

2. (a) Draw the graphs of the following equations on the same coordinate system: $y = \frac{2x+3}{4}$ and $12y = 6x + 10$
   (b) Compare the graphs and write down your comments.
   (c) Use any method that you may know to solve the following system of equations:

$$\begin{cases} 2x - 4y = -3 \\ -6x + 12y = 10 \end{cases}$$

3. Solve the following system of equations:

$$\begin{cases} hx + y = 1 \\ 4x - 2y = k \end{cases}$$

4. Given the following graphical representation of three lines, how many solutions does the system represented by these graphs have?

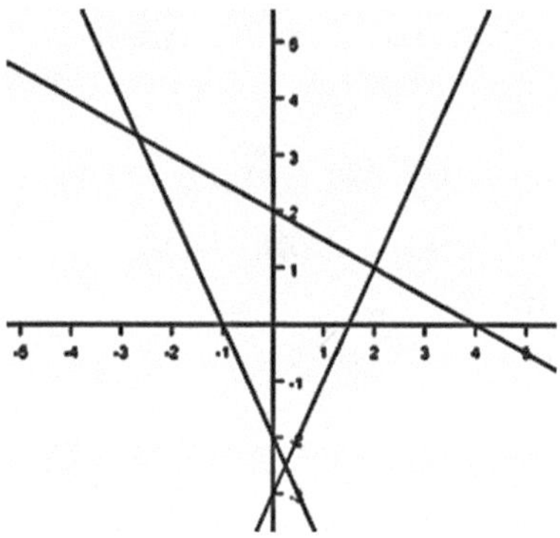

5. Solve the following system of equations:

$$\begin{cases} 4x - 3y = -1 \\ 2x + y = 4 \\ x - 3y = 3 \end{cases}$$

## Appendix 2: Questionnaire Applied to the Second Group of University Students

*Part I.* Question 1. Considering that in the first figure there are 2 lines in the plane and in all the other figures there are 3 lines, determine the number of solutions of the system of equations represented by each graph, what those solutions are and explain how you found them.

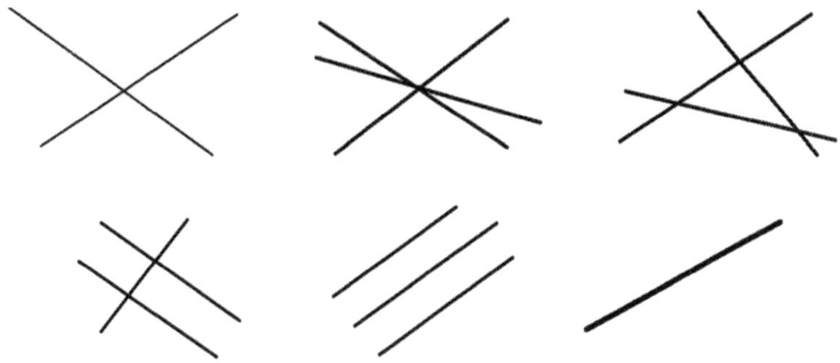

Question 2. Considering that in Fig. 7 there are 2 planes in the space and in all the other figures there are 3 planes, determine the number of solutions of the system of equations represented by each graph, what those solutions are and explain how you found them.

Figure 7

Figure 8

Figure 9

Figure 10

Figure 11

Figure 12

Figure 13

*Part II.* Question 3. For one of the previous figures write a system of equations that might represent it. (Although in the figures the coordinate axes were not included, you should have in mind the positions of the lines and planes with respect to each other).

*Part III.* Question 4. If it is possible, write a system of two equations in two unknowns so that it has:

(a) a unique solution (b) no solution (c) more than one solution

If it is not possible, explain why.

Question 5. If it is possible, write a system of three equations in three unknowns so that it has:

(a) a unique solution (b) no solution (c) more than one solution

If it is not possible, explain why.

Question 6. If it is possible, write a system of three equations in two unknowns so that it has:

(a) unique solution (b) no solution (c) more than one solution

If it is not possible, explain why.

Question 7. If it is possible, write a system of two equations in three unknowns so that it has:

(a) a unique solution (b) no solution (c) more than one solution

If it is not possible, explain why.

*Part IV*. Question 8. Solve the following system of equations:

$$\begin{cases} 4x - 3y = -12 \\ 2x + y = 4 \\ x - 3y = 3 \end{cases}$$

## Appendix 3: Questionnaire Applied to the Middle and High School Students

(1) The following figure shows lines associated to a system of three first degree equations in two unknowns. How many solutions does the system have? Why?

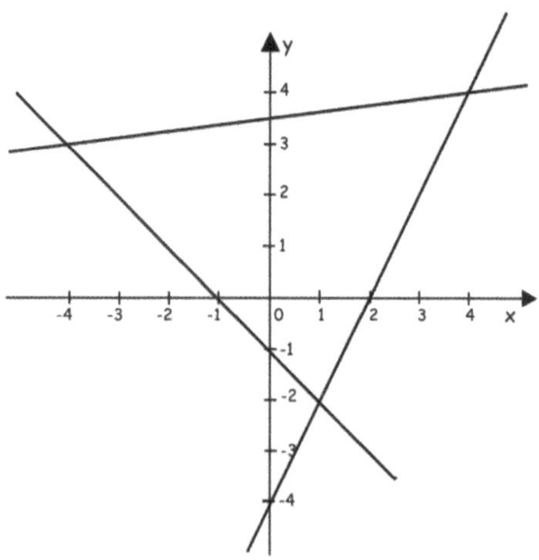

(2) The following figure shows lines associated to a system of four first degree equations in two unknowns. How many solutions does the system have? Why?

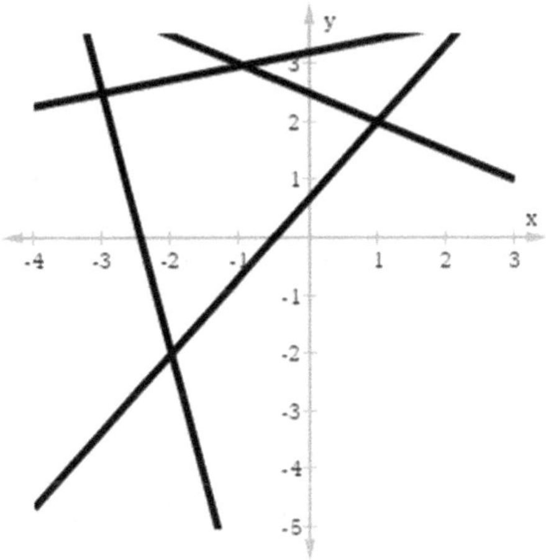


(3) The following figure shows lines associated to a system of three first degree equations in two unknowns. How many solutions does the system have? Why?

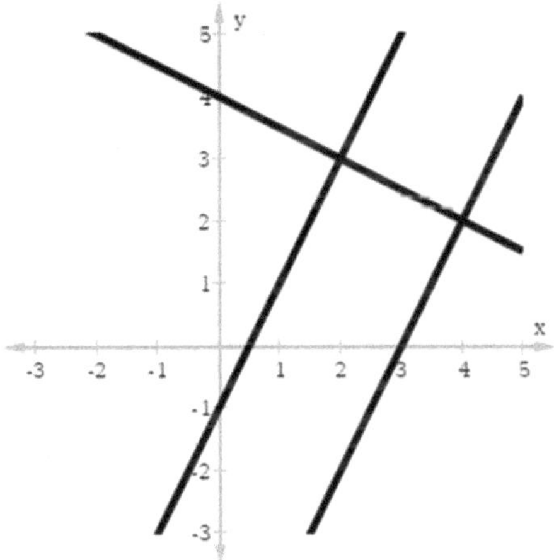


(4) Solve the following system of equations using the method you wish. Does the system have a solution? If your response is negative explain why and if it is affirmative indicate how many solutions there are and what they are.

$$\begin{aligned} x+y&=2\\ x-y&=8\\ x+2y&=4 \end{aligned}$$

(5) Can a system of three first degree equations in two unknowns have

(a) a unique solution?
(b) exactly two solutions?
(c) and exactly three?
(d) Can it have infinitely many solutions?
(e) And no solution?

Explain each one of your answers and illustrate it by means of a graphical representation.

(6) Can you put another line in the following figure so that the system of equations associated to all the lines has no solution? Explain your answer.

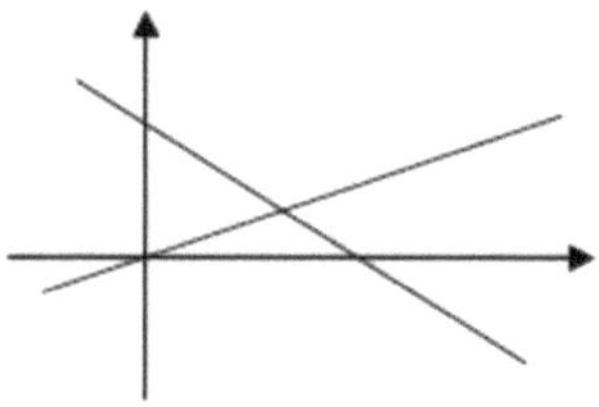

(7) Can you put three more lines in the following figure so that the system of equations associated to all the lines has a unique solution? Explain your answer.

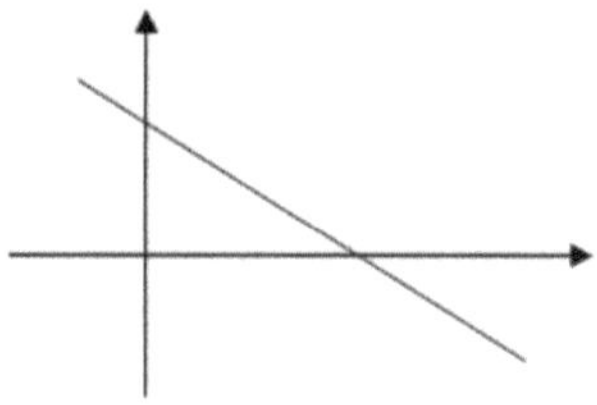

(8) Can you put another line in the following figure so that the system of equations associated to two lines has only two solutions? Explain your answer.

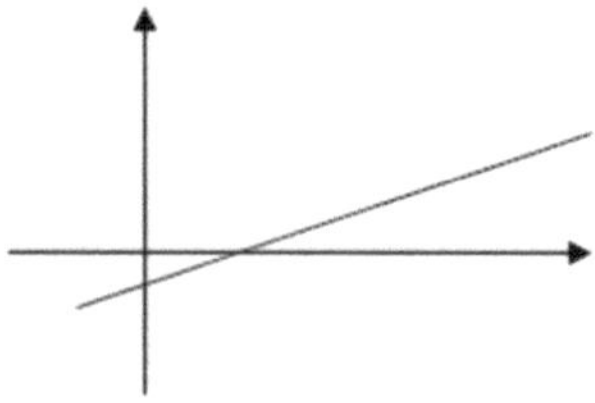

(9) Can you put another line in the following figure so that the system of equations associated to all the lines has infinitely many solutions? Explain your answer.

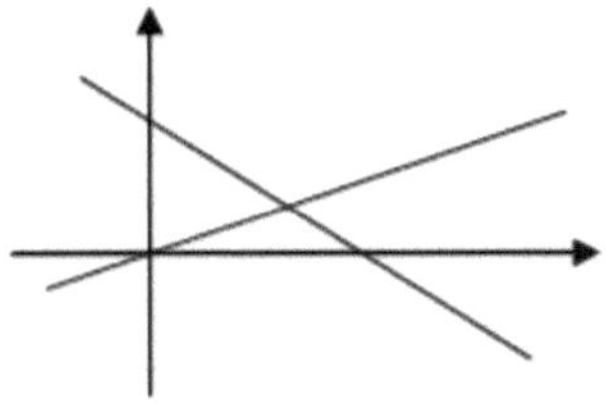

(10) Can you put two more lines in the following figure so that the system of equations associated to all the lines has infinitely many solutions? Explain your answer.

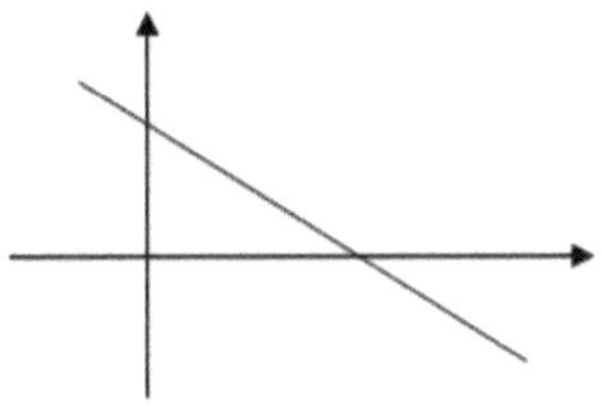

(11) Give a system of first degree equations whose unique solution is the ordered pair $(2, 1)$. Explain how you did it.

(12) Can a system of first degree equations have as a solution the ordered pair $(2, 1)$ and also other solutions? If your answer is negative explain why it is not possible and if it is affirmative give an example explaining how you obtain it.

(13) Can you put another line in the following figure so that the system of equations associated to them has the solution the ordered pair $(3, 4)$?

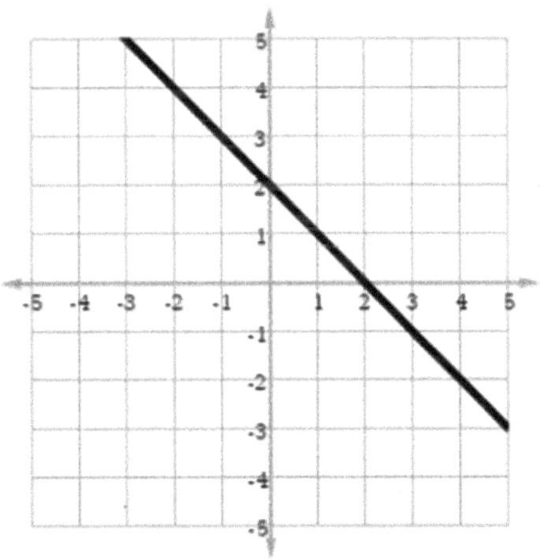

(14) Can you put another line in the following figure so that the system of equations associated to them has as its solutions only the ordered pairs $(-3, 2)$ and $(2, -1)$? Explain your answer.

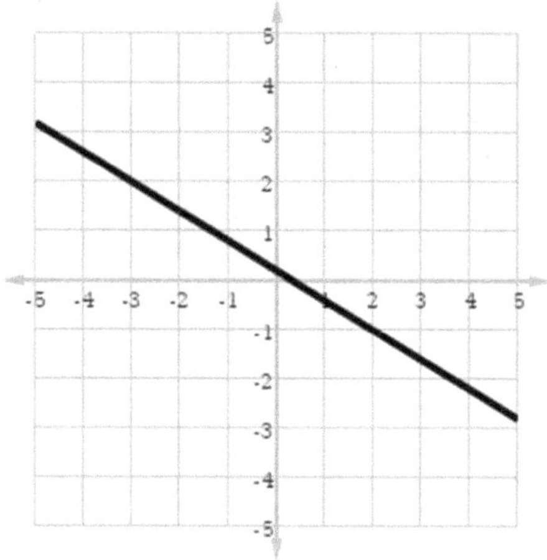


(15) Can you put another line in the following figure so that the system of equations associated to them has among its solutions the ordered pairs $(-3, 2)$ and $(2, -1)$? Explain your answer.

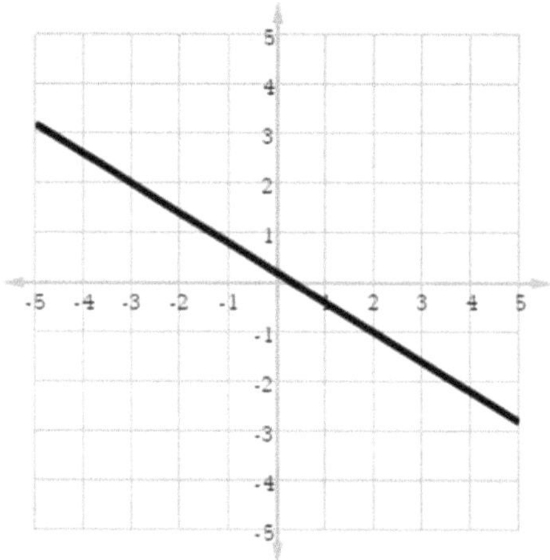


(16) Can a system of three first degree equations in two unknowns have

(f) a unique solution?
(g) exactly two solutions?
(h) and exactly three?
(i) Can it have infinitely many solutions?
(j) And no solution?

Explain each one of your answers and illustrate it by means of a graphical representation.[1]

(17) Explain what a system of equations is for you.
(18) Explain what a solution of a system of equations is for you.

## References

Borja-Tecuatl, I., Trigueros, M. & Oktaç, A. (2013). Difficulties in Using Variables—A Tertiary Transition Study. In S. Brown, G. Karakok, K. Hah Roh & M. Oehrtman (eds.), *Proceedings of the 16th Annual Conference on Research in Undergraduate Mathematics Education*, vol. 1, (pp. 80–94). Denver, Colorado.

Cutz Kantún, B. M. (2005). *Un estudio acerca de las concepciones de estudiantes de licenciatura sobre los sistemas de ecuaciones y su solución*. Unpublished masters' thesis. Cinvestav-IPN, Mexico.

DeVries, D. & Arnon, I. (2004). Solution-What does it mean? Helping linear algebra students develop the concept while improving research tools. *Proceedings of the 28th Conference of the International Group for the Psychology of Mathematics Education*, 2, 55–62.

Duval, D. (2006). A cognitive analysis of problems of comprehension in a learning of mathematics. *Educational Studies in Mathematics*, 61, 103–131.

Eslava, M. & Villegas, M. (1998). *Análisis de los modos de pensar sintético y analítico en la representación de las categorías de tres rectas en el plano*. Unpublished certification course thesis, Universidad Autónoma del Estado de Hidalgo, Mexico.

Kieran, C. (1981). Concepts Associated with the Equality Symbol. *Educational Studies in Mathematics*, 12, 317–326.

Mora Rodríguez, B. (2001). *Modos de pensamiento en la interpretación de la solución de sistemas de ecuaciones lineales*. Unpublished masters' thesis. Cinvestav-IPN, Mexico.

Ochoviet Filgueiras, T. C. (2009). *Sobre el concepto de solución de un sistema de ecuaciones lineales con dos incógnitas*. Unpublished doctoral thesis. Cicata-IPN, Mexico.

Panizza, M., Sadovsky, P. & Sessa, C. (1999). La ecuación lineal con dos variables: Entre la unicidad y el infinito. *Enseñanza de las Ciencias*, 17(3), 453–461.

Sfard, A. & Linchevski, L. (1994a). Between arithmetic and algebra: In the search of a missing link. The case of equations and inequalities. *Rend. Sem. Mat. Univ. Pol. Torino*, 52(3), 279–307.

Sfard, A. & Linchevski, L. (1994b). The gains and pitfalls of reification—The case of algebra. *Educational Studies of Mathematics*, 26, 191–228.

Sierpinska, A. (2000). On some aspects of students' thinking in linear algebra. In J.-L. Dorier (ed.), *On the teaching of linear algebra* (pp. 209–246). Dordrecht: Kluwer Academic Publishers.

Stadler, E. (2011). The same but different—Novice university students solve a textbook exercise. In M. Pytlak, T. Rowland & E. Swoboda (eds.), *Proceedings of the 7th Conference of European Researchers in Mathematics Education* (pp. 2083–2092). Rzeszow, Poland.

[1]Note that this question is repeated to see if the sequence of questions has made an effect on the students' conceptions.

# Rationale for Matrix Multiplication in Linear Algebra Textbooks

**John Paul Cook, Dov Zazkis and Adam Estrup**

**Abstract** Although matrix multiplication is simple enough to perform, there is reason to believe that it presents conceptual challenges for undergraduate students. For example, it differs from forms of multiplication students with which Linear Algebra students have experience because it is not commutative and does not involve scaling one quantity by another. Rather, matrix multiplication is a multiplication in the sense of abstract algebra: it is associative and distributes over matrix addition. Exposure to abstract algebra's general treatment of multiplication, however, usually occurs after students have taken Linear Algebra. This elicits the following question: How is matrix multiplication being presented in introductory linear algebra courses? In response, we analyzed the rationale provided for matrix multiplication in 24 introductory Linear Algebra textbooks. We found the ways in which matrix multiplication was explained and justified to be quite varied. In particular, two commonly employed rationalizations are somewhat contradictory, with one approach (isomorphization) suggesting that matrix multiplication can be understood from an early stage, while another (postponement) suggesting that it can only be understood upon consideration of more advanced concepts. We also coordinate these findings with the literature on student thinking in Linear Algebra.

**Keywords** Linear algebra · Matrix multiplication · Textbook analysis

## 1 Introduction

Matrix multiplication is foundational to many of the core concepts in introductory Linear Algebra. Indeed, results concerning systems of linear equations, span, linear (in)dependence, and linear transformations can all be (and often are) formulated in

J. P. Cook (✉) · A. Estrup
Oklahoma State University, Stillwater, USA
e-mail: cookjp@okstate.edu

D. Zazkis
Arizona State University, Tempe, USA



S. Stewart et al. (eds.), *Challenges and Strategies in Teaching Linear Algebra*, ICME-13 Monographs, https://doi.org/10.1007/978-3-319-66811-6_5

terms of matrix multiplication. This means that how conceptually coherent students understanding of matrix multiplication is has broad implications for their potential understanding of much of the Linear Algebra curriculum. Harel (1997), in his commentary on the undergraduate Linear Algebra curriculum, argued that "understanding must imply knowing *why*, not just *how*" (p. 111, emphasis ours). Matrix multiplication presents some conceptual challenges for students in this regard. For example, matrix multiplication is fundamentally different than the familiar multiplication of integers because it does not 'multiply' in the literal sense; moreover, it is generally not commutative. Instead, matrix multiplication is a 'multiplication' in the sense of abstract algebra: it distributes over (matrix) addition and is associative. This is an unfamiliar and non-intuitive idea for students who have had no prior exposure to abstract algebra. These conceptual peculiarities—along with the importance of matrix multiplication—bring to the fore the question of what approaches are available for motivating and explaining matrix multiplication to students. In other words, what rationale for matrix multiplication can be provided to students to help them overcome the challenges associated with this unfamiliar and non-intuitive operation?

One avenue of insight is to examine the rationale provided for matrix multiplication in introductory Linear Algebra textbooks. A quick glance at two popular textbooks suggests that there is substantial variation in rationale. For example, Kolman and Hill's (2007) initial presentation of matrix multiplication defined the matrix-matrix product AB using dot products of rows of A with columns of B. Noting the differences between matrix addition and matrix multiplication, the authors offered the following explanation:

> One might ask why matrix equality and matrix addition are defined in such a natural way, while matrix multiplication appears to be much more complicated. Only a thorough understanding of the composition of functions and the relationship that exists between matrices and what are called linear transformations would show that the definition of multiplication given previously is a natural one. These topics are covered later in the book. (Kolman & Hill, 2007, p. 24)

Kolman and Hill's (2007) connection to linear transformation is made 5 chapters after the above quote. Lay, Lay, and McDonald (2015), on the other hand, first introduced matrix multiplication in terms of the matrix-vector product Ax, which they defined as a linear combination of the columns of A. The authors justify the need for, and importance of, this definition in the following way: "a system of linear equations may now be viewed in three different but equivalent ways: as a matrix equation, as a vector equation, or as a system of linear equations. … you are free to choose whichever viewpoint is more natural" (p. 36). The connection to the matrix-matrix product is made several sections later in the same chapter and leverages the composition of linear transformations.

These two textbooks illuminate substantial differences in how matrix multiplication can be introduced, both in terms of the order of presentation and the rationale that is provided. Lay et al. (2015) go to considerable lengths to emphasize the utility of the matrix-vector product in terms of linear combinations of vectors, then use linear transformations to mathematically justify the matrix-matrix product in the first chapter of their text. On the other hand, Kolman and Hill (2007) define the matrix-matrix product outright with a cursory reference to linear transformations, the full treatment of which appears in Chap. 6. Moreover, there are drastic differences in the rationale these authors provided in their initial presentations of matrix multiplication. Kolman and Hill asserted that the ability to view their initial definition as natural depends on a thorough understanding of concepts that occur much later in the text and are presumably unfamiliar to students. In contrast, Lay et al. (2015) explained their initial definition in terms of concepts that occur earlier in the text and are presumably already familiar to students.

The contrast between the above examples provides impetus for the research questions that guided this study: are other introductory linear textbooks just as different? How else might matrix multiplication be motivated, justified, and explained? To these ends, we documented and analyzed the rationale for the definition(s) of matrix multiplication given in 24 introductory Linear Algebra textbooks. Our analysis revealed that there is indeed substantial variation in how matrix multiplication is rationalized, and that these rationalizations (as in the excerpts above) are not always entirely compatible. We conclude with a discussion of the potential pedagogical implications by coordinating our findings with research on student learning of matrix multiplication and, more generally, Linear Algebra.

## 2 Literature and Theory

Given that we are studying the presentation of matrix multiplication in textbooks, we first review literature pertaining to textbook analyses, including Harel's (1987) analysis of Linear Algebra textbooks. In particular, we detail Harel's framework for classifying the means by which textbook authors bridge the gap between students' existing knowledge and new content, which we operationalize in this study as a means to classify and analyze the rationale that the textbook authors provide for matrix multiplication. Research on student thinking in Linear Algebra appears later in the discussion section, in which we coordinate the results of our investigation with this body of literature.

### 2.1 *Textbook Analysis as an Avenue of Insight into Instruction*

Researchers have argued that textbook analysis can provide insight into how particular content is presented in mathematics classrooms (e.g. Reys, Reys, & Chavez, 2004; Robitaille & Travers, 1992). Though there is no guarantee that classroom instruction mirrors textbooks' content presentation, we note that textbooks "help teachers identify content to be taught [and] instructional strategies" (Thompson, Senk, & Johnson, 2012). Textbook analysis can thus be an efficient (though not comprehensive) means of gathering insight into classroom instruction. Harel (1987), for example, examined the differences in content presentation in Linear Algebra textbooks in order to "examine existing approaches to teaching" (p. 29); other textbook analyses that have been conducted at the undergraduate level include calculus (Weinberg & Weisener, 2011), combinatorics (Lockwood, Reed, & Caughman, 2016), and abstract algebra (Capaldi, 2012). Regarding the specific goals of our study, differences in content presentation are particularly revealing because of their potential influence on how students understand particular concepts (Bierhoff, 1996). Such differences also reflect "how experts in the field ... define and frame foundational concepts" (Lockwood et al., 2016, p. 9) and can therefore be useful for identifying key components of understanding these concepts.

## 2.2 Harel's Framework for Textbook Analysis

Harel (1987) conducted an analysis of Linear Algebra textbooks and reported differences in several respects; those particularly relevant to this study are differences in the sequencing of content and the justification provided for the introductory content. Our current study is distinct in two ways. First, Harel's analysis occurred nearly three decades ago (at the time of this writing), a period of time in which impactful attempts at nationwide Linear Algebra curriculum reform—such as the Linear Algebra Curriculum Study Group (Carlson, Johnson, Lay, & Porter, 1993; Harel, 1997)—were made. Presumably, these attempts, the broader changes in mathematics education, and the passage of time precipitated a different landscape of Linear Algebra textbooks than those studied by Harel. Second, Harel's study focused on general trends throughout entire textbooks (what might be called a *macro*analysis) and did not focus specifically on the different presentations of matrix multiplication (though he did affirm the status of matrix arithmetic in introductory Linear Algebra texts). Our current study, in contrast, focuses on the presentation of one specific concept (a *micro*analysis).

Harel's (1987) findings regarding sequencing of content and introductory content provide a useful framework with which to frame our analysis. Specifically, in this paper we document the ways in which the different forms of matrix multiplication are sequenced in order to gain insight into the structure of each textbook. Additionally, we used his characterization of the means by which textbooks rationalized their introductory content to classify the rationale for matrix multiplication for the textbooks in our study. He termed his classifications *isomorphization*, *postponement*, *analogy*, and *abstraction*.

- *Isomorphization* involves "[imposing] an isomorphism on two mathematical structures where one of these structures is familiar to the student" (p. 31). In other words, isomorphization introduces students to a new concept in a way that highlights how the new concept preserves the mathematical structure of a familiar concept. Because the current study focuses on introductory textbooks, we note that the term 'isomorphism' itself is not likely to be explicitly mentioned. As an example, Harel specified how some textbook authors motivated matrix multiplication by showing how it preserves the composition of linear transformations (we documented several cases of this exact use of isomorphization to justify the matrix-matrix product). Lay et al.'s (2015) rationale for the matrix-vector product (from our introduction) is another example of isomorphization because it highlights the structural equivalencies between linear systems, vector equations, and matrix equations.

- *Postponement* involves remarking on the necessity for and magnitude of a particular concept for which, in the authors estimation, such considerations are not yet clear to the student. Uses of postponement include cases for which the authors are depending on future, currently unfamiliar, ideas to temporarily justify the concept at hand. Kolman and Hill's (2007) justification for the matrix-matrix product (from our introduction) is an example of postponement because they argue that understanding matrix multiplication depends on understanding a future concept (linear transformations).
- *Analogy* is a technique used to demonstrate connections "between new ideas to be learned and familiar ones that are outside the content area of immediate interest" (Harel, 1987, p. 30). There are two types of analogies: mathematical, in which connections are made with a familiar mathematical concept, and real-world, in which connections are drawn between the new mathematical concept and an application to a real-world problem where the concept is relevant. Note that analogy is quite similar to isomorphization. Though we acknowledge that these classifications are certainly not disjoint, we reserved 'isomorphization' for instances of literal mathematical isomorphism that emphasize the preservation of mathematical structure; we used analogy for all other comparisons.
- *Abstraction* is a very similar strategy in which students are first introduced to general ideas in specific, familiar, and more concrete, situations. The most common example of abstraction, Harel noted, is when an entire concept is motivated by a small number of examples of that same general concept. For example, demonstrating the utility of and justification for a result related to matrix multiplication by initially focusing on specific cases with $2 \times 2$ matrices would be a use of abstraction. The distinction between abstraction and analogy is that abstraction invokes a specific example of the general concept to be learned, whereas an analogy involves the comparison of two different (albeit similar) concepts or situations.

In the following study being presented in this paper, we operationalized the framework above to classify the rationale textbook authors used to explain and motivate matrix multiplication.

## 3 Methods

### *3.1 Textbook Selection*

We narrowed our focus to introductory Linear Algebra textbooks because, as Harel (1987) reported, introductory textbooks have a strong, foundational emphasis on matrix arithmetic. We also restricted our search to textbooks written in English. To ensure that we did not omit any English-language textbooks currently in widespread use, we examined syllabi available online for introductory Linear Algebra courses at more than 106 Research-1 universities around the United States, conducted online searches of textbook provider websites, and examined the textbooks in our own respective university libraries. Notably, among the 106 Research-1 Linear Algebra syllabi that we examined, the most frequently appearing were Lay et al. (2015) (43), Bretscher (2012) (10), Leon (2014) (8), Strang (6), Kolman and Hill (5), Edwards and Penney (5), and Poole (4) (or previous versions of these textbooks). All other textbooks in our sample appeared 0, 1, or 2 times in this list. The texts appearing 0 times were those that we included in our sample via searches of online textbook provider websites or our own university libraries.

Furthermore, we focused on textbooks published within the past decade (at the time of this writing, since 2006) to obtain a more accurate snapshot of how matrix multiplication is being presented in today's Linear Algebra classrooms (though we did not exclude a textbook outside this range if we found evidence that it was in widespread use). Additionally, we omitted textbooks for which Linear Algebra was not the sole focus, such as those designed for courses in Linear Algebra and differential equations. Due to their propensity for introducing topics in very similar (if not identical) ways, we included at most one textbook from each author in our sample. Similarly, whenever possible, we examined the most recently published editions of textbooks, omitting all other releases. In some cases, however, we were not able to obtain access to the most recent edition and thus opted for the most recent edition available (e.g. Andrilli & Hecker, 2010). Overall, our sample included 24 introductory Linear Algebra textbooks. A complete list of the textbooks in our sample can be found at the beginning of Sect. 4, and their corresponding bibliographic information can be found in the "Bibliography of Textbooks" section following the references.

### *3.2 Procedure for Data Collection and Analysis*

In order to explore and contextualize the entirety of the rationale that textbooks presented for matrix multiplication, we decided that it was necessary to document the various forms of matrix multiplication in each text along with how they were sequenced before identifying, recording, and analyzing the associated rationale and justification. The data collection process began with using the table of contents and the index to identify the places where matrix multiplication appeared in each textbook, and then photocopying (or printing out) all pages in any section of a textbook in which matrix multiplication was defined, exemplified, or discussed. To ensure that our sample included all relevant presentations and discussions of matrix multiplication, this process was independently repeated by another researcher contributing to this project.

Next, we documented all forms of matrix multiplication featured in each textbook, the order in which they appeared, and any rationale the authors provided for the given forms. Our operational definition of rationale was broadly interpreted as any explicit attempt by the textbook author(s) to mathematically or pedagogically explain, justify, or demonstrate the purpose(s) or derivation of matrix multiplication. Each instance was classified using Harel's (1987) framework (isomorphization, postponement, analogy, abstraction); additional categories were created as necessary for rationale that did not conform to these four classifications (we adapted Harel's framework to include one additional category, detailed below: *computational efficiency*). We note that use of one strategy did not preclude use of another: as matrix multiplication is such a rich and connected concept, we allowed for the possibility (or, perhaps, probability) that textbooks would employ multiple strategies to communicate their rationale for this important concept. We also must acknowledge that, because we did not examine every page of the textbooks in our study, we cannot discount the possibility that we inadvertently omitted particular subtleties related to the sequencing of and rationale for matrix multiplication in certain textbooks. As such, the documentation in this paper should be regarded only as affirmation that these types of rationale do indeed appear in the textbooks in which they are cited and referenced. The absence of attribution of a type of rationale to a particular textbook does not necessarily imply that the textbook in question does not employ that type of rationale. For example, we will often use parenthetical citations to provide examples of textbooks employing the rationale in question; a citation like (e.g. Bretscher, 2012; Shifrin & Adams, 2010) means only that we documented such rationale in Bretscher's, and Shifrin and Adams's respective textbooks, not that these were the only texts in which this form of rationale appeared.

We used constant comparison (Creswell, 2007, 2008) of textbook materials to identify common themes across the data set, including common sequences for the forms of matrix multiplication and commonalities in rationale both within and across sequences. The final stage of our analysis—appearing in Sect. 5—was to triangulate the rationale that textbooks provided for matrix multiplication with the relevant literature on teaching and learning of Linear Algebra.

## 4 Results

There are many different, yet equivalent, forms of matrix multiplication (see, for example, Carlson, 1993). During our analysis we used *constant comparison* (Creswell, 2007, 2008) across textbooks to identify four primary forms of matrix multiplication that emerged prominently across the textbooks in our sample. In the descriptions below,[1] $A$ is a $m \times n$ matrix with column vectors $\mathrm{col}_1(A), \ldots, \mathrm{col}_n(A)$ in $\mathbb{R}^m$ and row vectors $\mathrm{row}_1(A), \ldots, \mathrm{row}_m(A)$ in $\mathbb{R}^n$; $x$ is a vector in $\mathbb{R}^n$ with components $x_1, \ldots, x_n$. Additionally, $B$ is an $n \times p$ matrix with column vectors $\mathrm{col}_1(B), \ldots, \mathrm{col}_p(B)$ in $\mathbb{R}^n$ and row vectors $\mathrm{row}_1(B), \ldots, \mathrm{row}_n(B)$ in $\mathbb{R}^p$. Up to mathematical equivalence, these forms of matrix multiplication are:

- ***Ax as LCC***: The matrix-vector product Ax as a linear combination of the columns of A: $Ax = x_1\mathrm{col}_1(A) + \cdots + x_n\mathrm{col}_n(A)$.
- ***Ax as DP***: The matrix-vector product Ax as a dot product of the rows of A with the column vector x: $Ax = \begin{bmatrix} \mathrm{row}_1(A) \cdot x \\ \vdots \\ \mathrm{row}_m(A) \cdot x \end{bmatrix}$.
- ***AB as DP***: The matrix-matrix product AB determined by dot products of row/column vectors: AB is the matrix in which the entry in row $i$, column $j$ (where $1 \leq i \leq m, 1 \leq j \leq p$) is given by: $\mathrm{row}_i(A) \cdot \mathrm{col}_j(B)$.
- ***AB as [Acol(B)]***: The matrix-matrix product[2] AB as a matrix whose columns are determined by the action of A on the columns of B: $AB = \left[A\mathrm{col}_1(B)|\cdots|A\mathrm{col}_p(B)\right]$.

Early in our analysis, we noticed that the form of matrix multiplication that appears first in a textbook affords necessary context for the rationale that the authors provide (both for the initial form and those that follow). Thus, we sorted the textbooks into three categories—based upon the form of matrix multiplication that was first introduced—as a means of contextualizing the rationale:

- Sequence 1: initiating with Ax as LCC (Ax as a linear combination of the columns of A);
- Sequence 2: initiating with Ax as DP (Ax as a dot product of the rows of A with the column vector x); and
- Sequence 3: initiating with AB as DP (AB as a dot product of the rows of A with the columns of B).

We found no evidence of any textbooks introducing matrix multiplication with the remaining form of matrix multiplication (AB as Acol(B)), though it often appeared after matrix multiplication had been presented in another way. Along these lines, our analysis revealed additional variations in sequencing beyond these textbooks' initial forms of matrix multiplication. For example, though 6 of the 9 textbooks in Sequence 1 proceeded in a similar way, there were two additional variations with respect to the order of the remaining forms, which we call Sequence 1a, Sequence 1b, and Sequence 1c (see Table 1). In Sequence 1b, Ricardo (2009), for example,

[1]Our descriptions given here are not necessarily identical to those given in each textbook but are instead offered as summaries of these methods that are mathematically equivalent.

[2]Methods for multiplying using block/partitioned matrices, if formally addressed in a textbook, typically appeared along with this form of the matrix-matrix product.

**Table 1** Textbooks in which Ax as LCC is the initial form of matrix multiplication

| First | Second | Third | Fourth | Textbooks exhibiting this sequence |
|---|---|---|---|---|
| Ax as LCC | Ax as DP | AB as [Acol(B)] | AB as DP | *Sequence 1a*<br>Cheney and Kincaid (2012)<br>Lay et al. (2015)<br>Nicholson (2013)<br>Solomon (2014)<br>Spence, Insel, and Friedberg (2007)<br>Strang (2009) |
| | | AB as DP | AB as [Acol (B)] | *Sequence 1b*<br>Ricardo (2009) |
| | AB as [Acol (B)] | AB as DP | – | *Sequence 1c*<br>Beezer (2015)<br>Holt (2012) |

**Table 2** Textbooks in which Ax as DP is the initial form of matrix multiplication

| First | Second | Third | Fourth | Textbooks exhibiting this sequence |
|---|---|---|---|---|
| Ax as DP | Ax as LCC | AB as [Acol(B)] | AB as DP | *Sequence 2a*<br>Bretscher (2012) |
| | | AB as DP | AB as [Acol(B)] | *Sequence 2b*<br>Hefferon (2008)<br>Leon (2014)[a] |
| | AB as DP | AB as [Acol(B)] | – | *Sequence 2c*<br>Shifrin and Adams (2010) |
| | | Ax as LCC | AB as [Acol(B)] | *Sequence 2d*<br>DeFranza and Gagliardi (2015) |

[a]Leon (2014) makes brief reference in one sentence to AB as [Acol(B)] before presenting it in terms of dot products, but we classify this textbook in Sequence 2b because AB as [Acol(B)] does not feature prominently until much later

proceeds in the following order (left to right in the table): (1) Ax as LCC, (2) Ax as DP, (3) AB as DP, (4) AB as [Acol(B)]. In Sequence 1c, Beezer (2015) proceeds in the following order: (1) Ax as LCC, AB as [Acol(B)], and (3) AB as DP. Tables 2 and 3 for Sequences 2 and 3 can be read similarly.

The 5 textbooks following Sequence 2 (initiating with Ax as DP) exhibited considerably more variation.

The textbooks in Sequence 3 were also quite varied.

We now shift to documenting and analyzing the rationale—both mathematical and pedagogical—that these textbooks offered for matrix multiplication and the way in which they presented it. We found examples of all four of Harel's (1987) classifications of rationale: *isomorphization*, *postponement*, *analogy* (both mathematical and real-world), and *abstraction*. We also documented examples in which a form of matrix multiplication was introduced for the purposes of *computational*

**Table 3** Textbooks in which AB as DP is the initial form of matrix multiplication

| First | Second | Third | Fourth | Textbooks exhibiting this sequence |
|---|---|---|---|---|
| AB as DP | Ax as DP | Ax as LCC | – | *Sequence 3a*<br>Larson (2016) |
| | | | AB as [Acol(B)] | *Sequence 3b*<br>Edwards and Penney (1988)<br>Kolman and Hill (2007)<br>Poole (2014) |
| | | – | – | *Sequence 3c*<br>Robinson (1991)<br>Venit, Bishop, and Brown (2013) |
| | Ax as LCC | Ax as DP | – | *Sequence 3d*<br>Anthony and Harvey (2012) |
| | AB as [Acol(B)] | Ax as DP | Ax as LCC | *Sequence 3e*<br>Anton and Rorres (2014)<br>Williams (2012) |
| | | Ax as LCC | Ax as DP | *Sequence 3f*<br>Andrilli and Hecker (2010) |

*efficiency*, a classification occurring frequently enough amongst our sample to warrant adapting Harel's framework.

## 4.1 *Isomorphization*

We documented two distinct examples of isomorphization: (1) rationalizing the matrix-vector product by identifying the advantages of equivalently reformulating linear systems and/or vector equations as matrix equations, and (2) framing the matrix-matrix product as an operation on matrices that preserves the composition of the corresponding linear transformations. Each of these is explained in detail below.

### 4.1.1 Reformulating Linear Systems and/or Vector Equations as Matrix Equations

Highlighting equivalencies between the matrix equation Ax = b and both systems of linear equations and vector equations was particularly prominent across each of the textbooks in Sequence 1; it also appeared in textbooks in Sequences 2 (e.g. Bretscher, 2012; Leon, 2014; Shifrin & Adams, 2010) and 3 (e.g. Edwards & Penney, 1988; Larson, 2016; Poole, 2014). It is important to note that the textbooks in Sequences 1 and 2 typically invoked isomorphization to accompany their initial definition of matrix multiplication, whereas the textbooks in Sequence 3 invoked isomorphization for forms of matrix multiplication presented after their initial treatment.

Indeed, a matrix equation preserves the algebraic structure of its corresponding linear system and vector equation, allowing these authors to link the matrix-vector product (in the form of a matrix equation) with a familiar idea (systems of linear equations). The following excerpt (adapted from Lay et al. (2015, p. 36) typifies this approach:

For example, the system

$$\begin{aligned} x_1 + 2x_2 - x_3 &= 4 \\ -5x_2 + 3x_3 &= 1 \end{aligned} \tag{1}$$

is equivalent to

$$x_1\begin{bmatrix}1\\0\end{bmatrix} + x_2\begin{bmatrix}2\\-5\end{bmatrix} + x_3\begin{bmatrix}-1\\3\end{bmatrix} = \begin{bmatrix}4\\1\end{bmatrix} \tag{2}$$

As in Example 2, the linear combination on the left side is a matrix times a vector, so that (2) becomes

$$\begin{bmatrix}1 & 2 & -1\\0 & -5 & 3\end{bmatrix}\begin{bmatrix}x_1\\x_2\\x_3\end{bmatrix} = \begin{bmatrix}4\\1\end{bmatrix} \tag{3}$$

Equation (3) has the form $A\mathbf{x} = \mathbf{b}$. Such an equation is called a **matrix equation**, to distinguish it from the vector equation such as is shown in (2).

Many textbooks called explicit attention to these equivalencies. Beezer (2015), for instance, formulated this equivalence in terms of the solution(s), stating that "every solution to a system of linear equations gives rise to a linear combination of the column vectors of the coefficient matrix that equals the vector of constants" (p. 175). Other such instances include the following:

- "By now we are comfortable with translating back and forth between vector equations and linear systems. … Ax = b is a compact form of the vector equation $x_1a_1 + x_2a_2 = b$, which in turn is equivalent to [a] linear system" (Holt, 2012, p. 63).
- "We can use these new concepts to understand a system of equations Ax = b. If A and b are given, such a system challenges us to determine whether b is in the span of the columns of A and, if so, to find the coefficients needed to express b as a linear combination of the columns of A" (Cheney & Kincaid, 2012, p. 42).

- "For a linear equation with $n$ unknowns of the form $a_1x_1 + a_2x_2 + \cdots + a_nx_n = b$ if we let $A = [a_1a_2 \ldots a_n]$ and $x = \begin{bmatrix} x_1 \\ x_2 \\ \vdots \\ x_n \end{bmatrix}$ and define the product Ax by $Ax = a_1x_1 + a_2x_2 + \cdots + a_nx_n$ then the system can be written in the form Ax = b" (Leon, 2014, p. 31).
- "We reiterate that a solution x of the system of equations Ax = b is a vector having the requisite dot products with the row vectors $A_i$" (Shifrin & Adams, 2010, p. 39).
- "The initial purpose of matrix multiplication is to simplify the notation for systems of linear equations" (Edwards & Penney, 1988, p. 35).
- "Then the matrix equation $AX = B$ is equivalent to the linear system … Here is further evidence that we got the definition of the matrix product right" (Robinson, 1991, p. 10).

Many of these texts also argued that multiple representations afford flexibility with respect to selecting a problem-solving approach. For example, Nicholson (2013) stated that "a change in perspective is useful because one approach or the other may be better in a particular situation … there is a choice" (p. 45). Strang (2009), noting that a linear systems (rows) approach is easy to visualize for a $2 \times 2$ case but exceptionally difficult (if not impossible) to visualize for higher dimensions, stated that his "own preference is to combine column vectors. It is a lot easier to see a combination of four column vectors in four-dimensional space, than to visualize how four hyperplanes might possibly meet at a point. (*Even one hyperplane is hard enough …*)" (p. 33, emphasis in original). Interestingly, these arguments align with the arguments in the literature about the importance of being able to move flexibly between multiple representations in Linear Algebra (e.g. Dorier, 2000; Harel, 1997; Larson & Zandieh, 2013), which we discuss further in Sect. 5.

### 4.1.2 Framing the Matrix-Matrix Product in Terms of Preserving the Composition of Linear Transformations

The second example of isomorphization involved framing the matrix-matrix product in terms of preserving the composition of the corresponding linear transformations. Textbooks in Sequences 1 (e.g. Cheney & Kincaid, 2012; Holt, 2012) and 2 (e.g. Bretscher, 2012; Hefferon, 2008) invoked this rationale. Textbooks in Sequence 3, on the other hand, typically treated linear transformations as tangential, rather than interrelated, at this early stage, opting to delay more formal treatments until later in the text.

The isomorphization approach in Sequences 1 and 2 typically proceeded as follows: if $T_A$: $R^n \rightarrow R^m$ is a linear transformation with $m \times n$ matrix $A$ and $T_B$: $R^p \rightarrow R^n$ is a linear transformation with $n \times p$ matrix $B$, then the $m \times p$ matrix $AB$

is defined so that it is the matrix of the linear transformation ( $|A \circ T_B)^T$: $R^p \to R^m$; that is, so that $(T_A \circ T_B)(\vec{x}) = (AB)\vec{x}$, where $\vec{x}$ is a vector in $R^p$. In their accompanying explanations, the textbook authors attempted to make the equivalence between composition of transformations and matrix multiplication clear:

- "When a matrix B multiplies a vector x, it transforms x into the vector Bx. If this vector is then multiplied in turn by a matrix A, the resulting vector is A(Bx). … Thus A(Bx) is produced from x by a *composition* of mappings—the linear transformations studied [previously]. Our goal is to represent this composite mapping as multiplication by a single matrix, denoted by AB, so that A (Bx) = (AB)x" (Lay, Lay, and McDonald, 2015, p. 96).
- "Because x was an arbitrary vector in $R^n$, this shows that $T_A \circ T_B$ is the matrix transformation induced by the matrix $[Ab_1, Ab_2, \ldots, Ab_k]$. This motivates the following definition" (Nicholson, 2013, p. 57).
- "The definition of matrix multiplication was framed precisely to make this equation valid" (Cheney & Kincaid, 2012, p. 152).
- "The matrix of the linear transformation $T(x) = B(Ax)$ is called the *product* of the matrices $B$ and $A$, written as $BA$" (Bretscher, 2012, p. 77, emphasis in original).
- "The matrix representing $g \circ h$ has the rows of $G$ combined with the columns of $H$" (Hefferon, 2008, p. 226).

It is worth noting that, because matrix multiplication appeared early in most of these textbooks, such an approach necessitated that linear transformations also be treated early. Another approach centered on the similar task of finding one matrix $C$ such that $A(Bx) = C(x)$ but without formally treating linear transformations first (e.g. DeFranza & Gagliardi, 2015; Shifrin & Adams, 2010; Strang, 2009). This alternative approach either avoided invoking linear transformations or made minimal references to them in passing (often with a note that a full treatment of linear transformations would follow in a subsequent chapter/section). For example, DeFranza and Gagliardi (2015) motivated matrix multiplication in a way that strongly suggested the relevance of linear transformations, asking "is there a single matrix which can then be used to transform the original vector $\begin{bmatrix} 1 \\ 3 \end{bmatrix}$ to $\begin{bmatrix} 4 \\ 1 \end{bmatrix}$?" (p. 30). Shortly thereafter, they remarked that "the notion of matrices as transformations is taken up again in Chap. 4" (p. 31). Several other textbooks—particularly those in Sequence 3—also opted for this tangential reference to the importance of linear transformations, presumably to provide some insight into the mathematical structure of matrices and linear transformations without committing to a formal treatment so early in the text (e.g. Anton & Rorres, 2014; Shifrin & Adams, 2010; Williams, 2012). The methods that such textbooks employed for justifying the associative law—which the above textbooks achieved by leveraging the associativity of composing linear transformations—utilized different strategies, including analogy and abstraction (and are thus detailed in subsequent sections).

## 4.2 Postponement

We found examples of postponement across all three sequences. The textbooks employing postponement in Sequences 1 (e.g. Beezer, 2015; Spence, Insel, & Friedberg, 2007) and 2 (e.g. Bretscher, 2012; Shifrin & Adams, 2010) primarily did so to explain the future utility of examining vector equations and linear combinations of vectors (by means of characterizing the matrix-vector product as a linear combination of column vectors). For example:

- "[This definition] is frequently the most useful for its connections with deeper ideas like the null space and the upcoming column space" (Beezer, 2015, p. 182).
- "We now make an observation that will be crucial in our future work: the matrix product Ax can also be written as [a linear combination of the columns of A]" (Shifrin & Adams, 2010, p. 53).
- "Note that the product Ax is the linear combination of the columns of A with the components of $\vec{x}$ as the coefficients … Take a good look at this equation, because it is the most frequently used formula in this text. Particularly in theoretical work, it will often be useful" (Bretscher, 2012, p. 31).

Note that many of these textbooks also employed isomorphization, similar to what we discussed in the previous section, in order to motivate linear combinations; thus, these texts are attempting to justify the present importance of linear combinations and vector equations (as an alternative viewpoint on linear systems) while also emphasizing their future importance.

Postponement was particularly prominent amongst textbooks in Sequence 3 (e.g. Anthony & Harvey, 2012; Kolman & Hill, 2007; Larson, 2016) as a means to rationalize the matrix-matrix product AB in terms of the vector dot product (which, for these textbooks, was the first appearing form of matrix multiplication). Many of these textbooks contrasted the intuitive, component-wise approach of matrix addition with matrix multiplication, the understanding or utility of which, they asserted, depended on subsequent concepts:

- "Since matrices are added by adding corresponding entries and subtracted by subtracting corresponding entries, it would seem natural to define multiplication of matrices by multiplying corresponding entries. However, it turns out that such a definition would not be very useful for most problems. Experience has led mathematicians to the following more useful definition of matrix multiplication" (Anton & Rorres, 2014, p. 29).
- "The most natural way of multiplying two matrices might seem to be to multiply corresponding elements when the matrices are of the same size, and to say that

the product does not exist if they are of different size. However, mathematicians have introduced an alternative rule that is more useful. It involves multiplying the rows of the first matrix times the columns of the second matrix in a systematic manner" (Williams, 2012, p. 71).

- "We now define the product of two matrices. From the way the other matrix operations have been defined, you might guess that we obtain the *product* of two matrices by simply multiplying corresponding entries. The definition of *product* given below is much more complicated than this but also considerably more useful in applications" (Venit, Bishop, & Brown, 2013, p. 90).

Considering the sequencing of these textbooks, the widespread use of postponement makes a certain amount of sense, as textbooks that have not yet discussed the matrix-vector product or linear transformations upon the presentation of the matrix-matrix product have fewer familiar concepts with which to justify their definition. It should be noted, though, that postponement was only used as a temporary (and not permanent) strategy: all of these textbooks eventually connected matrix multiplication to linear transformations.

## 4.3 Analogy

We documented uses of both *mathematical* and *real-world* analogies. The mathematical analogies focused on relating aspects of matrix multiplication to familiar arithmetic domains (notably the real numbers and the integers). The *real-world* analogies involved use of a practical real-world scenario to justify or explain the formula for the matrix-matrix product AB.

### 4.3.1 Mathematical Analogy

Strang (2009) motivated the matrix-matrix product by expressing the desire for a single matrix $C$ such that $A(Bx) = Cx$, which was fairly common. What distinguishes his approach, however, is that he does not formally treat linear transformations until near the end of the textbook (Chap. 6), and thus is unable to use linear transformations to justify the associativity of matrix multiplication. Instead, he used the associativity of integer multiplication as an analogy: "When multiplying $EAC$, you can do AC first or EA first. This is the point of an "associative law" like $3 \times (4 \times 5) = (3 \times 4) \times 5$. Multiply 3 times 20, or multiply 12 times 5. Both answers are 60. That law seems so clear that it is hard to imagine it could be false" (p. 58). The other instance of mathematical analogy involved comparison of the matrix equation $Ax = b$ to the real number equation $ax = b$. Edwards and Penney (1988), for instance, wrote that "[the matrix equation $Ax = b$] is analogous in notation to the single scalar equation $ax = b$ in a single variable $x$" (p. 38). This technique was typically used to justify the importance of the matrix equation $Ax = b$ (and thus the matrix-vector product) while also setting the stage for the importance of the inverse of a matrix (as a common method for solving both equations involves multiplication on the left by the appropriate inverse element). Accordingly, some textbooks (e.g. Bretscher, 2012) used the scalar equation $ax = b$ in order to accentuate the importance of inverse matrices.

### 4.3.2 Real-World Analogy

We found examples of textbooks leveraging real-world scenarios to justify the formula for the matrix-matrix product AB across Sequence 1 (e.g. Ricardo, 2009), Sequence 2 (e.g. Bretscher 2012), and Sequence 3 (e.g. Andrilli & Hecker, 2010; Larson, 2016; Williams, 2012). The scenario in Ricardo's (2009) presentation was typical: the textbook described two hypothetical universities, Alpha College and Beta University, that plan to purchase the same computer equipment (in different quantities). The following information about quantity and price of equipment is provided in the following tables (adapted from Ricardo, 2009, p. 182):

**Quantities**

| | System Unit (incl. 17" monitor) | Laser Printer | Surge Protector |
|---|---|---|---|
| Alpha College | 25 | 5 | 20 |
| Beta University | 35 | 3 | 15 |

**Unit Prices in Dollars**

| | Vendor 1 | Vendor 2 |
|---|---|---|
| System unit (incl. 17" monitor) | 1286 | 1349 |
| Laser printer | 399 | 380 |
| Surge protector | 39 | 37 |

Ricardo then demonstrated how to calculate the amount that Alpha College and Beta University would spend if they purchase their equipment from Vendor 1, which can be expressed as dot products of the respective rows (in the "Quantities" table) with the Vendor 1 column (in the "Unit Prices in Dollars" table); for example, $25(1286) + 5(399) + 20(39) = \$34,925$. These calculations, Ricardo noted, take the same form as the matrix-vector product Ax. The same calculations are repeated for Vendor 2 and are then summarized as a matrix-matrix product:

$$\begin{bmatrix} 25 & 5 & 20 \\ 35 & 3 & 15 \end{bmatrix} \begin{bmatrix} 1286 & 1349 \\ 399 & 380 \\ 39 & 37 \end{bmatrix} = \begin{bmatrix} 34925 & 36365 \\ 46792 & 48910 \end{bmatrix}$$

Ricardo followed this scenario with a remark on the matrix-matrix product, noting that "we can generalize this row-by-column operation in a natural way" (p. 183).

As with use of postponement, the use of real-world analogy seems well-situated for Sequence 3 because the sequencing of these textbooks afforded few mathematical footholds to introduce the matrix-matrix product. The textbooks following the other sequences, on the other hand, were able to leverage the matrix-vector product en route to developing their characterizations of the matrix-matrix product, lessening the need for real-world comparisons for justification.

## 4.4 Abstraction

We documented two distinct uses of textbooks explicitly employing abstraction, a classification that we reserved for cases in which textbooks made direct comments about the relationship between a specific example and its associated general concept, representation, or formula. First, DeFranza and Gagliardi (2015), in the absence of a formal treatment of linear transformations, used an argument with $2 \times 2$ matrices to justify that matrix multiplication is associative (i.e. that $A(Bx) = (AB)x$). This is a use of abstraction because it introduces students to and justifies associativity in a particular situation in order to justify the associativity of general matrix multiplication. Second, Shifrin and Adams's (2010) presentation focused explicitly on the matrix-vector product as a special case of the matrix-matrix product, emphasizing that the matrix-matrix product "is a generalization of multiplication of matrices by vectors" (p. ix). In a similar way, Andrilli and Hecker (2010) characterized the matrix-vector product as "a generalization of the dot product of vectors" (p. 59). Though the focus of abstraction is different in each case (the matrix-vector product and the vector dot product, respectively), both of these examples frame the matrix-matrix product in terms of versions of matrix multiplication that are more concrete and familiar. One possible reason for the lack of documented cases of abstraction is that, as we noted in the introduction, matrix multiplication is a notion for which students have little experiential basis, and thus the capacity for connecting general concepts to their more familiar, concrete instantiations is limited.

## 4.5 Computational Efficiency

We documented widespread use of the strategy of *computational efficiency*, which we characterize as motivating a concept by explicitly highlighting its potential to simplify procedures or calculations. Nearly all textbooks in Sequence 1—that had initially defined the matrix-vector product in terms of linear combinations—cast the dot product method as a means to expedite computing matrix multiplication (e.g. Holt, 2012; Nicholson, 2013; Strang, 2009); we also documented this strategy in Sequence 2 (e.g. Bretscher, 2012). Moreover, several textbooks in Sequence 1 coupled statements about computational efficiency with the theoretical importance of a linear combinations view (e.g. Beezer, 2015; Lay et al., 2015; Solomon, 2014). For example:

- "We can define AB using dot products and 'fast' matrix-vector multiplication … the column-wise description above is usually the best way to *understand* matrix multiplication, but the dot-product formula gives a convenient way to *compute* matrix products" (Solomon, 2014, p. 1.10).

Another argument appearing in favor of computational efficiency is the ability to use this to calculate "an individual entry of the product, without calculating the entire column that contains it" (Spence et al., 2007, p. 99). Other examples include:

- This formula enables us to compute any element in the dot product with one simple dot product" (Cheney & Kinkaid, 2012, p. 192).
- "In some applications, we only need a single entry of the matrix product AB" (Holt, 2012, p. 99).
- "It is useful to have a formula for the *ij*th entry of the product …" (Bretscher, 2012, p. 79).

Several textbooks—particularly those in Sequence 3—employed computational efficiency to motivate their introduction of the matrix-matrix product as an action of A on the columns of B (AB as [Acol(B)]) (e.g. Andrilli & Hecker, 2010; Anton & Rorres, 2014; Kolman & Hill, 2007). Anton and Rorres (2014), for instance, stated that that this form of matrix multiplication "has many uses, one of which is for finding particular rows or columns of a matrix product AB without computing the entire product" (p. 31). This recasting of certain forms of matrix multiplication purely in terms of their capacity to streamline calculations certainly seems to insinuate (or, in Solomon's case, explicitly assert) that the dot product methods are less conceptually illuminating for students, further demarcating the contrast in rationale around which we framed this analysis.

## 5 Pedagogical Implications and Future Research

In this section we coordinate the results of our analysis with the literature on student thinking in Linear Algebra in order to hypothesize which approaches might be advantageous (or disadvantageous) for student learning. We acknowledge, however, that we are not positioned to comment definitively on the relative pedagogical effectiveness of any particular content presentation, and thus any hypotheses resulting from this coordination are offered tentatively as avenues for future research.

We used an example of isomorphization (from Lay et al., 2015) and postponement (from Kolman & Hill, 2007) in the introduction to highlight the potential for variation in rationale regarding a central concept like matrix multiplication. As we noted before, there is a subtle tension between these explanations for the initially presented form of matrix multiplication. On one hand, textbooks invoking isomorphization (to vector equations and linear systems) for their initial form of matrix multiplication are attempting to frame matrix multiplication in terms of familiar concepts, procedures, and ideas. On the other hand, textbooks invoking postponement are (oftentimes explicitly) stating that the rationale for matrix multiplication are currently unable to be easily understood. Informally, we might characterize these two approaches as "this can be reasonably understood now using familiar ideas" and "this can only be understood later using more advanced ideas." Our analysis indicates that this tension is indeed reflected amongst a substantial number of other textbooks in our sample as well, particularly between textbooks in Sequence 1 (Ax as LCC) and Sequence 3 (AB as DP). Some of these other textbooks also clearly delineated this tension. For example, recall Solomon's (2014) comment that "the column-wise description above is usually the best way to *understand* matrix multiplication, but the dot-product formula gives a convenient way to *compute* matrix products" (p. 1.10) seems to suggest that the dot product approach in Sequence 3 is less conceptually enlightening and should be used purely for computation.

In addition to connecting matrix arithmetic to familiar notions like linear systems, the textbooks motivating matrix multiplication via isomorphization also did so to emphasize the importance of viewing matrix multiplication in various ways. We should note that it is possible that other textbooks outside of Sequence 1 similarly emphasized the importance of multiple representations; we are simply noting that the textbooks in Sequence 1 made this focus explicit (see Sect. 4.1). Such an approach seems to align with suggestions in the literature regarding the importance of mastering multiple forms of representation in Linear Algebra (e.g. Dorier, 2000). Harel (1997), for example, noted that "one of the most appealing aspects of Linear Algebra—yet a serious source of difficulty for students—is the 'endless' number of mathematical connections one can (must) create in studying it. Relationships between systems of linear equations, matrices, and linear transformations can be built in numerous ways, and problems about systems of linear equations are equivalent to problems about matrices which, in turn, are equivalent to problems about linear transformations" (pp. 111--112). Specific to matrix multiplication, Larson and Zandieh (2013) furthered this idea by identifying three productive ways in which students should understand the matrix-vector product Ax = b, each of which correspond to ways of understanding that emerged amongst textbooks in our sample:

- viewing b as a linear combination of the columns of A (i.e. Ax as LCC),
- viewing the rows of Ax = b as the equations in a linear system (i.e. Ax as DP), and
- viewing b as a linear transformation of the vector x.

Larson and Zandieh reiterated the importance of understanding all three viewpoints, citing the Invertible Matrix Theorem—a set of equivalent conditions to determine the invertibility of a matrix—as an example. The textbooks in Sequence 1 (and some in Sequence 2) appear to be well-positioned to emphasize and foster such flexibility right from the initial definition of matrix multiplication. How alternative emphases in the early stages regarding matrix multiplication might affect subsequent students' understanding of subsequent concepts remains an unexplored question in the literature. We also note that initiating with the matrix-vector product and its different characterizations naturally extends to the matrix-matrix product and can lead to flexible ways of conceptualizing the matrix-matrix product. For example, understanding the matrix-vector product as an isomorphization of a linear system can support understanding the matrix-matrix product as the structure needed to support a change of variables (substitution) in the linear system. Similarly, understanding the matrix-vector product as a linear transformation can support viewing the matrix-matrix product as the composition of linear transformations. In the same way that Harel (1997) and Larson and Zandieh (2013) argue in favor of multiple ways of understanding the matrix-vector product, we propose that it is similarly valuable to understand the matrix-matrix product in different ways. Specific affordances of these ways of understanding could be explored in future research.

Textbooks invoking postponement, particularly those in Sequence 3, typically supplemented their rationale with (1) a real-world example, and/or (2) a brief, tangential reference to linear transformations in order to justify the formula for matrix multiplication. Regarding real-world examples, Harel (1987) argued that such scenarios require the student to "distinguish between relevant and irrelevant features," which, in turn, might "weaken the anticipated motivational effect" (p. 30). We argue that one particularly relevant feature of such scenarios is potentially underemphasized: the appropriate arrangement of the arrays of information (from which the matrices are constructed). Indeed, most textbooks presenting a real-world scenario conveniently pre-arranged these arrays so that the formula for the matrix-matrix product was immediately clear. While we acknowledge that such application problems are intended as an introduction (and not as examples that comprehensively embody the structure of matrix multiplication), we question the effects of this convenient pre-arrangement because it partially sidesteps the potential for students to identify and abstract the interplay between the row and column vectors of the matrices in the product. Future research could explore how students might be able to construct the relevant features of the mathematical structure of matrix multiplication via such application problems.

Much of this discussion has focused on Sequences 1 and 3, largely because they espoused drastically different initial approaches to matrix multiplication (isomorphization vs. postponement, respectively). We found far less consistency amongst the 5 textbooks in Sequence 2. Their rationale for the matrix-matrix product, for example, spanned all categories of rationale (isomorphization, postponement, analogy, abstraction, and computational efficiency). Sequence 2 was, in some sense, a hybrid of textbooks that were similar in approach to the other two sequences. Bretscher (2012) and Leon (2014), for example, seemed to have more in common with the textbooks in Sequence 1 (due to their early emphasis on Ax as a linear combination of the columns of A, even though this was not their initial definition), and Shifrin and Adams (2010) and DeFranza and Gagliardi (2015) seemed to have more in common with Sequence 3 (due to their limited emphasis in the early stages on viewing Ax as a linear combination). Perhaps what Sequence 2 reveals most, then, is that the approaches of Sequences 1 and 3 are not altogether pedagogically incompatible. Indeed, a textbook making use of a definition in terms of dot products first could very well afford ample focus to the linear combinations definition (e.g. Bretscher, 2012; Leon, 2014). Thus, significant conclusions drawn from the sequencing of matrix multiplication alone should be regarded cautiously and in need of additional study. On a more general level, though, the fact that there are apparent contradictions in the two most conspicuous methods of rationale for these respective approaches does highlight the need for research to examine which approach might be more pedagogically effective.

## 6 Conclusions

This study contributes to the literature in three important ways. First, the primary contribution of this paper is our documentation and analysis of the ways in which introductory, English-language linear algebra textbooks conceptualize and sequence matrix multiplication. This analysis provided very specific information about the four characterizations of matrix multiplication that expert mathematicians believe to be the most important (Ax as LCC, Ax as DP, AB as DP, and AB as [Acol(B)]). Particularly, we noticed that experts value fluency amongst these multiple characterizations of matrix multiplication and often discussed unique insights or abilities that each one offered (for example, Ax as LCC allows one to reformulate the concept of span in terms of linear systems, and Ax as DP enables one to calculate individual entries in a product). The order in which these characterizations appeared often had significant implications for the way in which they were rationalized. The textbook authors collectively employed a wide variety of techniques to rationalize and explain matrix multiplication: each category of rationale in Harel's (1987) framework (in addition to computational efficiency) appeared in our sample.

Second, we coordinated these findings with the literature on the teaching and learning of Linear Algebra to hypothesize about ways of understanding matrix multiplication that might be advantageous (or disadvantageous) for students to have. In addition to highlighting productive avenues for future research, this information can be used to inform a *conceptual analysis*, a description of "what students might understand when they know an idea in various ways" (Thompson, 2008, p. 57). Conceptual analyses are particularly important in research on student learning because, as noted by Thompson (2008), their uses include (1) devising ways of understanding a particular concept that might be powerful for students, and (2) characterizing the nature of student struggles with that concept. Conceptual analyses can also be used to design and analyze student thinking in the context of instructional sequences that aim to develop these powerful ways of understanding.

Third, we have adapted Harel's (1987) framework for analyzing rationale to include a category for *computational efficiency*. Additionally, though it has been used exclusively to study rationale for Linear Algebra concepts in Linear Algebra textbooks, we suggest that its potential scope is much broader and extends beyond textbook analysis. For example, the different types of rationale in this framework—isomorphization, postponement, abstraction, analogy, and computational efficiency—could also be used to analyze the rationale that an instructor provides for the introduction of a new concept in a lecture setting. In this way, we anticipate that our adaptation of Harel's framework can be useful for future research.

## References


Bierhoff, H. (1996). Laying the Foundations of Numeracy. A Comparison of Primary School Textbooks in Britain, Germany and Switzerland. *Teaching Mathematics and its Applications, 15*(4), 141–60.

Carlson, D. (1993). Teaching Linear Algebra: Must the Fog Always Roll In? *College Mathematics Journal, 24*(1), 29–40.

Carlson, D., Johnson, C. R., Lay, D. C., & Porter, A. D. (1993). The Linear Algebra Curriculum Study Group recommendations for the first course in Linear Algebra. *The College Mathematics Journal, 24*(1), 41–46.

Capaldi, M. (2012, February). A study of abstract algebra textbooks. In *Proceedings of the 15th Annual Conference on Research in Undergraduate Mathematics Education* (pp. 364–368).

Creswell, J. W. (2007). *Qualitative inquiry and research design: Choosing among five Approaches* (2nd Edition). California: Sage Publications.

Creswell, J. W. (2008). *Educational research: Planning, conducting, and evaluating quantitative and qualitative research* (3rd Edition). Upper Saddle River, NJ: Pearson.

Dorier, J. L. (Ed.). (2000). *On the teaching of Linear Algebra* (Vol. 23). Springer Science & Business Media.

Harel, G. (1987). Variations in Linear Algebra content presentations. *For the learning of mathematics, 7*(3), 29–32.

Harel, G. (1997). The Linear Algebra curriculum study group recommendations: Moving beyond concept definition. *MAA NOTES*, 107–126.

Larson, C., & Zandieh, M. (2013). Three interpretations of the matrix equation Ax = b. *For the Learning of Mathematics, 33*(2), 11–17.

Lockwood, E., Reed, Z., & Caughman, J. S. (2016). An Analysis of Statements of the Multiplication Principle in Combinatorics, Discrete, and Finite Mathematics Textbooks. *International Journal of Research in Undergraduate Mathematics Education*, 1–36.

Reys, B. J., Reys, R. E., & Chavez, O. (2004). Why Mathematics Textbooks Matter. *Educational Leadership, 61*(5), 61–66.

Robitaille, D. F., & Travers, K. J. (1992). International studies of achievement in mathematics.

Thompson, P. W. (2008). Conceptual analysis of mathematical ideas: Some spadework at the foundations of mathematics education. In Proceedings of the annual meeting of the International Group for the Psychology of Mathematics Education (Vol. 1, pp. 45-64). PME Morelia, Mexico.

Thompson, D. R., Senk, S. L., & Johnson, G. J. (2012). Opportunities to learn reasoning and proof in high school mathematics textbooks. *Journal for Research in Mathematics Education, 43*(3), 253–295.

Weinberg, A., & Wiesner, E. (2011). Understanding mathematics textbooks through reader-oriented theory. *Educational Studies in Mathematics, 76*(1), 49–63.

## Bibliography of Textbooks

Andrilli, S., & Hecker, D. (2010). *Elementary Linear Algebra* (4th ed.). Massachusetts: Academic Press.

Anthony, M., & Harvey, M. (2012). *Linear Algebra: concepts and methods*. Cambridge University Press.

Anton, H. & Rorres, C. (2014). *Elementary Linear Algebra: Applications version* (11th ed.). Hoboken, NJ: Wiley.

Beezer, R. A. (2015). *A first course in Linear Algebra* (version 3.50). Retrieved from http://linear.ups.edu/download/fcla-3.50-tablet.pdf.

Bretscher, O. (2012). *Linear Algebra with applications* (5th ed.). New Jersey: Pearson.

Cheney, W. & Kinkaid, D. (2012). *Linear Algebra: Theory and applications* (2nd ed.). Massachusetts: Jones & Bartlett Learning.

DeFranza, J. & Gagliardi, D. (2015). *Introduction to Linear Algebra with applications*. Illinois: Waveland Press.

Edwards, C. H., & Penney, D. E. (1988). *Elementary Linear Algebra: Custom Edition for Arizona State University*. Pearson College Div.

Hefferon, J. (2008). *Linear Algebra*. Available online.

Holt, J. (2012). *Linear Algebra with applications*. New York: W.H. Freeman.

Kolman, B., & Hill, D. (2007). *Introductory Linear Algebra* (9th ed.). New Jersey: Pearson.

Larson, R. (2016). *Elementary Linear Algebra* (8th ed.). Massachusetts: Houghton Mifflin.

Lay, D. C., Lay, S., & McDonald, J. (2015). *Linear Algebra and its applications* (5th ed.). Pearson.

Leon, S. (2014). *Linear Algebra with applications* (9th ed.). New Jersey: Pearson.

Nicholson, K. (2013). *Linear Algebra with applications* (7th ed.). New York: McGraw Hill.

Poole, D. (2014). *Linear Algebra: A modern introduction* (4th ed.). Massachusetts: Houghton Mifflin.

Ricardo, H. (2009). *A modern introduction to Linear Algebra*. CRC Press.

Robinson, D. J. S. (1991). *A course in Linear Algebra with applications* (pp. I–XIII). Singapore: World Scientific.

Shifrin, T. & Adams, M. (2010). *Linear Algebra: A geometric approach* (2nd ed.). New York: W. H. Freeman.

Solomon, B. (2014). *Linear Algebra, geometry and transformation*. CRC Press.

Spence, L., Insel, A. & Friedberg, S. (2007). *Elementary Linear Algebra: A matrix approach* (2nd ed.). New Jersey: Pearson.

Strang, G. (2009). *Introduction to Linear Algebra* (4th ed.). Massachusetts: Wellesley Cambridge Press.

Venit, S., Bishop, W., & Brown, J. (2013). *Elementary Linear Algebra* (2nd ed.). Ontario: Nelson Education.

Williams, G. (2012). *Linear Algebra with applications* (8th ed.). Massachusetts: Jones & Bartlett Learning.

# Misconceptions About Determinants

**Cathrine Kazunga and Sarah Bansilal**

**Abstract** In Zimbabwe, the topic determinant of matrices is usually covered as part of the first-year linear algebra courses. In this study, we focused on Zimbabwean teachers who were studying the topic at university while also teaching the topic to their high school pupils at a different level. The study explored the misconceptions displayed by 116 in-service mathematics teachers, with respect to determinants of matrices. The participants responded to tasks based on determinants of matrices and their applications. More than half of the participants struggled with finding the determinant of the inverse of a matrix, transpose of matrices, and the application of properties of determinants. The teachers exhibited many misconceptions, which were mainly a result of the incorrect application of rules outside the domain in which they were defined. The study suggests possible ways of teaching the concept of determinant to reduce the possible misconceptions among the mathematics teachers and their students. It is recommended that future course outlines of in-service teachers' programmes should include more formal learning opportunities for teachers to develop a more conceptual understanding of the concept of determinant of a matrix.

**Keywords** Linear algebra · Determinant of matrices
Undergraduate mathematics · In-service teachers · Misconceptions

## 1 Introduction

Globally, many mathematics education researchers have been concerned with students' difficulties related to the undergraduate linear algebra course (Dorier & Sierpinska, 2001). There is agreement that teaching this course is a frustrating experience for both teachers and students, and despite all the efforts to improve the curriculum, the learning of linear algebra remains challenging for many students

C. Kazunga (✉) · S. Bansilal
University of KwaZulu-Natal, Durban, South Africa
e-mail: kathytembo@gmail.com



S. Stewart et al. (eds.), *Challenges and Strategies in Teaching Linear Algebra*,
ICME-13 Monographs, https://doi.org/10.1007/978-3-319-66811-6_6

(Dorier & Sierpinska, 2001). In addition, Stewart and Thomas (2007), in their study of university students on vector space, argued that students may cope with procedural aspects of the module where they apply algorithms and procedures when solving systems of equations and manipulating matrices, but have difficulties with the crucial ideas underpinning them.

Mathematics lecturers may benefit from learning about the common problems experienced by students in learning matrix algebra concepts. By becoming more informed about students' problems, they could find ways of adjusting their instructional techniques so that the learning opportunities could be improved. Research on the demands of developing pedagogic content knowledge amongst mathematics and science lecturers in higher education is limited (Bansilal, 2014; Fraser, 2016; Major & Palmer, 2006). In her study with science lecturers, Fraser (2016) reported that experienced science lecturers affirmed the importance of anticipating students' difficulties in their instructional practices, while Major and Palmer (2006) noted that student learning was an important aspect reported by faculty staff engaged on a professional development programme. Lecturers who are interested in developing their understanding of the learning experiences of their students would benefit from knowing more about misconceptions that are developed during the learning process. Furthermore, few studies have reported specific misconceptions developed by students about the concept of determinants of matrices (Aygor & Ozdag, 2012). This apparent dearth of research motivated us to explore the misconceptions exhibited by in-service teachers in solving problems involving determinants. Thus, the purpose of the study was to explore the misconceptions of a sample of 116 mathematics undergraduate in-service teachers with respect to determinants of matrices. It is hoped that the study will help mathematics lecturers better understand the misconceptions encountered by students so that they can anticipate student difficulties in matrix algebra regarding determinants of matrices.

## 2 Literature Review

### *2.1 Conceptualising Determinant*

The concept of determinant plays a vital role in linear algebra. The determinant can be used to find area and volume of polygons and shapes in two dimensions and three dimensions respectively, to find solution of systems of equations with $n$ unknowns and $n$ equations, as one of the properties of invertible matrix theorem, and it can be used in vector space to determine whether the vectors are linearly independent or not. The determinant also can be used in the calculation of eigenvalues and eigenvectors which play a greater role in multivariate statistics and non-linear differential equations. The determinant is used to find eigenvalues and eigenvectors. Larson et al. (2008) and Rasmussen and Blumenfield (2007) call this way of finding eigenvalues and eigenvectors an eigenvector first approach. Hence, determinants are an essential part of matrix algebra for solving systems of equations and as part of curricula for Ordinary and Advanced level physics and Computer Science (Todorova, 2012). In our study we explored misconceptions exhibited on determinants when mathematics in-service teachers are taught it as a numerical value of square matrix. Todorova (2012) points out that the study of determinants may sometimes cause problems if the concept of determinant is interpreted solely as a numerical characteristic (value) of a square matrix.

## 2.2 *Origin of the Theory of Determinant*

The theory of determinant in linear algebra emerged as a way of solving systems of linear equations (Andrews-Larson, 2015; Dorier, 2000). Andrews-Larson (2015) examined the possibility of using history to inform instruction considering linear systems of equations and their solutions. The author asserted that theory of determinant emerged separately in the 17th and 18th centuries in both Japan and Europe after the ancient Chinese method of solving systems of equations. In 1693, a Japanese mathematician Seki Kowa developed a version of determinant as part of a method of solution nonlinear systems of equations. In 1750, a Swiss mathematician Gabriel Cramer independently developed a way of specifying the solution to a system of linear equations. Cramer generalised a method for their computation by leveraging the combinatorics of cleverly superscripted but unspecified coefficients. Andrews-Larson (2015) points out that Cramer obtained a general way of solving system of equations with $n$ linear equations and $n$ unknowns which he framed using combinatorics of the superscripts. He discovered that the $n$ by $n$ system of equations is solved by forming $n$ fractions, each of which has $n!$ terms in both the numerator and the denominator. Cramer discovered the conditions necessary for which the system of equations has a unique solution or no solution (Andrews-Larson, 2015). The way in which Cramer structured the coefficients in the notational system shaped the way in which the determinant was specified. However, today we have Cramer's rule, which is a familiar method used to solve system of equations named after Cramer to recognise the credibility of his work. An English mathematician James Joseph Sylvester coined the term 'matrix' in 1850, and made use of determinants. Thus, determinants can be used as a tool to determine whether a system of equations is consistent or inconsistent. Andrews-Larson (2015) argues that the inclusion of history to inform instruction when teaching linear systems of equations and their solutions can enhance understanding.

Furthermore, Dorier (2000) provides a historical account and explains the relationship between concepts of determinants and rank. He states that from around 1840 to 1879, within the theory of determinants, the concept of rank took shape. He further points out that the concept of inclusive dependence was rapidly renamed after Euler's work connected to way of finding determinant of the augment matrix of system of linear equations. Between 1840 and 1879 the concept of rank was, therefore, implicitly central to the description of the system of linear equations. The theory of determinants made an analogous treatment of equation of dependence of equations and $n$-tuples possible. Dorier stresses that a knowledge of historical development provides the teacher and the researcher with a field of investigation from which they can better understand the students' difficulties and, more generally, from which they can put some 'meat' onto the (bare bones) of the axiomatic approach.

Selinski et al. (2014) used adjacency matrices to analyse students' interpretation of and connections between linear algebra concepts such as linear independence, determinants, span, invertibility and null space. They asked students certain questions which required them to say whether it is true or false that an invertible matrix has a determinant zero (singular matrix). The invertible matrix can also be described in terms of determinant. The above-mentioned authors observed that the students could connect some of the properties of the invertible matrix, including the determinant issue. However, in a study conducted among school students, Todorova (2012) found that they failed to make connections between the determinant and the area of non-regular polygons using the dimensional GeoGebra system. There is a connection between the invertible matrix and the determinant. The determinant is also connected to vectors. If the determinant of a $3 \times 3$ invertible matrix say A is zero, then the column vectors of A are independent.

Todorova (2012) examined conceptual difficulties of students with respect to the concept definition and various images of a determinant. As Todorova (2012) puts it, most of the textbooks introduce the term 'determinant' clearly and students usually have little or no difficulty in understanding the meaning of determinant (that it is actually a number) and the procedure of obtaining this number. However, it is in applying the concept to solve problems that they experience challenges. She offers an approach for understanding the concept of determinant with visualisation in GeoGebra. Todorova (2012) argues that it is possible to develop the concept of a determinant using the dimensional GeoGebra system. Students could visualise the connection between the determinant and area of plane shapes which could be at any position in the Cartesian plane. She points out that the concept of determinant of a $3 \times 3$ matrix can be defined using plane geometry figures and also as volume of solids. The results of her study showed that students are able to think about the concept of a determinant as a value of area of plane shapes and volume of solid shapes but they have challenges with the concept in applying its definition to solve problems involving irregular geometric figures. However, the students who participated in her study were able to solve simple problems after the formal definition was introduced but had challenges when solving problems that involved irregular polygons.

Wawro (2014) adapted and extended the work of Selinski et al. (2014) using adjacency matrices by emphasising that the centrality of 'the determinant of A is zero' was higher than 'the determinant of A is non-zero'. A deeper examination of the involved argument revealed that the Selinski's class reasoned about how determinants connected to other concepts by focusing more on matrices with zero determinants rather than non-zero determinants. Although the invertible matrix theorem has a determinant not equal to zero, students used 'determinant equal to zero' more often throughout the semester as they reasoned about novel problems. Thus, students in this study preferred the use of 'singular matrix' to 'non-singular matrix'.

Donevska-Todorova (2014) categorised thinking of determinants into three modes of description, namely the geometric language or synthetic-geometric mode of thought, the arithmetic language or analytic-arithmetic mode of thought, and the algebraic language or analytic-structural mode of thought. In the first mode (synthetic-geometric) determinants are viewed as oriented volumes and areas of parallelepiped and parallelograms respectively spanned by vectors. In the arithmetic language or analytic-arithmetic mode of thought determinants are viewed as the sum of permutations. In the algebraic language or analytic-structural mode of thought, determinants are viewed as functions satisfying three axioms (multilinear form, norm and two equal rows in a matrix give a zero value for its determinant). She elaborates that university- and further-level students viewed the axioms of determinants according to the abstract or analytical structure mode of thinking. For our study, the mathematics undergraduate in-service teachers were given tasks based on the analytic-arithmetic mode of thinking.

Sierpinska (2000), who also conducted a study on students' thinking in linear algebra, concluded that students tend to think in practical rather than theoretical ways. The study revealed that students are reluctant to move flexibly among the three modes of reasoning in linear algebra 'languages': 'visual geometric' language, 'arithmetic' language and the 'structural' language. Sierpinska avoided the development of the obstacle of formalism in students by assigning tutors to different groups of students, instead of having one teacher teaching the whole class. This author states that the development of linear algebra started as a process of thinking analytically about the geometric space and argues that the familiar objects of analytic thinking in linear algebra are vectors and matrices which have lost their numeric substance. They are no more "boxes with numbers", but units whose internal structure is not of much interest in the reasoning. The author points out that the theory of determinant and techniques for their calculation lose the prominent position they hold in analytical-arithmetic thinking. She stresses that by teaching the concept of determinant geometrically, the structure and properties of the linear algebra objects can be illustrated (Sierpinska, 2000). The analytical-structural mode is when students conceptualising matrices as either having or not having a determinant equal to zero, and when they make use of properties of determinant, while calculation determinant of square matrices as the analytical -arithmetic (Sierpinska, 2000). Sierpinska (2000) pointed out that in analytical-arithmetic thinking students make use of formula to compute the determinant. Sierpinska concluded that it is difficult to avoid the analytical-arithmetic thinking when solving determinants problems.

Ndlovu and Brijlall (2016) point out that mathematics lecturers and teachers need to focus on their students' understanding of interrelationships between concepts, rather than carrying out procedures. In schools, students construct mathematical knowledge as isolated facts and assimilate rules cognitively as a list of unconnected actions. Ndlovu and Brijlall (2016) carried out an APOS (Action–Process–Object–Schema) study with 31 pre-service teachers, five of whom were interviewed. The researchers observed that most of the pre-service teachers did not display evidence of conceptual understanding. They also observed that the pre-service teachers understood the procedure of evaluating determinants but could not explain the connection made between general statement of evaluating determinants and its applicability to other contexts. Our study went a step further by exploring the misconceptions exhibited by in-service teachers when solving problems involving determinants and possible strategies that can be used to improve the understanding of determinants.

The difficulties students encounter in linear algebra has been the focus of many researchers (Stewart & Thomas, 2008). In our study, we focused on misconceptions encountered by non-traditional mathematics undergraduate students. These are students who carry out their studies through blocks during school holidays or vacations. They differ from conventional students who pursue their studies full–time. We consider a misconception as a conceptual construction, created by the students, which is logical in relation to her/his current knowledge, but which is not in line with specific mathematical knowledge (Confrey, 1990; Siyepu, 2013). Aygor and Ozdag (2012) investigated misconceptions exhibited by 60 undergraduate students while solving problems on matrices and determinants. Their results revealed many misconceptions which are related to confusion between matrices and the determinant of matrices. For example, some students took the relationship $\det A = -\det B$ (that is the determinant of matrix $A$ equal to minus one times the determinant of matrix $B$) to mean $A = -B$ (that is, matrix $A$ equals minus one times matrix $B$). Some students also took the relationship $\det A = k \det B$ (that is, determinant of matrix $A$ is equal to $k$ times determinant of matrix $B$) to mean $A = kB$ (that is matrix $A$ equals $k$ times matrix $B$). Again some students took the relationship $\det A + \det B$ (determinant of matrix $A$ plus determinant of matrix $B$) to mean $A + B$ (matrix $A$ plus matrix $B$). Aygor and Ozdag (2012) point out that misconceptions might cause students to perform badly in mathematics; hence, curriculum materials should be made available to prevent, halt or reduce misconceptions (Aygor and Ozdag, 2012).

## 2.3 A Framework to Understand Matrix Algebra Misconceptions

Errors, according to Lannin et al. (2007, pp. 44–45), should be taken as "opportunities for deepening one's understanding and as important components of the learning process". The view that errors or misconceptions are part of the process of learning, rather than something that should be eliminated, is supported by many mathematics education researchers (Mahlabela, 2012). Smith, diSessa and Roschelle (1993a, b) explain that misconceptions develop as part of the process of constructing productive knowledge and caution that conceptions are embedded in complex systems, and are not single units of knowledge. This perspective makes it "easier to understand how some conceptions can fail in some contexts and play productive roles" in others. The crucial issue is where the conceptions are used and how they are used, since that marks the difference between whether it is a productive conception or a misconception. Smith et al. (1993a, b) argue that persistent misconceptions can actually be understood as efforts made by learners to extend their existing conceptions to other instructional contexts in which the conceptions are not productive. Mahlabela (2012) explains that misconceptions often occur when learners overgeneralise a concept from one domain to another, as put forward by Nesher (1987, p. 38): "Misconceptions are usually an outgrowth of an already acquired system of concepts and beliefs wrongly applied to an extended domain." Naidoo (2009) avers that misconceptions are fundamental in learning because learners can create misconceptions in the sense-making process of knowledge acquisition. In this chapter, we consider matrix algebra misconceptions as conceptual constructions, created by the learners, which are logical in relation to their current knowledge, but which are not in line with specific mathematical knowledge (Confrey, 1990; Siyepu, 2013).

## 3 Methodology

A qualitative research approach was adopted to answer the research question. Qualitative research methods involved the systematic collection, organisation and interpretation of textual or verbal data (Cohen, Manion, & Morrison, 2011). The participants of the study were 116 in-service teachers who were enrolled in a course on a part-time basis at a Zimbabwean university. The participants were non-traditional students in the sense that they were mature teachers who were categorised as unqualified mathematics teachers because their initial training was no longer considered sufficient. As part of a large-scale programme (funded by international aid organisations in partnership with the Zimbabwean government), the unqualified teachers attended in-service courses at local universities which were designed to upgrade their qualifications. The design of the course was such that the participants would complete the work usually done in an undergraduate three-year degree programme except that the lectures were offered in two intensive block sessions for each semester. The 116 in-service mathematics teachers who were studying matrix operations concepts in a linear algebra course agreed to participate in this study.

The data from the study was generated from the written responses of the 116 participant teachers to tasks based on determinants of matrices and its application collected through activity sheets and interviews. The participants who were willing to participate in the study signed the consent letter and they were given one and a half hours to answer the six questions in an activity sheet. The purpose of the research was to identify teachers' misconceptions when answering questions involving determinants. The actual tasks are presented in the results section.

The participants were coded using tags ‘S1’, ‘S2’ and so forth up to ‘S116’. The order did not have any significance, so that while enabling organisation of the data the responses could not be linked in publications to the original participants.

We employed “interpretation analysis” to guide us in the data analysis process (Gall, Borg, & Taylor, 2003, p. 562). According to Gall et al., interpretation analysis is “the process of examining case study data closely in order to find constructs, themes, and patterns that can be used to describe and explain the phenomenon being studied” (2003, p. 562). In this setting, the process involved finding patterns and themes within data to identify the misconceptions, and constituted thematic analysis (Given, 2008). Hence, the data collected was arranged and sorted to identify common themes, patterns, differences, and similarities that were used to organise the presentation of the results (Cohen et al., 2011).

## 4 Results

We present the results for the different questions, followed by a description of misconceptions or errors that were identified in the particular items.

### *4.1 Determinants of 3 × 3 and 4 × 4 Matrices*

There were six sub-questions based on calculating the determinants of $3 \times 3$ and $4 \times 4$ matrices as well as the determinant of the transpose of a matrix. These are shown in Table 1, together with the percentage of correct responses.

Table 1 shows that most participants were able to calculate the determinants of $3 \times 3$ matrices correctly, with few presenting incorrect or blank responses. It was in the calculation of the determinants for the $4 \times 4$ matrix, D, that some misconceptions emerged.

#### 4.1.1 Incorrect Procedures for Calculating the Determinants of 4 × 4 Matrices

One common misconception about calculating the determinant of the $4 \times 4$ matrix related to the inappropriate application of the Sarrus rule (which is only used for $3 \times 3$ matrices). There were 20 teachers who obtained the answer of zero when they applied the Sarrus rule for $3 \times 3$ matrices to an extended domain including $4 \times 4$ matrices.

Another incorrect strategy used by 10 participants was to extend the method used to find the determinant of a $2 \times 2$ matrix, (that is when $A = \begin{pmatrix} a & b \\ c & d \end{pmatrix}$ then

**Table 1** Results for items on finding determinants of matrices of different order

1. Consider the following matrices and find their determinants:

$$A = \begin{pmatrix} 1 & 2 & 3 \\ 2 & 5 & 3 \\ 1 & 0 & 8 \end{pmatrix} \quad B = \begin{pmatrix} 1 & 2 & 0 \\ -1 & 1 & 0 \\ 2 & 5 & 1 \end{pmatrix}$$

$$C = \begin{pmatrix} 2 & 3 & 1 \\ 5 & 3 & 4 \\ -1 & 2 & 5 \end{pmatrix} \quad D = \begin{pmatrix} 1 & 2 & 1 & 2 \\ 3 & 0 & 1 & 2 \\ 2 & 1 & 2 & 1 \\ 1 & 0 & 2 & 2 \end{pmatrix}$$

| | Percentage with correct answer (%) |
|---|---|
| **1.1** A | 64.7 |
| **1.2** B | 74.1 |
| **1.3** C | 61.2 |
| **1.4** D | 8.6 |
| **1.5** $C^T$ | 24.1 |
| **1.6** $A^T$ | 24.1 |

$|A| = ad - bc$). These participants ‘generalised’ the method, applying it to both $3 \times 3$ and $4 \times 4$ matrices by multiplying the entries of the diagonals and then subtracting the results. This incorrect strategy is illustrated in Fig. 1, which shows the response of S14.

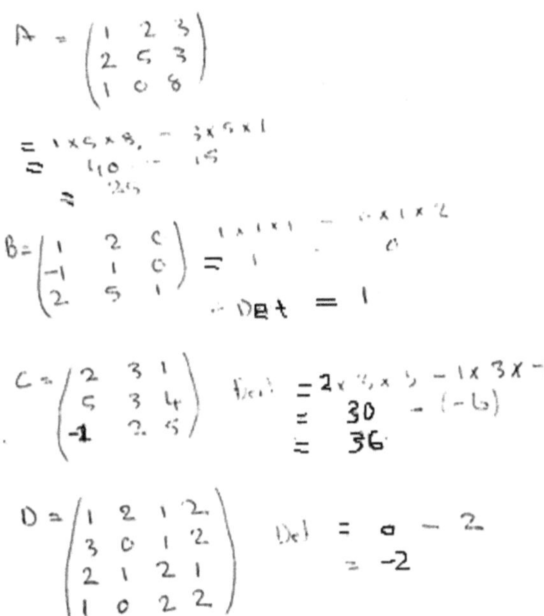


**Fig. 1** The response of S14 who found the difference between the products of entries on the diagonals

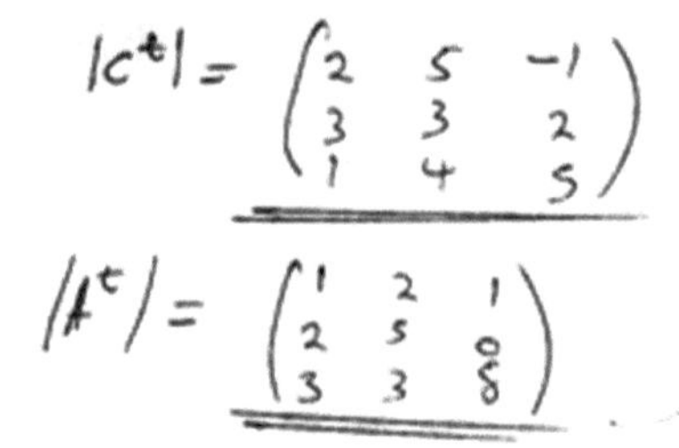

**Fig. 2** The response of S1 who interpreted the determinant of the transpose of a matrix as being equivalent to the transpose matrix

### 4.1.2 Confusion Between the Transpose of a Matrix and Its Determinant

From Table 1, it is seen that only 28 participants correctly calculated the determinant of the transpose matrix (Q 1.5 and Q 1.6). The responses reveal a widespread misconception (by 65 participants) that the determinant of a transpose matrix is the same as the transpose matrix, that is, $|A^T| = A^T$, as illustrated in Fig. 2, through the response of S1.

The response of S1 presents the transpose of the matrix as the determinant of the transpose matrix, suggesting that he sees the two as being equivalent.

## *4.2 Misconceptions About Properties of Determinants*

There were 4 sub-questions where teachers were expected to apply properties of determinants to work out the determinants. The results for these items appear in Table 2.

Table 2 shows that 77 participants correctly calculated the determinant of the square of a matrix (Q 2.1). However, some teachers took the determinant $|A^2| = 2|A|$ leading to the answer of 2(8) = 16.

For Question 2.2, 53 participants managed to find $|A^{-1}|$ correctly. There were 25 participants who incorrectly applied the rule $A^{-1} = \frac{1}{|A|}A$ to this situation. That is, they presented the question as if they were asked to find $A^{-1}$ which is equal to $\frac{1}{|A|}A$.

**Table 2** Results for questions based on determinants of related matrices

| 2 Suppose that A, B are matrices where $\|A\| = 8$ and $\|B\| = 2$. Find | |
|---|---|
| | Percentage with correct answer (%) |
| **2.1** $\|A^2\|$ | 66.4 |
| **2.2** $\|A^{-1}\|$ | 45.7 |
| **2.3** $\|B^T\|$ | 47.4 |
| **2.4** $\|A^{-1}B^T\|$ | 25.0 |

Furthermore, instead of using the matrix A, they substituted the value of $|A| = 8$. Hence their answer to $|A^{-1}|$ was written as $\frac{1}{8}A = \frac{1}{8} \times 8 = 1$. Therefore, these 25 participants showed confusion between the determinant of the matrix and the matrix itself (for A as well as $A^{-1}$); that is, they took $A^{-1} = |A^{-1}|$ and $A = |A|$. Figure 3 illustrates this misconception from the work of S14.

There were a large number of participants (32) who incorrectly took the determinant of $A^{-1}$ to be equal to the determinant of A, that is $|A^{-1}| = |A|$, as illustrated in the response of S19 which appears in Fig. 4.

For Question 2.3 there were 55 teachers who managed to find the determinant of $B^t$ given the determinant of B. Most of the participants with incorrect answers gave the response $2^T$, which is actually not defined, since transpose as an operation can only be carried out on a matrix and not on a number. The underlying misconception is taking the determinant of the transpose of matrix as equal to the transpose of the determinant of the matrix, $|B^T| = |B|^T$, illustrated in Fig. 5, using the response of S2.

For Question 2.4 only 29 teachers managed to find $|A^{-1}B^T|$. It was found that the incorrect responses for Questions 2.2 and 2.3 influenced the teachers in their approach to Question 2.4; that is their incorrect approach to working out the determinant of the inverse and the determinant of the transpose matrix led to incorrect answers for Question 2.4 as well. There were many responses where the equal sign was used inappropriately as an operator symbol or a "do something" symbol (Kieran, 1981), instead of a sign denoting equivalence between two expressions. This misuse of the equal sign is illustrated in Fig. 6 through the response of S3. However, it is important to note that the misconception was not confined to this question only but appeared in responses to other questions as well.

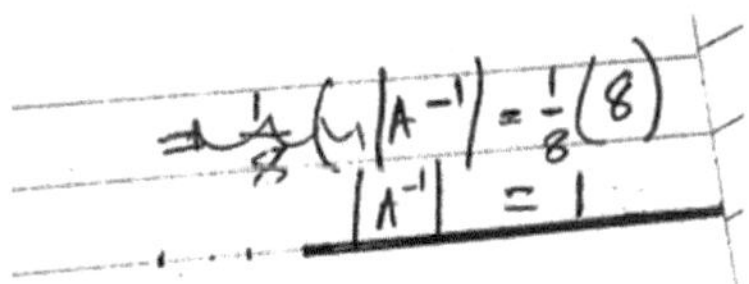

**Fig. 3** Response of S14 showing a misinterpretation of $|A^{-1}|$

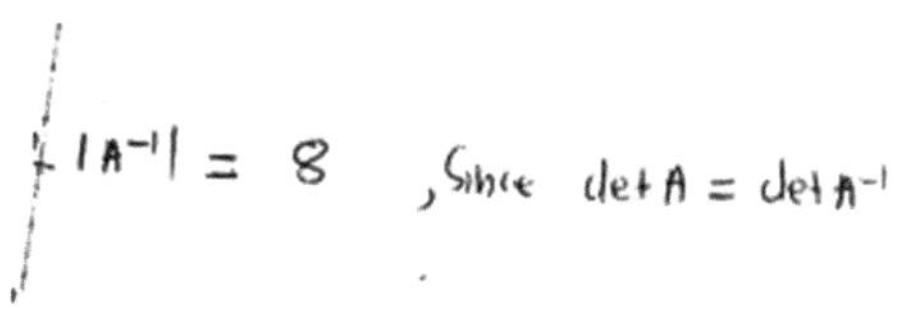


**Fig. 4** Response of S19 who took the determinant of the inverse of a matrix as equal to the determinant of the matrix

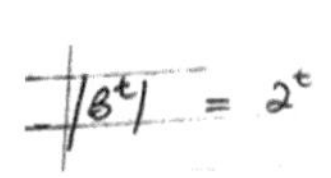

**Fig. 5** The response of S2 to finding the determinant of the transpose matrix

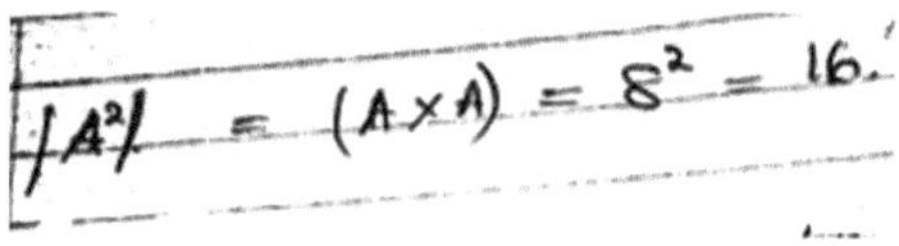


**Fig. 6** The response of S3 to finding the determinant of the square of a matrix

## 4.3 Difficulties with Deducing Relationships Between Determinants of Related Matrices

Teachers were presented with the determinant of a matrix and asked to find the determinant when the columns were swopped (Q 3.1), of the matrix transpose (Q 3.2) and to apply the multilinear property of determinants (Q 3.3). These sub-questions appear below, together with the percentage of participants who got each answer correct.

Table 3 shows that less than 40% gave the correct responses for the three items, indicating that they did not know the properties of the determinant of matrices. Most participants could not deduce relationships such as det C = 3 det A, which arises from the multilinear property of the determinant function (Donevska-Todorova, 2014). Instead, many tried to find the determinant of matrices B, C and D from scratch, using the variable entries as illustrated in Fig. 7.

There were 10 participants who incorrectly extended the method used to find the determinant of a $2 \times 2$ matrix by multiplying the entries of the diagonals and then subtracting the results as shown in Fig. 8. These participants had the same misconception illustrated earlier in Fig. 1.

We now consider the teachers' responses to items which required working with algebraic relationships arising from determinants of matrices with algebraic entries, as shown in Table 4.

Table 4 shows that only 48 and 33 participants provided correct answers for Questions 4.1 and 4.2 respectively. It is of interest that some participants had managed to evaluate the determinant of the $3 \times 3$ matrix in Question 2, yet they were unable to evaluate the determinant of a matrix of the same size but one whose entries included variables. The responses included some basic algebraic errors: some participants (9) were unable to simplify the algebraic expression representing the determinant; 18 wrote $x^2 = \frac{3}{2} \rightarrow x = \sqrt{\frac{3}{2}}$; that is, they left out the negative value of the square root $\left(-\sqrt{\frac{3}{2}}\right)$; 13 participants left out the negative value of the square root $\left(-\sqrt{\frac{6}{4}}\right)$; that is, they took $\sqrt{x^2} = x$; three participants stopped at the equation $4x^2 - 6 = 0$; that is, they did not solve for $x$. One wrote $x^2 = \frac{3}{2} \rightarrow x = 1\frac{1}{2}$, indicating that $x$ and $x$ squared were seen as the same $x$ ($x^2 = x$) which is not linear to mathematical knowledge. Figure 9 includes the response of participant S5, illustrating this misconception.

For Question 4.2, there were 28 participants who could not simplify the algebraic expression representing the determinant, while 12 participants obtained the

**Table 3** Questions based on determinants of matrices with algebraic entries and the percentage of correct responses

3. Suppose the matrix A is such that $|A| = \begin{vmatrix} a_1 & a_2 & a_3 \\ b_1 & b_2 & b_3 \\ c_1 & c_2 & c_3 \end{vmatrix} = -2.$

Given the following matrices:

$$B = \begin{pmatrix} a_3 & a_2 & a_1 \\ b_3 & b_2 & b_1 \\ c_3 & c_2 & c_1 \end{pmatrix} \quad C = \begin{pmatrix} a_1 & a_2 & a_3 \\ b_1 & b_2 & b_3 \\ 3c_1 & 3c_2 & 3c_3 \end{pmatrix}.$$

$$D = \begin{pmatrix} a_1 & b_1 & c_1 \\ a_2 & b_2 & c_2 \\ a_3 & b_3 & c_3 \end{pmatrix}$$

Find their determinant:

| | Percentage with correct answer (%) |
|---|---|
| **3.1** B | 25.0 |
| **3.2**. C | 32.8 |
| **3.3** D | 37.9 |

correct expression for both determinants but did not continue to solve for $x$. They left the answer as a quadratic expression $2x^2 - 3x - 3 = 0$.

## 5 Discussion

The results showed that many of the teachers held various misconceptions about determinants of matrices and their application. It was surprising that even though these teachers teach the topic of matrix operations at school, 34 out of the 116 participants calculated the determinants of $2 \times 2$ matrices incorrectly. Eighty-eight participants were not able to find the determinant of a $3 \times 3$ matrix. A further concern is that many participants had problems with basic algebra. For example, when working with determinants of matrix to find the unknown value, some participants took $\sqrt{x^2} = x$ and $x^2 = x$. These findings concur with those from Kazunga and Bansilal's (2015) study which found that many pre-service teachers have low level engagement with most matrix algebra concepts. The participants also displayed misconception concerning the use of square and square root function. Many participants used the equal sign inappropriately as an operator symbol or a 'do something' symbol, which is a misconception often encountered in children learning early algebra (Kieran, 1981).

Many of the misconceptions that were identified were related to the incorrect application of rules outside the domain in which they were defined. Most of the teachers applied the method of finding the determinant of $2 \times 2$ matrices which involve subtraction of the product of diagonal entries, to find the determinants of

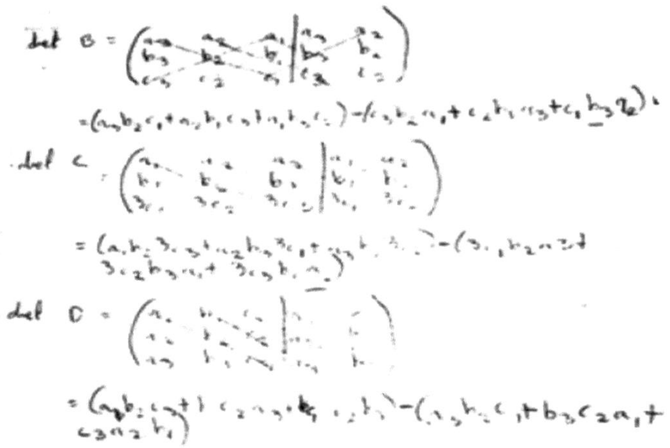

Fig. 7 The response of S4 to finding the determinant of the square matrix

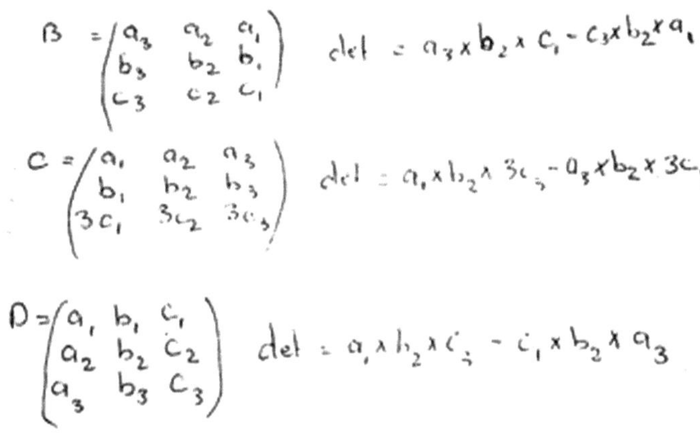

Fig. 8 The response of S14 to show misconception 1 on determinant of square of matrix with algebraic entries

$3 \times 3$ and $4 \times 4$ matrices. This misconception was evident in the responses to Question 1 as well as in Question 3. This tendency to overgeneralise a concept from one domain to another has been observed in research about misconceptions. Nesher's (1987) view of misconceptions is that these are "usually an outgrowth of an already acquired system of concepts and beliefs wrongly applied to an extended domain" (p. 38). Smith et al. (1993a, b) argue similarly that misconceptions, especially the widespread ones, are associated with some contexts where they are used successfully and persistent misconceptions may be viewed as "efforts to extend their existing useful conceptions to instructional contexts where they turn out to be unproductive" (p. 61). In this context of calculating determinants, the

**Table 4** Questions based on application of determinant of a matrix and percentage of correct responses

4.1 Suppose that $B = \begin{pmatrix} 1 & x & x \\ -x & -2 & x \\ x & x & 3 \end{pmatrix}$, work out what the value of $x$ is, if $\det(B) = 0$

4.2 Suppose that $\begin{vmatrix} x & -1 \\ 3 & 1-x \end{vmatrix} = \begin{vmatrix} 1 & 0 & -3 \\ 2 & x & -6 \\ 1 & 3 & x-5 \end{vmatrix}$.

Now solve for $x$

| | Percentage with correct answer (%) |
|---|---|
| **4.1** | 39.7 |
| **4.2** | 28.4 |

**Fig. 9** The response of S5 who considered $x^2 = x$

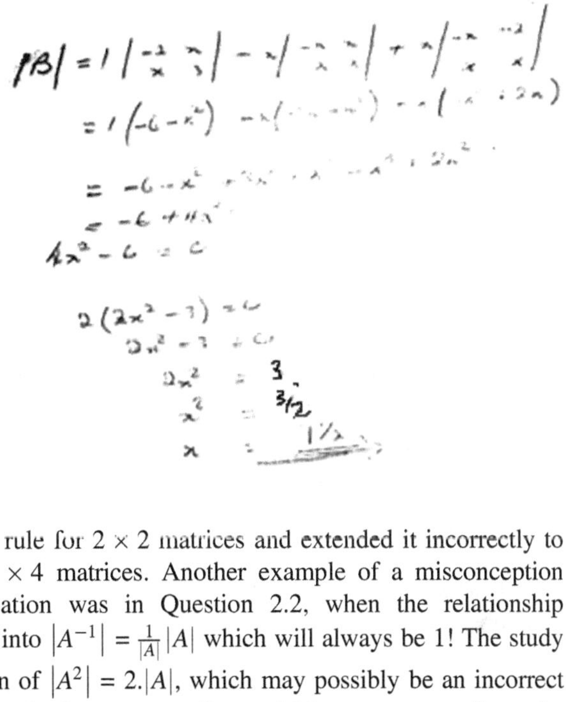

teachers overgeneralised the rule for 2 × 2 matrices and extended it incorrectly to the domain of 3 × 3 and 4 × 4 matrices. Another example of a misconception related to an overgeneralisation was in Question 2.2, when the relationship $A^{-1} = \frac{1}{|A|}A$ was transformed into $|A^{-1}| = \frac{1}{|A|}|A|$ which will always be 1! The study also found the misconception of $|A^2| = 2.|A|$, which may possibly be an incorrect extension of the derivative rule for powers of a variable to powers of matrix. Researchers (Lannin et al., 2007; Naidoo, 2009; Smith et al., 1993a, b) have emphasised that the development of misconceptions is fundamental in learning. Instructors could design tasks that explore common misconceptions to help students engage more deeply with the concepts. Interrogation of the misconceptions can provide rich opportunities for students to question the domain under which certain procedures can be applied and why they cannot apply in domains outside their definition. It is hoped that the identification of these misconceptions can help instructors to be more aware of them so that it can help inform the design of their instructional programmes.

We also found that teachers were confused about operations of matrices and the relationship with determinants. Some teachers assumed that the inverse of a matrix A was the same as the determinant of A. The result of the inverse operation on a matrix is another matrix while the result of the determinant operation is a number. There was also confusion about whether the order of the operations on a matrix made a difference. For example, many teachers took the determinant of the transpose of a matrix to be equal to the transpose of the determinant of the matrix; that is, raised to the power (exponent) $T$ (i.e. det $(A^T)$ = $(\det A)^T$ where the second expression is not defined since the transpose of a number does not exist). Many participants exhibited combinations of many misconceptions. For example, 25 teachers took $A^{-1}$ as being equal to $\frac{1}{|A|}A$, which they simplified to $\frac{1}{8}A$, and then took A as $|A| = 8$, hence getting $\frac{1}{8} \times 8 = 1$. The confusion indicates that the participants had problems with their conceptual understanding and did not view the determinant as more than a set of steps that lead to an answer. However, the determinant is more than just a set of steps; it is actually a function with particular properties (Donevska-Todorova, 2014). Students who have not moved past an operational perspective to a functional perspective of a determinant will be limited to just carrying out a set of calculations and will not be able to easily solve problems based on properties of the determinant function.

A common finding that emerged across various questions, was that many teachers considered the determinant of a matrix A as being the same as the matrix A (i.e., det A = A). This was found in Questions 1.5 and 1.6 where participants took the determinant of the transpose matrix as the transpose itself; that is, det $A^T$ = $A^T$ Similarly, in Question 2.2, many participants assumed that det A and A were equal, as well as that det $A^{-1}$ = $A^{-1}$. This confusion actually conflates a matrix which is an array of numbers with a number (value of the determinant). Similar misunderstandings were reported by Aygor and Ozdag (2012) who found that many students took the relationship $\det A = k \det B$, to mean $A = kB$ while some students took the relationship $\det A +$ $\det B$ to mean $A + B$. Hence this confusion may be quite widespread and instructors should be aware of this possibility when teaching the concept.

As noted in an earlier section, the participants in this study were introduced to determinants as a numerical value associated with a square matrix, which may have limited their development of the concept. The review of the literature provided some innovative ways in which students' understanding can be improved. Andrews-Larson (2015) suggests that including the history of determinant when teaching determinants will enhance understanding. Students can also develop the concept of a determinant using the GeoGebra or other dynamic geometry software, which can support connections between geometric and algebraic modes of description (Donevska-Todorova, 2014). The use of visualisation approaches will enable students to discover the multilinear properties of determinants as they double or multiply by a scalar the $x$ or $y$ or both coordinates. In addition to the use of visualisation and history in the linear algebra class, Dorier (2000) suggests the use of the 'practical' approach rather than a theoretical treatment. He especially proposes a 'structural' approach because most university students are practical-minded. In a 'practical' approach students will be working on the applications of linear algebra. It is important to note that it is not sufficient to use new methods on their own to ensure that students grasp the properties of determinants of a matrix. Linear algebra instructors will need to apply a variety of approaches such as using history, visualisation and the practical approach as they teach determinants, while also taking care to identify and deal with students' misconceptions as they emerge during the course.

## 6 Conclusion

In this study we set out to identify misunderstandings of the participant teachers with respect to finding and applying determinants of matrices, an area which is under-reported in the literature. We identified many misconceptions which do not appear to have been reported in other studies. It is hoped that the identification of such misconceptions can help mathematics educators who teach the topic of matrix operations. If educators become more aware of their students' misconceptions, they will be in a stronger position to help their students. By anticipating such difficulties, mathematics educators could plan instruction that could help the students to move past such epistemological obstacles. They could also design rich exploratory tasks which could use the misconceptions as a basis for in-depth discussion and interrogation of the students' understanding of the concept of determinant. This may help to shift students' understanding of determinant from seeing it as a set of steps towards seeing it as a function which has various properties. Students' misconceptions and the errors they make can be used productively as pedagogic tools to help move students towards this deeper conceptual understanding. Furthermore, it is hoped that this chapter has provided pertinent advice on how the current in-service course within which our study was located can be improved. This will ensure that future teacher participants can gain a deeper understanding of the determinant and other concepts in linear algebra.

**Acknowledgements** This study was made possible by a fellowship received from Organisation for Women in Science for the Developing World (OWSD).

## References


Andrews-Larson, C. (2015). Roots of Linear Algebra: An Historical Exploration of LineSystems. *PRIMUS: Problems, Resources, and Issues in Mathematics Undergraduate Studies, 25*(6), 507–528.

Aygor, N., & Ozdag, H. (2012). Misconceptions in linear algebra: The case of undergraduate students. *Procedia Social and Behavioural Sciences, 46*, 2989–2994.

Bansilal, S. (2014). Examining the invisible loop: Tutors in large scale teacher development programmes. *Africa Education Review, 11*(3), 1–19.

Cohen, L., Manion, L., & Morrison, K. (2011). *Research methods in education* (7th ed.). USA: Routledge.

Confrey, J. (1990). A review of the research on student conceptions in mathematics, science, and programming. In C. Cazden (Ed.), *Review of Research in Education, 16*, 3–56.

Donevska-Todorova, A. (2014). Three Modes of Description and Thinking of Linear Algebra Concepts at Upper Secondary Education. *Beiträge zum Mathematikunterricht, 48*(1), 305–308.

Dorier, J.-L. (2000). Epistemological analysis of the genesis of the vector spaces. In J.-L. Dorier (Ed.), *On the teaching of linear algebra* (pp. 3–81). Dordrecht, The Netherlands: Kluwer Academic Publishers.

Dorier, J.-L., & Sierpinska, A. (2001). Research into the teaching and learning of linear algebra. In D. Holton (Ed.), *The teaching and learning of mathematics at university level: An ICMI study* (pp. 255–273). Dordrecht, The Netherlands: Kluwer Academic Publishers.

Fraser, S. P. (2016). Pedagogical content knowledge (PCK): Exploring its usefulness for science lecturers in higher education. *Research in Science Education, 46*(1), 141–161.

Gall, M. D., Borg, W. R., & Taylor, T. (2003). *Educational research: An introduction.* Oregon: Longman.

Given, L. M. (2008). Qualitative research methods. In N. J. Salkind (Ed.), *The encyclopedia of educational psychology* (pp. 827–831). Thousand Oaks: Sage Publications.

Kazunga C., & Bansilal, S. (2015). Zimbabwean pre-service teachers' responses to matrix algebra assessment item. *Proceedings of 10th Southern Hemisphere Conference on the Teaching and Learning of Undergraduate Mathematics and Statistics*, Port Elizabeth, South Africa.

Kieran, C. (1981). Concepts associated with the equality symbol. *Educational Studies in Mathematics, 12*(3):317–326.

Lannin, J. K., Barker, D. D., & Townsend, B. E. (2007). How students view the general nature of their errors. *Educational Studies in Mathematics, 66*(1), 43–59.

Larson, C., Zandieh, M. & Rasmussen, C. (2008). A trip through eigen-land: Where most roads lead to the direction associated with the largest eigenvalue. *Proceedings of the 11th Annual Conference for Research in Undergraduate Mathematics Education*, San Diego, CA.

Mahlabela, P. (2012). *Learner errors and misconceptions in ratio and proportion. A case study of grade 9 learners from a rural KwaZulu-Natal school* (Unpublished master's thesis). University of KwaZulu-Natal, Durban.

Major, C. H., & Palmer, B. (2006). Reshaping teaching and learning: The transformation of faculty pedagogical content knowledge. *Higher Education, 51*(4), 619–647.

Naidoo, K. S. K. (2009). *An investigation of learners' symbol sense and interpretation of letters in early algebraic learning* (Unpublished master's thesis). University of Witwatersrand, Edenvale.

Ndlovu, Z. & Brijlall, D. (2016). Pre-service mathematics teachers' mental constructions of the determinant concept. *International Journal of Education Science, 14*(2), 145–156.

Nesher, P. (1987). Towards an instructional theory: The role of student misconceptions. *For the Learning of Mathematics, 7*(3), 33–40.

Rasmussen, C. Blumenfield, H. (2007). Reinventing solutions to systems of linear differential equations: A case of emergent models involving analytic expressions. *The Journal of Mathematical Behavior, 26*(3), 195–210.

Selinski, N. E., Rasmussen, C., Wawro, M. & Zandieh, M. (2014). A method for using adjacency matrices to analyze the connections students make within and between concepts: The case of linear algebra. *Journal for Research in Mathematics Education*, *45*(5), 550–583.

Sierpinska, A. (2000). On some aspects of students' thinking in linear algebra. In J.-L. Dorier (Ed.), *On the teaching of linear algebra* (pp. 209–246). Netherlands: Springer.

Siyepu, S. W. (2013). An exploration of students' errors in derivatives in a university of technology. *The Journal of Mathematical Behavior, 32*(3), 577–592.

Smith, J. P., diSessa, A. A., & Roschelle, J. (1993). Misconceptions Reconceived: A Constructivist Analysis of Knowledge in Transition. *The Journal of the learning Science, 3*(2), 115–163.

Stewart, S., & Thomas, M. O. J. (2007). Embodied, symbolic and formal aspects in linear algebra. *International Journal of Mathematics Education in Science and Technology*, *38*(7), 927–937.

Stewart, S., & Thomas, M. O. J. (2008). Linear algebra thinking: Embodied, symbolic and formal aspects of linear independence. *The 11th Conference of the Special Interest Group of the Mathematical Association of America on Research in Undergraduate Mathematics Education (RUME)*, San Diego, California.

Todorova, A. D. (2012). Developing the Concept of a Determinant Using DGS. *Electronic Journal of Mathematics & Technology*, *6*(1), 114–125.

Wawro, M. (2014). Student reasoning about the invertible matrix theorem in linear algebra. *ZDM —The International Journal on Mathematics Education*, *46*(3), 389–406.

# Dealing with the Abstraction of Vector Space Concepts

**Lillias H. N. Mutambara and Sarah Bansilal**

**Abstract** University mathematics students often find the content of linear algebra difficult because of the abstract and highly theoretical nature of the subject as well as the formal logic required to carry out proofs. This chapter explores some specific difficulties experienced by students when negotiating vector space and subspace concepts. Seventy-three in-service mathematics teachers' responses to two items testing the ability to prove that a given set is not a subspace and that a given set is a subspace of a vector space were studied in detail. Follow-up interviews on the written work were conducted to identify the participants' ways of understanding. The action–process–object–schema (APOS) theory was used to unpack the structure of the concepts. Findings reveal that the teachers struggled with the vector sub-space concepts mainly because of prior non-encapsulation of prerequisite concepts of sets and binary operations and difficulties with understanding the role of counter-examples in showing that a set is not a vector subspace.

**Keywords** APOS · Vector subspace · Binary operations · Counter-example · Vector space

## 1 Introduction

Linear algebra is considered to be one of the most widely applicable subjects for students in the field of mathematics in that it can be applied to many different content areas, such as engineering and statistics, and can be studied for mathematical abstraction. Salgado and Trigueros (2015) note that linear algebra has become a compulsory course in many undergraduate degree programmes because of its wide application in the different disciplines. However, when the students take their first linear algebra course, they seem to encounter cognitive barriers. Almost two decades ago, Dorier, Robert, Robinet, and Rogalski (2000) noted that the

L. H. N. Mutambara · S. Bansilal (✉)
School of Education, University of KwaZulu-Natal, Durban, South Africa
e-mail: Bansilals@ukzn.ac.za



S. Stewart et al. (eds.), *Challenges and Strategies in Teaching Linear Algebra*,
ICME-13 Monographs, https://doi.org/10.1007/978-3-319-66811-6_7

teaching of vector spaces had completely disappeared in the secondary schools and the teaching had become less formal with no studies on algebraic structures. Some criticisms voiced by students about linear algebra concern the use of formalism and the lack of connections with what they already know, since this is not done at secondary level. Dorier et al. (2000) elaborated that the formalism is experienced when students need to learn new definitions, symbols, words and theorems. Stewart and Thomas (2010) noted that many students in the first years cope well with the procedural aspects of solving systems of linear equations but struggle to understand the crucial concepts underpinning the material involving the study of vector space concepts such as subspace, linear independence and spanning. Teachers often complain that the students have limited skills in elementary Cartesian geometry, and display an inconsistent use of the basic tools of logic or set theory (Dorier et al., 2000). The authors further argued that the lack of prior knowledge in logic and elementary set theory contribute much to the creation of errors in linear algebra.

Testing whether the axioms of vector spaces are satisfied or not requires an understanding of the formalism of setting out proofs and arguments in mathematics. Students need to make judgements about when to use a specific example and when to use a general description in setting out proofs. An argument that one of the axioms of a vector space is not satisfied requires a different kind of setup than showing that a set satisfies the vector space axioms. Similarly, showing that a subset W of a vector space V qualifies to be a subspace of V is qualitatively different from the process of identifying a counter-example to show that W is not a subspace of V. In the study on which this chapter draws, we explored some of these demands associated with the concepts of vector spaces and subspaces as experienced by a sample of 73 participants who may be considered as non-traditional undergraduate students. The participants were practising teachers who had not attained an undergraduate degree and were therefore considered as unqualified mathematics teachers. They had enrolled in the undergraduate degree programme in order to attain the qualification; hence this research was concerned with the difficulties experienced by these teachers in developing an understanding of vector subspace.

The study was guided by the action–process–object–schema (APOS) theory advocated by Dubinsky (1991). The following research questions underpinned the study:

- How do the participant teachers respond to the conceptual demands of the abstraction associated with vector space concepts?
- What APOS mental constructions can be inferred from the teachers' written and verbal responses to items based on vector space concepts?

We hope that the insights gained from this study will contribute to a deeper understanding of some of the epistemological barriers faced by students when studying vector spaces and that the results can be used to improve the design and delivery of the in-service upgrading programme offered to unqualified Zimbabwean teachers.

## 2 A Review of the Literature

An APOS study was conducted by Ndlovu and Brijlall (2015) based on pre-service teachers' mental constructions of concepts when learning matrix algebra. The study found that most of the pre-service teachers were operating at the action and process level, with a few operating at the object level. The authors argue that the lack of background knowledge of basic algebra schema hampered the teachers from developing adequate schemas at the object level. Many pre-service teachers could not manipulate numbers correctly when multiplying matrices and some of them failed to use notation correctly. The goal of mathematics teaching is that students understand mathematical concepts that are introduced to them or information that they discover for themselves. Hiebert and Carpenter (1992) asserted that one of the most widely accepted ideas in mathematics education is that students should understand mathematics. In studying student's conceptual understanding of a subspace Britton and Henderson (2009) argued that the abstract 'obstacle of formalism' and the theoretical nature of linear of linear algebra are the root causes of the difficulties experienced. They believed that lecturers teach students for procedural rather than conceptual understanding and that students have poor background knowledge of the concepts on proofs, logic and set theory. Their research was conducted with 500 students who had completed a first-year course in linear algebra and two calculus courses. One of the questions required the students to show that $V = \{t(1,2,3)|t \in R\}$ is a subspace of $\mathbb{R}^3$. Results revealed that most of the students could show that the set was non-empty but many were not able to prove that the set was closed under addition. Some students chose particular vectors instead of arbitrary vectors while many of them worked out the addition and then stated that the sum belonged to the set V (without explicitly showing why it did). Some had some misconceptions regarding the definition of a subspace and wrote solutions of the form $t(1,0,0) + t(0,2,0) + t(0,0,3)$ or statements such as "V spans $\mathbf{R}^3$ and dim $\mathbf{R}^3 = 3$". This showed that the students were mixing up concepts and showing rote learning of the concepts on vector subspace. The researchers also noted that students had problems with logic and set theory and with moving from abstract to algebraic representation. The students also experienced problems with the logic required to make the necessary proofs for problems on set theory. Hillel (2000) argued that the different modes of representation used in linear algebra posed a conceptual difficulty that is peculiar to linear algebra. The three different modes of representation are the abstract, algebraic, and geometric mode. The abstract mode uses the language of vector space, subspace, and linear span. The algebraic mode uses the language and concepts relating to matrices, rank, solutions of systems of equations, etc., while the geometric mode uses the language and concepts 2 and 3 dimensional space points, lines planes, etc. Hillel noted that students have difficulties in moving from one mode to another, as well as within the same mode. For example, the students within the geometric mode experienced challenges with the description of a vector which emanates from seeing it as an arrow and a point.

Britton and Henderson (2009) also noted that students experienced difficulties in moving from the abstract representation in which the question was usually phrased, to the algebraic representation which is required for the proof.

Wawro (2014) views reasoning as a valuable skill and part of the practice of mathematics. Wawro, Sweeney, and Rabin (2011) noted that in order to attain such reasoning skills, individuals should engage in mathematical activities of defining mathematical concepts, problem solving, proving and making arguments with justifications, as well as example generation. Wawro's study (2011) focused on students' understanding of the concept of a subspace using Tall and Vinner's (1981) theory on concept image and concept definition. The authors found that the students had varied definitions of the term 'subspace', and identified common imagery for a subspace as a geometric object, part of a whole, and as an algebraic object. They also noted that students had incorrect conceptions that $\mathbf{R}^{\mathbf{k}}$ is a subspace of $\mathbf{R}^{\mathbf{n}}$ for $k<n$. They concluded that the students struggled to understand mathematical ideas—especially definitions—because of the cognitive conflicts between the concept 'image' and concept 'definition'. However in the study, the definition helped the students to understand concepts and supported the development of further mathematical ideas.

Dorier et al. (2000) maintain that the difficulties with proof and formalism in understanding linear algebra are content specific. Uhlig (2002) asserts that some of the problems can be reduced if the approach to the teaching of linear algebra is changed from starting with definitions and lemmas to one that is exploratory in nature and where proofs are built in from the beginning, for example, by starting with a discussion about the solvability of linear equations before actually teaching how to solve them.

Stewart and Thomas (2009) looked at students' learning of basic linear algebra concepts in terms of Tall's (2008) three worlds of mathematics framework. They identified a linguistic confusion between scalar multiple of a vector and scalar product (dot product of vectors)—students considered the scalar $k$ as a vector of the form $(k_1, k_2, k_3)$. This showed unfamiliarity with recognising a vector and a scalar in different representations.

Some difficulties in linear algebra are related to misconceptions and errors that students can make. Several studies (Brodie, 2010; Luneta & Makonye, 2010; Siyepu, 2013) observed that many students perform badly in their first years at university mainly due to errors and misconceptions they inherit from their prior knowledge, as part of their experience. Misconceptions are seen as incorrect structures that students build, which is normally done repeatedly, and an error can be a mistake (Luneta & Makonye, 2010). Cangelosi, Madrid, Cooper, Olson, and Hartter (2013) reported that students memorise algebraic rules with no conceptual understanding attached to the concepts. The students then have difficulties in keeping track and applying the rules appropriately. They also found that language and notation can also hinder or enhance students' mathematical development.

Problems with theoretical and abstract concepts have been studied in different areas of mathematics. Hazzan (1999) maintains that students attempt to reduce the level of abstraction of new concepts that they learn in an attempt to make the concepts more mentally accessible so that they can work with them. Hazzan (1999) distinguishes between three interpretations of abstraction level. First, there is the interpretation of abstraction level as the strength of the relationships between the mental object and the thinking person; that is, previous experiences and interactions with the concept render it more familiar to the person. This view is consistent with the definition given by Hershkowitz, Schwarz, and Dreyfus (2001) of abstraction as "an activity of vertically reorganizing previously constructed mathematics into a new mathematical structure". The process of this vertical reorganization results in a more abstract concept.

A second interpretation of abstraction level is the process–object cognitive development path described in APOS theory. Although Hazzan (1999) does not refer to action conceptions, her description of working with procedures canonically is similar to the action conception in APOS theory which is evident when a student's working on a procedure is prompted by an external trigger or step-by-step calculation.

A third interpretation of abstraction level is the degree of complexity of the mathematical concept (Hazzan, 1999). Often students try to reduce abstraction by considering a simpler entity; for example, they may replace a set by one of its elements or they may ignore some of the properties of the object.

One of the complexities associated with abstraction is the logic and deduction required to prove or disprove statements. Zaslavsky and Ron (1998) conducted a study with 150 high school students who were given six mathematical statements (four of which were false) and they were asked to identify the false statements and say why they were false. They found that only one student generated four correct counter-examples for the false statements. For each of the four false statements, at least 33% were not able to determine that the statement was false. Two-thirds of the students did not find it appropriate to use counter-examples to show that a statement was false. However, there were many students who accepted a counter-example as sufficient evidence for refuting a statement, but were unable to distinguish between an example that satisfies the condition of a counter-example and one that does not satisfy them. In an attempt to generate a counter-example, they either used a counter-example to the converse statement or they used an example that satisfies the statement. Bansilal's (2015) study conducted with 48 pre-service student teachers found similarly that most participants were unable to produce a counter-example to show that the statement 'Every real number is rational' is not true. Thirteen participants stated that it was not true, but only five were able to produce a relevant counter-example. More than half the students were convinced that the statement was true, which provided "confirming examples for their assertion by choosing real numbers which were also rational" (Bansilal, 2015, p. 46).

Zaslavsky and Peled (1996) conducted a study with the aim of identifying difficulties associated with the concept of binary operation regarding the associative and commutative properties. The 103 participants (36 experienced teachers and 67 student teachers) were given a false statement and asked to produce a counter-example to convince a student that the statement was false. The results showed that 41 participants produced some example, but only 15 produced at least one correct example. Of the 62 participants who did not produce any example, 13 believed that the statement was true. Zaslavsky and Peled (1996) suggest that the issue of order was a critical source of confusion. They maintain that both commutativity and associativity deal with some sort of change in order. The first property deals with change in order of the elements while the second (associativity) involves change in the order in execution of the operations.

## 3 Theoretical Framework

We used the APOS theory as a framework to make sense of the data. According to Arnon et al. (2014), APOS theory is based on the extension of Piaget's (1965, 1973) principle of reflective abstraction that an individual learns mathematics by applying certain mental mechanisms to build specific mental structures. According to the APOS theory, the main mental mechanisms for building the mental structures include interiorisation, coordination and encapsulation. The mental structures refer to the action, process, object and schema. As actions are repeated and reflected on, the student moves from relying on external cues to having internal control over them. This is characterised by an ability to imagine carrying out the steps without necessarily having to perform each one explicitly. Interiorisation is the mechanism that makes this mental shift possible. Encapsulation occurs when an individual becomes aware of a process as a totality upon which transformations can act. At this stage the student can analyse properties of the object and compare objects arising from the same process (Arnon et al., 2014).

Many actions, objects and processes are interconnected in the individual's mind and these will be organised to form a coherent framework called a schema. An object can be assimilated by an existing schema, thus extending the span of the schema. According to Piaget, schema development also passes through stages of development. The intra level is the preliminary level and is characterised by analysing particular events or objects in an isolated manner in terms of their properties, where explanations are local and not global and relationships between objects may not be perceived. At the inter level, comparison and reflection upon properties of objects lead to the establishment of relationships. The individual can coordinate two different interpretations of the concept to mean the same thing. During the trans stage, the student reflects upon and coordinates the relations and is aware of the complete structure. Using these definitions, we now present a genetic decomposition of the vector space concept.

## 3.1 Genetic Decomposition of the Vector Space Concept

We draw upon the work of Parraguez and Oktaç (2010) and Arnon et al. (2014) to present a summarised description of the genetic decomposition of the vector space concept. The construction of the vector space concept is developed as the coordination of the prerequisite concepts of set and binary operations. Hence we refer to the set and binary operations schema as components of the vector space concept.

*Set schema.* At an action level, an individual conceives of a set when given a specific listing or a particular condition of set membership. The action of gathering and putting objects together in a collection according to some condition is interiorised into a process. This is encapsulated into an object when an individual can apply actions or processes to the process, such as compare two sets, consider a set to be an element of another and analyse properties of the set (Arnon et al., 2014).

*Binary operation schema.* A binary operation is a function of two variables defined on a single set or on a Cartesian product of two sets. At an action level, given a binary operation, an individual can take two specific elements of the sets and apply the formula. The individual interiorises the action into a process that takes two objects (elements) and acts on these to produce a new object (element) that is the result of the binary operation. At the object level, an individual can distinguish between two binary operations, check whether a binary operation satisfies an axiom, and compare objects arising from two different binary operations (Arnon et al., 2014).

Parraguez and Oktaç (2010) describe how these two schemas can be drawn together to form the notion of vector space:

> The objects that are sets with two kinds of operations (addition and multiplication by a scalar) can be coordinated through the related processes and the vector space axioms that involve both operations, to give rise to a new object that can be called a vector space (Parraguez & Oktaç, 2010, p. 2116).

The concept of vector space further evolves in the individual's mind as connections are made between other actions, processes, objects and schemas. Initially, for example, the student can verify different sets as being vector spaces or not, but does not see the vector space structure inherent in all of them. Parraguez and Oktaç (2010) describe this stage as being the Intra level in the development of the vector space schema. As an individual's conception deepens, a further stage called the Inter level can be identified where the object of vector space starts having relationships with other concepts such as subspace, linear transformations, basis, etc. When the student reflects upon these relations, through synthesis they can be recognised as part of a whole structure that makes up a vector space schema. This implies that the student is reasoning at the Trans level of the vector space schema and the student can recognise and work with non-standard examples of vector spaces and can invoke her/his schema when needed.

This description emphasises the complexity of the construction of the vector space concept which is built upon layers of abstraction. APOS theory developed from the ideas of Piaget about reflective abstraction. Dubinsky (1991, p. 99) explains that the "first part of reflective abstraction consists of drawing properties from mental or physical actions at a particular level of thought". This abstraction of properties "is projected onto a higher plane of thought" where other actions can be performed on the mental construction by drawing upon more powerful nodes of thought (Ibid, p. 99). De Lima and Tall (2008, p. 4) explain that as concepts become progressively more abstract, a parallel development occurs to compress these concepts "to construct thinkable concepts". These authors maintain that compression of knowledge is at the heart of mathematical thinking and the "process of making links leads to a compression of knowledge from complicated phenomena to rich concepts with useable properties and coherent links to other concepts" (De Lima & Tall, 2008, p. 4). The construction of the vector subspace concept is dependent upon a robust understanding of some prerequisite concepts. First, the binary operation and set concepts are developed through to higher levels of abstraction via the action–process–object path. The vector space concept is then constructed by coordinating these layers and testing the different axioms, which leads to an even higher layer of abstraction. As the vector space schema develops, at each stage the previous layer is re-organised as increasing coordination and coherence develop across the objects and relationships. The vector subspace concept is built upon this schema—students will not be able to see the connections between a vector space and a vector subspace if they have not developed the vector space schema up to at least an inter level when they can see links and inter-relationships between the various processes and objects.

## 4 Methodology

This study was conducted with 73 in-service mathematics teachers who did not have degree qualifications and were therefore considered as unqualified in Zimbabwe. The teachers were enrolled in a part-time in-service course at a Zimbabwean university that was designed to upgrade their qualifications, so that they could attain degrees. At the time of the study, the teachers had already completed a first course in linear algebra and calculus and were engaged in a second course in linear algebra that included the concepts of vector spaces, subspace, linear combinations, linear independence, basis, dimension, linear transformation and diagonalisation, eigenvalues and eigenvectors as well as solving systems of linear differential equations. However, it was noted that this module was taught concurrently with a module on mathematical discourse and structures. This module introduces students to the concepts on sets and relations, operations and structures, logic, mathematical proofs, and numbers. The design of the programme was such that the teachers would complete the equivalent of an undergraduate three-year degree programme except that the lectures were offered in two intensive block sessions for each semester. These block release sessions, coinciding with the school and university holidays, were very intensive with classes being held from 08:00 to 18:00 every day. This was designed so that the teaching time over the block sessions was equivalent to that for the full-time undergraduate degree, except that it was packed into the holiday period.

The data was collected from the teachers' written responses to an activity sheet consisting of nine items which were intended to probe their understanding of vector spaces and vector sub-spaces. In this chapter we focus on the responses to two tasks that were based on the vector space consisting of $2 \times 2$ matrices over the real field $\mathbb{R}$, and which were considered as having different levels of demand: one task required the teachers to confirm that a subset is a vector space and the second required them to show that a given set was not a subspace. Semi-structured interviews were conducted thereafter, with seven teachers who displayed different levels of engagement with the written tasks. The purpose of the interviews was to develop a deeper understanding of the ways in which the teachers responded to the tasks. Some of the interview questions were common to all participants and probed their experiences of the topic, while the rest of the questions were a follow-up to the issues noted in their solutions. The interviews were audio-recorded and thereafter transcribed verbatim.

The written responses of the teachers were analysed using themes suggested by the genetic decomposition. These themes were then organised so that they could serve as descriptors of cognitive difficulties experienced by teachers.

The two tasks based on the vector space of $2 \times 2$ matrices appear in Table 1, together with comments.

**Table 1** Research tasks

| Item | Comments |
|---|---|
| 1. Let **V** be the vector space over of all $2 \times 2$ matrices over the real field $\mathbb{R}$. Show that **W** is not a subspace of **V**, where **W** consists of all $2 \times 2$ matrices which have a zero determinant | For this, teachers were expected to find a counter-example to show that the set W is not closed under vector addition |
| 2. Show that the set of all $M_{2\times2}$ matrices of the form $\begin{bmatrix} a & 0 \\ 0 & b \end{bmatrix}$ is a vector space | For, this teachers could argue that since $M_{2\times2}$ is already a vector space, then it was only required to show that the given subset formed a vector subspace of $M_{2\times2}$. Alternatively, teachers could show that the ten axioms for a vector space were satisfied |

# 5 Results

The results are presented for Task 1 followed by Task 2. Note that T4, for example, refers to Teacher 4 and that the terms ‘students’ and ‘teachers’ are used interchangeably in these descriptions since the participants were teachers who were also students.

## 5.1 Results for Task 1

Task 1 required the teachers to generate a counter-example to show that the set of $2 \times 2$ matrices with a zero determinant is not a subspace. Task 1 was intended to provide insight as to whether teachers were on the path of developing strong schema of vector spaces. However, most teachers’ responses to this question showed that they had considerable difficulties with the formal reasoning required to present an argument why the set W did not fulfil the condition of being a subspace. To do that they needed to understand how counter-examples function in the process of rejecting conjectures (Bansilal, 2015; Zaslavsky & Ron, 1998).

The overall analysis for Task 1 showed that 16 (22%) of the teachers did not attempt the question, while 15 (21%) had completely incorrect responses. These teachers were quite lost in the task, There were 27 out of 73 (37%) of the teachers who attempted to add two matrices with zero determinants and showed that the sum had a zero determinant. Most of them proceeded to show the closure property of multiplication, and made various conclusions, many of which were incorrect. We now present more detail about some of the issues that emerged from the analysis of the written responses and some interviews.

### 5.1.1 Engaged Comfortably at Higher Abstract Layers of Reasoning

Some teachers’ responses suggested that they were comfortable with the reasoning required at the higher abstract layers. Seven students were able to present an example that did not satisfy the closure condition for vector addition. An example of such a response was that given by T6, who chose the two matrices: $\mathbf{x} = \begin{bmatrix} 0 & 1 \\ 0 & 0 \end{bmatrix}$ and $\mathbf{y} = \begin{bmatrix} 1 & 0 \\ 1 & 0 \end{bmatrix}$, and wrote (ii) $\mathbf{x}+\mathbf{y} \in V$: $\mathbf{x}+\mathbf{y} = \begin{bmatrix} 0 & 1 \\ 0 & 0 \end{bmatrix} + \begin{bmatrix} 1 & 0 \\ 1 & 0 \end{bmatrix} = \begin{bmatrix} 1 & 1 \\ 1 & 0 \end{bmatrix} \notin V$.

She concluded that the determinant of the sum $\mathbf{x} + \mathbf{y}$ was not zero, and hence did not satisfy the condition, hence the set **V** was not a subspace. She demonstrated understanding of the concept as she displayed a clear understanding of the determinant of a matrix. The teachers’ approaches and coherent arguments showed that they were comfortable with coordinating the binary operation and set processes to present arguments about why the subspace criteria were not fulfilled. These seven teachers' responses are aligned to that required by object-level reasoning about vector spaces. The teachers were able to coordinate the binary operation process and the set process and seem to have encapsulated the process into an object when they presented the argument about why the subspace criteria were not fulfilled. This supports Dubinsky's (1991) contention that an individual operating at the object level is able to take the process as a whole and create clear linkages between the concepts. However, most other teachers were unable to demonstrate such ease with the abstractions of the vector subspace concept and displayed different degrees of uncertainty.

### 5.1.2 Tried to Show the Set Was a Subspace, Contrary to the Instruction

The analysis identified some teachers who used examples to show that the set was a subspace, despite the instruction to the contrary. For example, T4 did not attempt Task 1. When probed in the interview, she explained her reasoning about why she concluded that the set was a subspace. R stands for the researcher and T4 represents Teacher 4. Note that the dialogue is capture verbatim and language errors have been left unchanged.

**R**: I understand you did not attempt this question. What exactly does this question requires us to do

**T4**: To find a matrix that gives determinant zero.

R: Oh OK, can you give an example of such a matrix?

**T4**: (Writing down) $\begin{bmatrix} 1 & 0 \\ 1 & 0 \end{bmatrix}$ det = 0.

**R**: So we have shown that it is not a subspace of the vector space.

**T4**: No, no, no, we find another matrix; we can use a multiple $\begin{bmatrix} 2 & 0 \\ 2 & 0 \end{bmatrix}$. So the closure property u + v = $\begin{bmatrix} 3 & 0 \\ 3 & 0 \end{bmatrix}$ det = 0, satisfied therefore subspace.

R: What else? Are we done? The question said, "Show that it is not a subspace." Is the set not a subspace of the given set?

**T4**: No, No, No take positive scalar k = 2 to give $2\begin{bmatrix} 1 & 0 \\ 1 & 0 \end{bmatrix} = \begin{bmatrix} 2 & 0 \\ 2 & 0 \end{bmatrix}$ det = 0 therefore subspace.

The excerpt above shows that T4 chose two vector elements whose sum belonged to the set, and took the single example as evidence that the set was a subspace. When probed further, she considered an example of a scalar multiple of a vector which also satisfied the condition on the subset. She was aware of the conditions that are needed to be satisfied by a subspace that is showing the closure property for addition and scalar multiplication. However, she was confused about the role of examples in proving or disproving a statement. Taking an example which satisfies a condition is not evidence that the condition is always true (Bansilal, 2015). At first she may have thought that she was showing that it is a subspace, which is why she then wanted to go on to the scalar multiplication condition. The teacher was also not clear about the determinant of a matrix, and did not view it as a function whose input is a matrix (Donevska-Todorova, 2014) but seemed to take it as a detached calculation.

### 5.1.3 Used an Illustrative Example to Show that the Result of the Binary Operations Belonged to the Set, But Concluded It Was not a Subspace

T13 considered four matrices, **v**, **u**, **s** and **t**, each with zero determinants as shown in Fig. 1.

T13 attempted to use two vectors and tried to show that the closure property of addition and scalar multiplication was not satisfied. However, for the vector addition, the determinant of the sum that she obtained was not zero, but she concluded that the sum belonged to W. The teacher made a computational error when trying to add the two matrices. The scalar multiple had zero determinant and these results were interpreted to mean that the set was not a subspace. The written response of T13 reveals her difficulty in showing that the set is not a subspace—she was just following procedures without understanding, which indicates action-level reasoning. Teaching students to engage in mathematics by applying a set of memorised

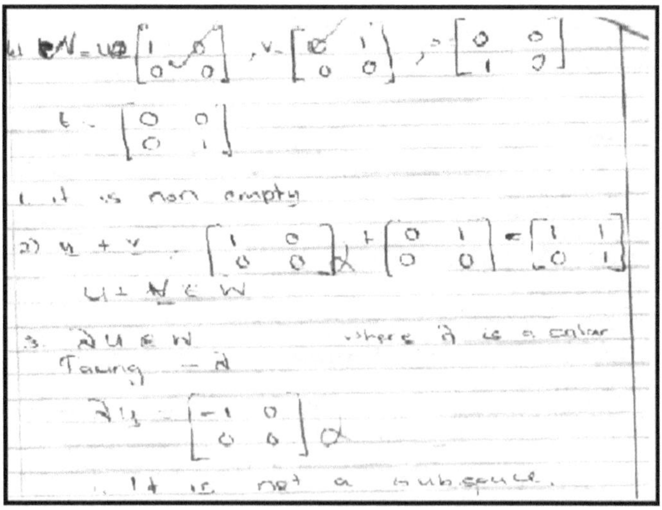

**Fig. 1** Response of T13 for Task 1

algorithms is seen as hindering their mathematical procedures (Foster, 2014). In an attempt to better understand why she was struggling, she was interviewed. The interview with T13 showed that she was still struggling to understand what the question really asked for:

R: You are talking of two vectors, **v** and **u**, so why did you choose three vectors for **v** and one vector for **s** in your solution?

T13: If the determinant is zero it is no longer a subspace. Maybe I confused myself because I see now that I must get a non-zero determinant. So I will choose matrix $\begin{bmatrix} 1 & 1 \\ 1 & 1 \end{bmatrix}$ and matrix $\begin{bmatrix} 2 & 1 \\ 1 & 2 \end{bmatrix}$. $A = \begin{bmatrix} 1 & 1 \\ 1 & 1 \end{bmatrix} + \begin{bmatrix} 2 & 1 \\ 1 & 2 \end{bmatrix} = \begin{bmatrix} 3 & 2 \\ 2 & 3 \end{bmatrix}$. Determinant is zero

R: So what can we do?

T13: I have another matrix A = $\begin{bmatrix} 4 & 1 \\ 1 & 2 \end{bmatrix}$, det = 8 − 1 = 7. Choose another one; I think so.
Then I will choose a negative scalar, will change for example $-2\begin{bmatrix} 1 & 1 \\ 1 & 1 \end{bmatrix} = \begin{bmatrix} -2 & -2 \\ -2 & -2 \end{bmatrix}$ and the determinant is zero. It doesn't, no it doesn't work—it is a scalar multiple.

The interview responses showed her struggle to show that the set is not a subspace. She stated that if the determinant [of the sum] is zero then it means that the set is not a subspace. Two matrices were then chosen, one of which did not belong to the set W since it did not have a zero determinant. She added the matrices and said the determinant was zero, which was not true. She then rushed to attempt to show the closure property of scalar multiplication, saying that a negative scalar should provide the counter-example. In her written response she also used $(-\lambda)$ as her scalar. Her ability to continue to use rules without reasoning is an indicator that she was still operating at the action level of understanding.

### 5.1.4 Confused About the Role of the Counter-Example

Some teachers knew they needed to produce a counter-example, but seemed not to know what the counter-example should show as shown in the written work of T39 appearing in Fig. 2.

He proceeded in a similar manner as explained in the interview by T4, except at the end T39 tried to twist the result to imply that closure property of addition was not satisfied on W. After adding the two matrices, he proceeded to find the determinant which was equal to zero. He then concluded that it showed that the resultant determinant was not equal to zero and concluded that it meant it could not

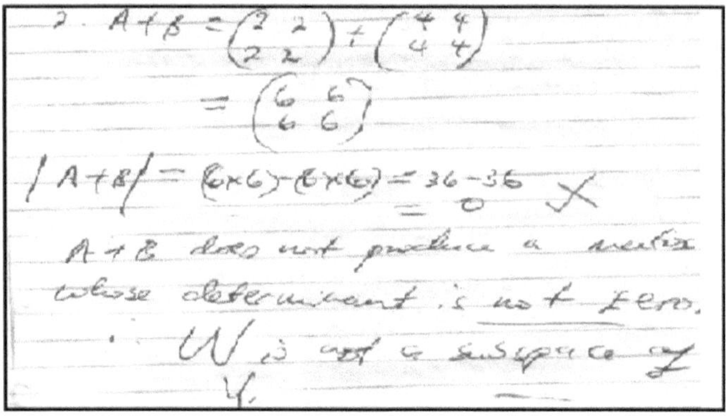


**Fig. 2** Written response of T39 showing confused argument

be a subspace. However his logic was misguided. He chose two elements belonging to the subset, added them and found that the determinant was 0, which does not indicate anything significant in this case. This suggests that he knew he was looking for a counter-example, but was not sure what the counter-example should show.

### 5.1.5 Not Able to Produce an Argument Around the Appropriate Counter-Example

There were some teachers who presented an appropriate counter-example but struggled to produce the accompanying argument about the counter-example as shown in Fig. 3.

The response of T46 shows a counter-example produced by the teacher because the determinant of the sum was not equal to zero but the teacher incorrectly concluded that the sum belonged to the subset.

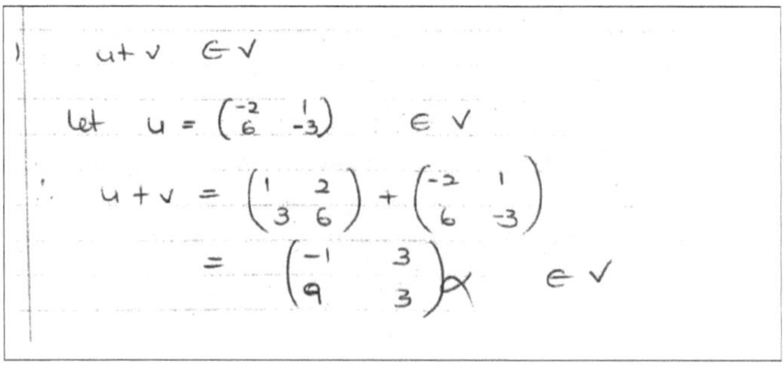


**Fig. 3** Response of T46 with a counter-example that was not recognised

### 5.1.6 Chose an Inappropriate Counter-Example

Unlike the case of teachers such as T46 who were able to identify suitable counter-examples, some teachers were unable to find an appropriate counter-example. For example, one teacher, T47, whose response appears in Fig. 4, elected two elements of W and assumed incorrectly that the determinant of the sum was not equal to 0. Hence she seemed to know what was required but could not identify the appropriate counter-example to fulfil her purpose. Note too that she did not mention explicitly the determinant of the sum, suggesting that she had a limited understanding of the determinant of a matrix and did not see it as a function which acts on a matrix.

The response R47 above shows that she added two matrices belonging to W that were made up of identical entries. The determinant of the sum was zero but T47 assumed that the determinant was not zero, allowing her to conclude that the set was not a subspace. This shows that T47 knew what she wanted from her counter-example but was not able to find the appropriate counter-example with the required property. However other teachers were not even clear about what they wanted to accomplish, as shown by T7 in Fig. 5.

### 5.1.7 Uncertain About What the Counterexample Should Do

Many teachers were not clear what the role of a counter-example is. The response from T7 shows a matrix with elements 1, 2, 3 and 4 and the teacher shows that the

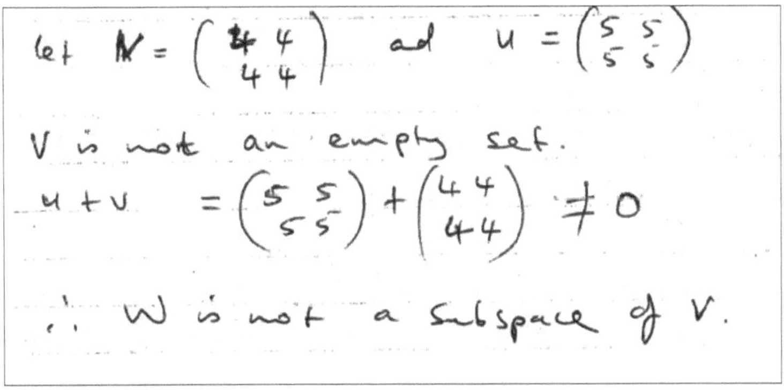


**Fig. 4** Response of T47 of matrices with zero determinants

**Fig. 5** Response of T7 showing uncertainty

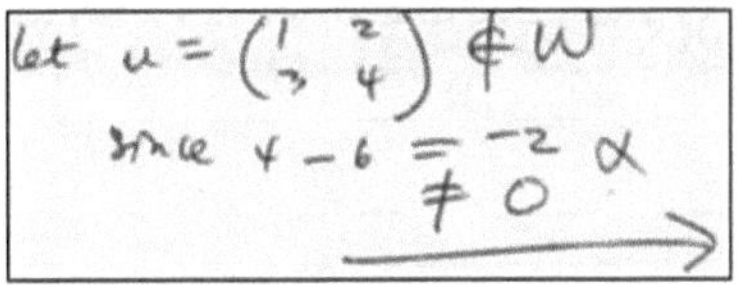

determinant of the matrix is not equal to zero. That is, he produces a $2 \times 2$ matrix, $\mathbf{u}$ which does not belong to the given set, and then shows that the determinant of $\mathbf{u} \neq 0$.

### 5.1.8 General Confusion

Some of the teachers, such as T69 and T12, produced responses which were unrelated to the questions, as shown in Figs. 6 and 7.

T69 incorrectly interpreted the problem and applied wrong procedures to solve the problem. He used the aspect of finding linear independence and concluded that it does not span the space. The answer given indicated that he saw some connection between subspaces, linear independence and spanning, but was not clear about how they are connected.

The response of T12 shows that he had considered a particular type of $2 \times 2$ matrices that have identical entries. These matrices belong to the given set, because their determinants are zero. The sum of the matrices does satisfy the condition of having a zero determinant, but the teacher was confused about what he was trying

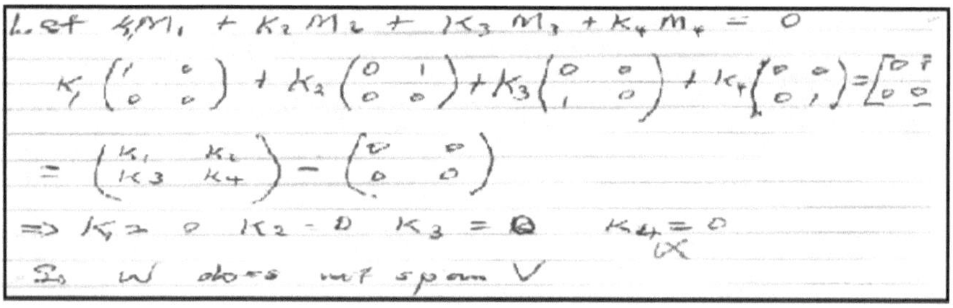

Fig. 6 Written response of T69 showing linear independence

Fig. 7 Response of T12 trying to use row echelon form

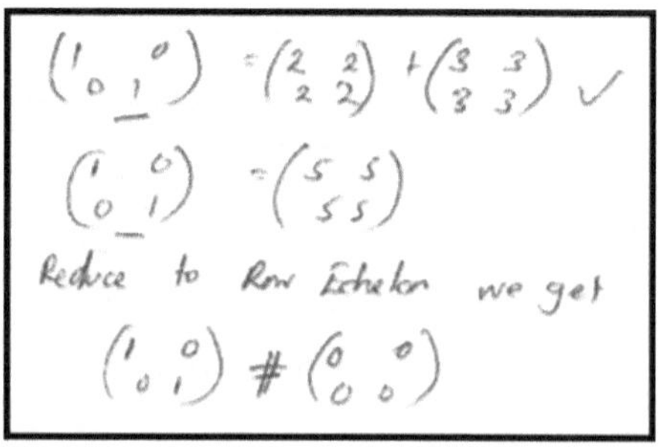

to do. He seemed to be trying to show that the sum should be equal to the identity matrix. He was also confused about equal matrices and brought in the aspect of reducing to row echelon form. These misconceptions had accumulated in a number of areas and emerged when the teachers were asked a question that required object-level reasoning about a vector space.

### 5.1.9 APOS Insights from the Responses to Task 1

Some teachers, such as T46, showed some progress towards interiorisation of the process of checking the subspace axioms, but struggled with articulating the arguments about why the set was not a subspace. However, most of the teachers had fundamental problems relating to the prerequisite schema for binary operations and set. The teachers did not seem to have developed object-level conceptions of those prerequisites and hence could not cope with the demands of the task which required the teachers to identify that the closure property of the binary operation on the set of $2 \times 2$ matrices was not fulfilled. Showing that a condition is not fulfilled requires sophisticated reasoning and arguments and this is not available to those who have not moved past a process conception of all the axioms. Many of these teachers were limited to carrying out procedures in a step-by-step manner. Dubinsky (1991) asserts that the ability to carry out procedures is at the action level of the APOS theory. The lack of the prerequisite construction of the set schema and binary operation schema hampered the teachers in developing a sufficiently strong schema for a subspace.

## 5.2 *Results for Task 2*

There were six teachers who went through each of the ten axioms and showed that they were satisfied by the elements of the given set. Three (3%) students presented totally incorrect responses, indicating that they had no idea of what was expected in this task. It seemed that these had not reached the action level using the genetic decomposition as they were still operating at the pre-action level. A further 12 (16%) of the students were able to identify and tried to prove some of the axioms, indicating action-level engagement with the set of $2 \times 2$ matrices. Fifty-three (73%) of the teachers attempted to show that the ten axioms for a vector space were satisfied; however, most of them had problems identifying exactly what they wanted to show for different axioms. There are some pertinent issues that emerged in the analysis related to how the teachers solved the problem. We identified five issues that emerged in the analysis of the responses to Task 2, and these issues are discussed below.

### 5.2.1 Difficulties in Recognizing What Needed to be Shown for Particular Axioms

Task 2 was intended to provide insight as to whether the teachers had developed a coherent vector space schema. However, some of them had problems identifying exactly what they wanted to show, for example the response by T11 indicates the teacher's attempt at showing that the set is closed under vector addition (Fig. 8).

The response of T11 shows that the teacher found the sum of an element of V and another arbitrary $2 \times 2$ matrix and then concluded that the sum belongs to V, without considering whether it satisfied the condition for elements to be in the set, similar to the finding reported by Britton and Henderson (2009) but which was for a different vector space. On the choice of the set V, this student seemed to be reproducing an example done in class. In terms of APOS, her actions of carrying out the binary operation for addition had not been interiorized into a process.

### 5.2.2 Used Specific Elements to Illustrate Axioms

There were many teachers who considered specific elements from the set, and showed that they satisfied the conditions of the axioms. The response from T8 in Fig. 9 shows such a response.

Figure 9 shows that the teacher had tested the axioms for specific elements of V, instead of considering generalised examples, an approach which was also identified in the study by Britton and Henderson (2009), using a different vector space. Furthermore, T11 was confused about the relationship between the property of commutativity of the binary operation and that of closure because he concluded that because $\mathbf{u} + \mathbf{v} = \mathbf{v} + \mathbf{u}$, it is true that $\mathbf{u} + \mathbf{v} \in \mathbf{V}$.

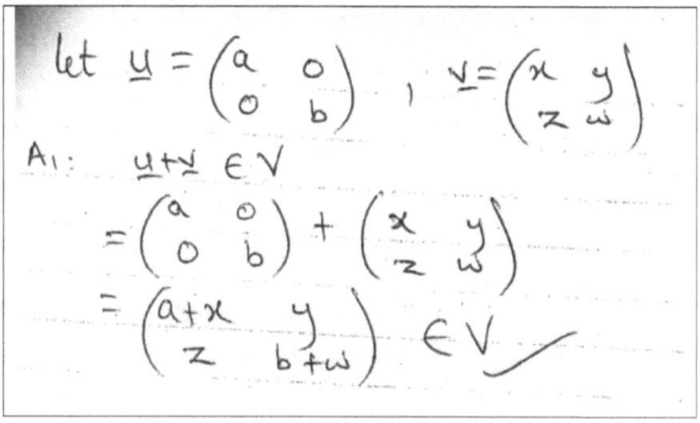


**Fig. 8** Response of T11 using one element from the set and another general $2 \times 2$ matrix

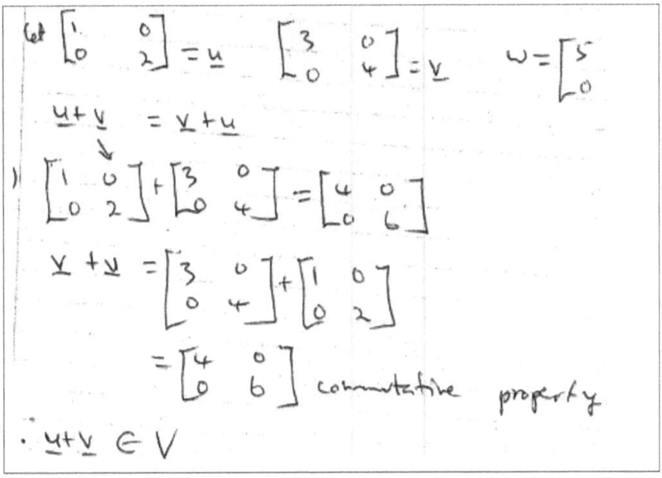

**Fig. 9** Response from T8 considering only specific elements

### 5.2.3 Confused About the Inverse Element for Vector Addition

Some teachers were evidently confused about what the identities for the different operations were. With respect to the additive inverse for vector addition, one teacher tried to show existence of the inverse of the 2 × 2 matrix as shown in Fig. 10.

Teacher 8, whose response appears in Fig. 10, showed that the determinant of the matrix is not zero which then implies that the inverse of the matrix exists. However, the required element was the inverse element for vector addition.

### 5.2.4 Confused About the Identity for Scalar Multiplication

The teachers' responses to Task 2 further showed that many of the teachers were able to state all the ten axioms but were unable to prove some of them. The responses showed that most of the teachers were unable to prove axiom ten which

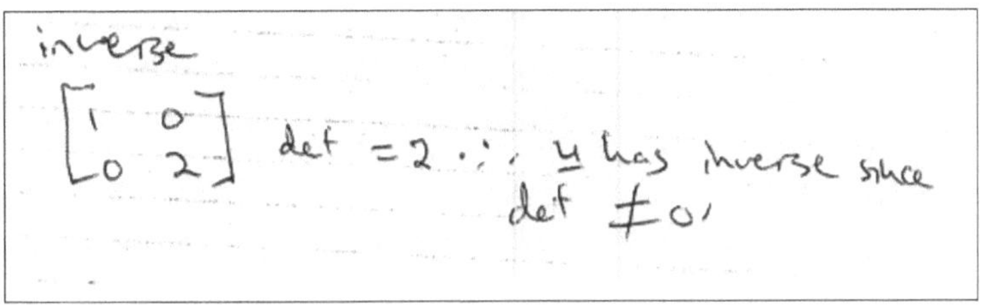

**Fig. 10** Response from T8 showing confusion between identity for vector addition and existence criterion for matrix inverse

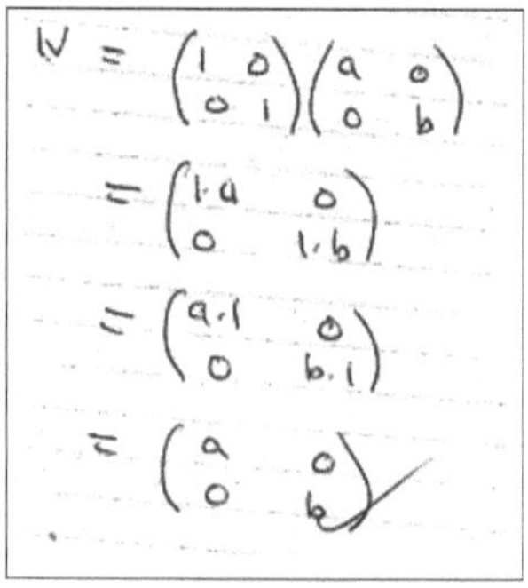

**Fig. 11** Response of T20 taking the identity matrix instead of the scalar identity

states that $\forall \mathbf{v} \in V$, $1.\mathbf{v} = \mathbf{v}$—only six teachers managed to prove that the axiom held. The teachers were not clear about what '1' in the axiom referred to, in the scalar multiplication $1.\mathbf{u}$ when the vector elements were matrices. The most common misconception was taking the scalar 1 as the identity matrix that is $\begin{bmatrix} 1 & 0 \\ 0 & 1 \end{bmatrix}$ which was shown by 44 teachers. One teacher, T20, took the identity matrix instead of the scalar value 1, and then carried out matrix multiplication, as shown in Fig. 11, hence he did not apply the scalar identity property.

In the interview, T7 revealed his confusion between the identity matrix and that of the identity for scalar multiplication:

**R**: [Referring to a question in the activity sheet]. How do you show axiom 10 that $1.\mathbf{v} = \mathbf{v}$?

**T7**: The 1 is represented by the identity matrix which is $\begin{bmatrix} 1 & 0 \\ 0 & 1 \end{bmatrix}$.

**R**: What if I just multiply by the scalar 1?

**T7**: Aah no, it is not still correct but this 1 is an identity we are multiplying with an identity, so this one must take the form of **v**.

**R**: So we cannot use the scalar 1?

**T7**: Yes 1 is a scalar. Here we are trying to show that eh … if we multiply a matrix with its identity.

**R**: But it is possible to multiply by 1?

**T7**: It is very possible if **v** is not a matrix.

The above excerpt shows that T7 had a misconception about scalar multiplication—he did not accept that it is possible to multiply a 2 × 2 matrix with a scalar and he therefore replaced the identity for scalar multiplication with the identity matrix $I_2$ so that multiplication of the matrix **v** by $I_2$ leaves **v** unchanged. This shows confusion between scalar multiplication and matrix multiplication—showing that the binary operation of scalar multiplication had not yet been interiorised.

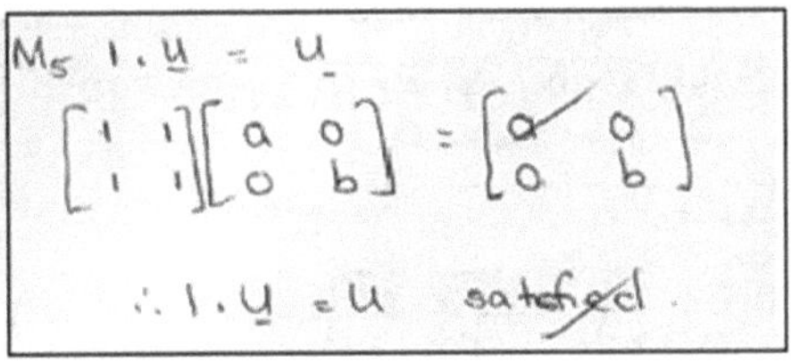


**Fig. 12** Response of T13 taking matrix with 1's as the scalar identity

Some teachers tried to use the identity matrix, but did not even identify it correctly. Two of them took $\begin{bmatrix} 1 & 1 \\ 1 & 1 \end{bmatrix}$ as an identity matrix, as shown in the response of T13 appearing in Fig. 12.

One teacher presented the matrix $\begin{bmatrix} 0 & 1 \\ 1 & 0 \end{bmatrix}$ as the identity, while another wrote $\begin{bmatrix} a & 0 \\ 0 & b \end{bmatrix} \cdot \begin{bmatrix} \frac{-1}{ab} \end{bmatrix}$, i.e. $\mathrm{v} \times v^{-1} = 1$. There were other teachers who took the identity element for scalar multiplication as the zero matrix. The widespread confusion and misconceptions related to the identity for scalar multiplication mainly arising from their weak background in set theory and the binary operations.

### 5.2.5 Confused About the Binary Operations

Confusion about the operation of vector addition was identified in the response by T14, who took vector addition as pairwise multiplication of corresponding elements.

The response in Fig. 13 shows that the teacher was aware that the property of addition needs to be satisfied, but instead of adding the two matrices, corresponding elements were multiplied. The teacher displayed the same misconception when he tried to prove the commutative as well as the associative property of addition, confirming that he had not developed even an action conception of the binary operation for addition and this did not allow him to develop the necessary

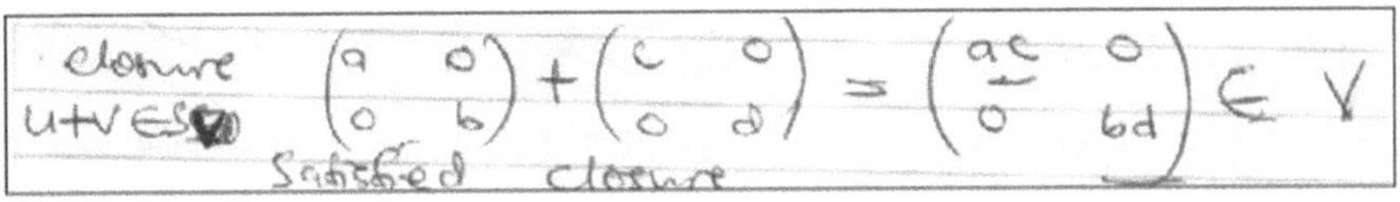


**Fig. 13** Response of T14 showing confusion between addition of matrices and pairwise multiplication of corresponding elements

construction for a strong vector space schema. Another teacher took the additive identity as the identity matrix and wrote:

$$\begin{bmatrix} a & 0 \\ 0 & b \end{bmatrix} + \begin{bmatrix} 1 & 0 \\ 0 & 1 \end{bmatrix} = \begin{bmatrix} a & 0 \\ 0 & b \end{bmatrix}.$$

Some of the students wrote u.0 = 1 for the additive identity.

### 5.2.6 APOS Insights

Only 6 (9%) of the students were able to move back and forth and managed to show correctly that the given set was a vector space. This required the coordination of the concepts of Set, and the two binary operations, to show that the axioms were satisfied. In terms of the genetic decomposition, it is only these teachers who had made progress towards the development of the vector space schema. Most other teachers were waylaid at many different places, indicating uneven interiorisation of some but not all concepts. Many students were able to state the axioms for the distributive and associative property of scalar multiplication, and had challenges in proving them, while many could not even state them. Teachers displayed a number of misconceptions peculiar to the different axioms. The widespread confusion between the identities and operations indicate that many of the teachers had not developed even process conceptions of the binary operations, because they were unable to carry out a binary operation if the two elements were not presented to them. It can be seen here that the teachers were just following procedure without understanding how some of the axioms are proved. In his research Harel (2000) noted that students struggle to understand the vector space axioms. The students whom he interviewed could not prove that for any A in a vector space V (−1) A = −A. The students could not articulate the argument indicated in the axiom, but Harel (2000) denoted that this proof was even done in class.

## 6 Discussion

The study attempted to uncover the conceptual difficulties that in-service teachers experience when learning the vector space concepts. We also made an attempt to understand some common misconceptions and errors that the students made. From the responses of the teachers, it was evident that many of the teachers were experiencing problems with the abstraction level of the vector space concepts with its axiomatic approach as well as formalism required for communicating the arguments.

Thomas (2011) posits that conceptual understanding, which is strictly related to the goal of becoming mathematically competent, is evident when one is able to solve problems in an unfamiliar situation. This is in line with Stewart and Thomas's

(2010) argument that many students experience difficulties when learning linear algebra because of its abstract and epistemological nature. This meant that the students in our study could not coordinate the set schema or the binary operation schema and so they could not carry out the process of checking all the axioms, and could not construct a robust proof. Some teachers were not clear about what the elements of the subset in Task 1 were. Students who were unable to identify elements of the set showed that they had not developed a conception of the set of matrices beyond an action level. Furthermore, some also displayed a weak algebraic background, as shown by teacher T13 in Fig. 1. This confirms some of the work reported in the literature that students find the abstract nature of concepts of a vector subspace challenging. The existence of different modes of representation contributes to their difficulties. The question on the subspace is represented in algebraic mode, whilst the definition of a subspace is phrased in abstract mode (Britton & Henderson, 2009). This is in agreement with Hillel (2000) who conducted a study on five experienced lecturers teaching concepts in linear algebra. He argued that the lecturers themselves confused the students because they persistently moved within the modes without explanations. The in-service teachers were unable to connect the algebraic and abstract mode of representation. Similarly, Hazzan and Zaski (2005) argue that abstractness of mathematics is complex because abstract concepts have many facets, with some concepts being more abstract than others.

We found that the teachers had some misconceptions emanating from previous concepts that they had encountered. Some teachers confused matrix addition with pairwise multiplication of the matrix elements. Others confused the identity elements for the binary operations with the identity matrix. It appeared as if many teachers did not understand that determinant is actually a function. This problem regarding determinants has led students to improper usage of the symbolic mathematical language as seen, for example, by T13 writing det $= 8 - 7 = 1$. The question is: determinant of what? It seems as if they do not understand that determinant is actually a function. The definition is substantiated by Donevska-Todorova (2014) who explains the determinant as a function. It may also be that the teacher's (T8) confusion about the existence of an inverse of a matrix (a non-zero determinant) in Fig. 10 arose from the teacher's misconception of the determinant.

In terms of APOS theory, the responses produced by many showed that they did not even have an action conception of binary operations. For example, the response of T14 in Fig. 13 showed that the teacher struggled to carry out a binary operation. It may be that for such teachers, the instructor may have moved too quickly to the more abstract treatments of binary operations which required process or object conceptions. However, APOS theory emphasises that a conception begins with an external action. The action level is a very important building block upon which other conceptions develop, and instructors must take care that enough attention is paid to practising the binary operation using various vector spaces before moving on to more complicated questions. Ndlovu and Brijlall (2015) argue similarly that when abstract algebra is introduced via definitions and axioms only, it can become a source of conceptual difficulty.

The analysis revealed that many teachers were confused about the identity elements for the binary operations. Teachers who struggled with identifying the identity elements but were able to carry out the binary operations also illustrated action conceptions, because they could only carry out operations on elements that were presented to them. Action-level conceptions were seen to be limiting because students could only carry out an operation in a step-by-step manner with the elements in front of them. In order to identify possible identity elements, then it is necessary for the student to have moved past looking at the binary operation as a step–by-step procedure (action) to one that has been interiorised (process) which allows the student to imagine the result of the operation. Those who resorted to the use of specific examples in Task 2 demonstrated action-level reasoning or to what Hazzan (1999) calls 'canonical procedures'. T8's use of specific elements, as shown in her response when attempting to show that the ten vector space axioms are satisfied, indicates that she has not moved past an action conception because she needs the comfort of the concrete matrices to carry out the vector addition operations.

For Task 2, some teachers showed that certain actions had been interiorised into processes as some of them were able to prove some of the nine axioms. However, it was clear that not even a process-level engagement with binary operations was sufficient to show that the axioms were satisfied. The response of T8 shows how the teacher confused the commutativity property and the closure property. Commutativity of operations and the distributive law requires that students are able to compare the results arising from different binary operations, which requires object-level reasoning.

Although some teachers demonstrated process-level engagement with the binary operations, this was not sufficient in providing a justification that the set W was not a vector subspace of V. The understanding of the role of examples in proving or disproving a statement was crucial in this task. If one wants to prove that a proposal does not hold, it is sufficient to produce one counter-example. However, if one produces an illustrative example of a proposal it is not sufficient to prove that the statement holds true. The responses of T39 (Fig. 2) and T4 (interview in Sect. 5.1.2) show that they produced illustrative examples which satisfied the condition (added two vector elements from a set W and showed that the sum belonged to W). The statements are that given any two elements of the non-empty W, the sum belongs to W and the scalar multiple of an element of W also belongs to W. To disprove this it is sufficient to produce a counter-example for any of the statements. However, to prove the statements, one would need to show that for any general elements each of the statements is true. The issue of the determinant being zero seemed to have caused some confusion in the minds of teachers such as T39 (in Fig. 2) who got entangled in the argument. Zaslavsky and Ron (1998) as well as Bansilal (2015) highlight the confusion experienced by students in distinguishing between an example that satisfies the condition of a statement and a counter-example that provides evidence that a statement does not hold. In Zaslavsky and Ron's study, the content was high school algebra and geometry and students struggled with using examples and counter-examples appropriately. In our study, the content was the highly abstract vector space concepts and providing counter-examples required a sound understanding of these concepts which explains why so many of the teachers struggled with the first task.

Some participants, such as T4 in this study, argued that W was a subspace of V contrary to what the question asked for. This was a similar response as those in the study by Zaslavsky and Ron (1998) where they found that almost a third of the students were not able to determine that certain statements were false. Similarly, Bansilal (2015) reported that more than half the participants took the statement that every real number is rational as true.

## 7 Conclusion

In this study we presented responses from 73 teachers who were enrolled in a linear algebra course at a Zimbabwean university. The study attempted to unpack some of the cognitive difficulties experienced by the teachers in negotiating the meaning of the various vector space concepts. The study showed that many problems were related to the teachers' understanding of the underlying concepts of binary operations and sets. We found that many teachers were confused about the identity elements for the binary operations and matrix operations. Further research is needed to help us understand why and how students become confused about these different operations, and to understand the extent of the difficulty in other vector spaces. Furthermore, most teachers struggled with explaining why a given set did not form a vector subspace because of the increased demand of using counter-examples appropriately. Further research may help us understand how to set out the teaching of proofs relating to when a subset of a vector space does not form a vector space. If the teaching of proof relating to this and other properties and relationships could be successfully scaffolded for students, they would be better prepared to deal with the abstractness of these concepts.

## References

Arnon, I., Cottrill, J., Dubinsky, E., Oktac, A., Fuentes, S. R., Trigueros, M., & Weller, K. (2014). *APOS Theory. A framework for research and curriculum development in mathematics education.* New York: Springer.

Bansilal, S. (2015). An exploration of students' conversions from a symbolic to a verbal representation. *Journal of Communication, 6*(1), 38–47.

Britton, S., & Henderson, J. (2009). Linear algebra revisited. An attempt to understand students' conceptual difficulties. *International Journal of Mathematics Education in Science and Technology, 40*(7), 963–974.

Brodie, K. (2010). *Teaching mathematical reasoning in secondary school classrooms.* London: Springer.

Cangelosi, R., Madrid, S., Cooper, S., Olson, J., & Hartter, B. (2013). The negative sign and exponential expressions: Unveiling students' persistent errors and misconceptions. *The Journal of Mathematical Behavior*, *32*(1), 69–82.

De Lima, R. N., & Tall, D. (2008). Procedural embodiment and magic in linear equations. *Educational Studies in Mathematics*, *67*(1), 3–18.

Donevska-Todorova, A. (2014). Three modes of description and thinking of linear algebra concepts at upper secondary education. In J. Roth & J. Ames (Eds.), *Beiträge zum Mathematikunterricht* (pp. 305–308). Munster: WTM-Verlag.

Dorier, J.-L., Robert, A., Robinet, J., & Rogalski, M. (2000). The obstacle of formalism in linear algebra. In J.-L. Dorier (Ed.), *On the teaching of linear algebra* (pp. 85–125). Dordrecht, The Netherlands: Kluwer Academic Publishers.

Dubinsky, E. (1991). Reflective abstraction in advanced mathematical thinking. In D.O. Tall (Ed). *Advanced mathematical thinking* (pp. 95–126). Springer: Netherlands.

Foster, C. (2014). "Can't you just tell us the rule?" Teaching procedures relationally. In S. Pope (Ed.), *Proceedings of the 8th British Congress of Mathematics Education* (pp. 151–158). Nottingham: University of Nottingham.

Harel, G. (2000). Principles of learning and teaching mathematics, with particular reference to the learning and teaching of linear algebra: Old and new observations. In J.-L. Dorier (Ed.), *On the teaching of linear algebra* (pp. 177–189). Dordrecht, The Netherlands: Kluwer Academic Publishers.

Hazzan, O. (1999). Reducing abstraction level when learning abstract algebra concepts. *Educational Studies in Mathematics*, *40*(1), 71–90.

Hazzan, O. & Zazski, R. (2005). Reducing abstraction: The case of school mathematics. *Educational Studies in Mathematics, 22*(1), 101–119.

Hershkowitz, R., Schwarz, B., & Dreyfus, T. (2001). Abstraction in context: Epistemic actions. *Journal for Research in Mathematics Education, 32*(2), 195–222. https://doi.org//10.2307/749673.

Hiebert, J., & Carpenter, T. P. (1992). *Learning and teaching with understanding*. New York: Macmillan.

Hillel, J. (2000). Modes of description and the problem of representation in linear algebra. In I. J. L. Dorier (Ed.), *The teaching of linear algebra in question* (pp. 191–207). Dordrecht, The Netherlands: Kluwer Academic Publishers.

Luneta, K. & Makonye, P. J. (2010). Learner errors and misconceptions in elementary analysis: A case study of a grade 12 class in South Africa. *Acta Didactica Napocensia, 3*(3), 35–46.

Ndlovu, D., & Brijlall, D. (2015). Pre-service teachers' mental constructions of concepts in matrix algebra. *African Journal of Research in Mathematics Science and Technology Education, 19* (2), 1–16.

Parraguez, M., & Oktaç A. (2010). Construction of the vector space concept from view point of APOS theory. *Linear Algebra and its Applications*, *432*(8), 2112–2124.

Piaget J. (1965). *The child's conception of number* (C. Gattegno & F. M. Hodgson, Trans.). New York: W.W. Norton (Original work published 1941).

Piaget, J. (1973). Comments on mathematical education. In A. G. Howson (Ed.), *Developments in mathematical education: Proceedings of the Second International Congress on Mathematical Education* (pp. 79–87). Cambridge, UK: Cambridge University Press.

Salgado, H., & Trigueros, M. (2015). Teaching eigenvalues and eigenvectors using models and APOS theory. *The Journal of Mathematical Behavior, 39*, 100–120.

Siyepu, S. W. (2013). Students' interpretations in learning derivatives in a university mathematics classroom. In Z. Davis & S. Jaffer (Eds.), *Proceedings of the 19th Annual Congress of the Association for Mathematics Education of South Africa, 1*, 183–193.

Stewart, S., & Thomas, M. O. J. (2009). A framework for mathematical thinking: The case of linear algebra. *International Journal of Mathematical Education in Science and Technology, 40*(7), 951–961.

Stewart, S., & Thomas, M. O. J. (2010). Student learning of basis, span and linear independence in linear algebra. *International Journal of Mathematical Education in Science and Technology, 41*(2), 173–188.

Tall, D., & Vinner, S. (1981). Concept image and concept definition in mathematics with particular reference to limits and continuity. *Educational studies in mathematics, 12*(2), 151–169.

Tall, D. O. (2008). The transition to formal thinking in mathematics. *Mathematics Education Research Journal, 20*(2), 5–24.

Thomas, S. (2011). *An activity theory analysis of linear algebra teaching within university mathematics* (Unpublished doctoral dissertation). University of Loughborough, UK.

Uhlig, F. (2002). The role of proof in comprehending and teaching elementary linear algebra. *Educational Studies in Mathematics, 50*(3), 335–346.

Wawro, M. (2014). Student reasoning about the invertible matrix theorem in linear algebra. *ZDM —The International Journal on Mathematics Education, 46*(3), 389–406.

Wawro, M., Sweeney, G., & Rabin, J. M. (2011). Subspace in linear algebra: Investigating students' concept images and interactions with the formal definition. *Educational Studies in Mathematics, 78*(1), 1–19.

Zaslavsky, O., & Ron, G. (1998). Students' understandings of the role of counterexamples. In A. Olivier, & K. Newstead (Eds.), *Proceedings of the 22nd Conference of the International Group for the Psychology of Mathematics Education* (Vol. 4). Stellenbosch: PME.

Zaslavsky, O., & Peled, I. (1996). Inhibiting factors in generating examples by mathematics teachers and student teachers: The case of binary operation. *Journal for Research in Mathematics Education, 27*(1), 67–78.

# Stretch Directions and Stretch Factors: A Sequence Intended to Support Guided Reinvention of Eigenvector and Eigenvalue

**David Plaxco, Michelle Zandieh and Megan Wawro**

**Abstract** In this chapter, we document the reasoning students exhibited when engaged in an instructional sequence designed to support student development of notions of eigenvectors, eigenvalues, and the characteristic polynomial. Rooted in the curriculum design theory of Realistic Mathematics Education (RME; Gravemeijer, 1999), the sequence builds on student solution strategies from each problem to the next. Students' used their knowledge of how matrix multiplication transforms space to engage in problems involving stretch factors and stretch directions. In working through these problems students reinvented general strategies for determining eigenvectors, eigenvalues, and the characteristic polynomial.

**Keywords** Linear algebra · Eigenvector · Eigenvalue
Realistic mathematics education · Inquiry oriented curriculum

## 1 Background

A number of researchers have studied various aspects of student conceptions of eigenvectors and eigenvalues (e.g., Gol Tabaghi & Sinclair, 2013; Salgado & Trigueros, 2015; Sinclair & Gol Tabaghi, 2010; Stewart & Thomas, 2006; Thomas & Stewart, 2011). This chapter focuses on aspects of student understanding relating to the equation $A\vec{x} = \lambda\vec{x}$. Specifically, we introduce an instructional sequence from the IOLA curriculum which is based on the instructional design theory of Realistic Mathematics Education (RME; Gravemeijer, 1999). We document existing student understanding and how it informs their approaches in this task sequence. These

D. Plaxco (✉)
Clayton State University, Morrow, GA, USA
e-mail: davidplaxco@clayton.edu

M. Zandieh
Arizona State University, Tempe, AZ, USA

M. Wawro
Virginia Tech, Blacksburg, VA, USA



S. Stewart et al. (eds.), *Challenges and Strategies in Teaching Linear Algebra*,
ICME-13 Monographs, https://doi.org/10.1007/978-3-319-66811-6_8

examples also demonstrate the types of student understanding the curriculum makes possible by engaging students in reflection on their own prior mathematical activity.

In previous work, members of our research team explored student understanding of the equation $A\vec{x} = \lambda\vec{x}$ or $A\begin{bmatrix} x \\ y \end{bmatrix} = 2\begin{bmatrix} x \\ y \end{bmatrix}$ in which the students were told that $A$ is a $2 \times 2$ matrix and $\vec{x}$ is a vector or $\begin{bmatrix} x \\ y \end{bmatrix}$ is a vector in $\mathbb{R}^2$ (Henderson, Rasmussen, Sweeney, Wawro, & Zandieh, 2010; Larson, Zandieh, Rasmussen, & Henderson, 2009). Students who were in a linear algebra class but had not yet studied eigentheory interpreted the equations in a variety of ways such as concluding the equation was only true if $A = 2$, concluding that $\det(A) = 2$, carrying out the multiplication to create a system of equations to solve for an $x, y$ pair (or pairs), and arguing that the way $A$ acts on the vector must be the same as what multiplication by 2 does to the vector. Students used a variety of symbolic, numeric, and geometric interpretations as they discussed the equation in terms of a system of equations, a linear transformation, or a vector equation. This is closely related to the framework of Larson and Zandieh (2013) who described a similar set of interpretations and representation used by students more broadly for the equation $A\vec{x} = \vec{b}$. Building on this research, our team developed an instructional sequence for learning eigenvalues and eigenvectors to mitigate issues that students might have with the equation $A\vec{x} = \lambda\vec{x}$. Rather than approaching eigentheory instruction by beginning with the equation $A\vec{x} = \lambda\vec{x}$, the sequence uses geometric notions of stretch factors and stretch directions of a linear transformation.

The eigentheory instructional sequence consists of four tasks and is the third of three units in the Inquiry-Oriented Linear Algebra curriculum (IOLA, iola.math.vt.edu). Each unit was developed from the perspective of Realistic Mathematics Education, which holds students' mathematical activity at the center of mathematical progress in the classroom (e.g., Freudenthal, 1991; Gravemeijer, 1999). Students work on tasks in small groups and explain their group's work to the rest of the class. A role of the instructor is to serve as a broker between students' mathematical activity and the mathematics of the mathematical community (Rasmussen, Zandieh, & Wawro, 2009; Zandieh, Wawro, & Rasmussen, 2017). One aspect of the role of the instructor is to introduce students to definitions and symbols used in the mathematics community that align with the mathematical activity in which students have already been engaged through their work on the tasks in the unit. In other words, in this curriculum definitions such as eigenvector and eigenvalue and symbols such as $A\vec{x} = \lambda\vec{x}$ are introduced only *after* the students have been working with the tasks in ways that experts would recognize as appropriate to symbolize with this expression.

In Units 1 and 2, the curriculum develops and explores various linear algebra concepts and how they relate to each other. These include: linear combination, span, linear independence, row reduction, systems of equations, linear transformations, and matrix operations. Unit 3 of the IOLA curriculum develops diagonalization and eigentheory. The first two tasks of Unit 3 are discussed in detail in

Zandieh, Wawro, and Rasmussen (2017). We summarize that student activity in the next section to help frame the story of this chapter, in which we share the third task of the sequence. Task 3 focuses on student exploration of the relationships that an expert would think of as being summarized by the equation $A\vec{x} = \lambda\vec{x}$. Our discussion of Unit 3 Task 3 centers on examples of typical student responses from small group discussions in two classes. We collected these examples from students' work during semester long implementations of the IOLA curriculum at two different universities. Students working on this Task drew on their mathematical experience with Tasks 1 and 2 of Unit 3 as well as their work in prior units. In order to provide a sense of how students in these two classes produced their responses, we briefly outline the IOLA curriculum prior to this task and the types of activity in which students had been engaging.

## 2 Students' Prior Mathematical Activity

In general, the IOLA materials provide students with early and frequent opportunities to interpret problem situations using systems of equations, vector equations, and matrix equations, as well as to translate between these representations and explain connections between them. Specifically, in Unit 1, which is about span and linear independence, students have opportunities to represent travel scenarios (involving vectors representing travel on a magic carpet and a hover board to particular locations) as vector equations (Wawro, Rasmussen, Zandieh, & Larson, 2013; Wawro, Rasmussen, Zandieh, Sweeney, & Larson, 2012). Many students convert these vector equations to systems of equations, and some who have previous experience in linear algebra represent the equations using an augmented matrix or a matrix equation.

In Unit 2 of the IOLA curriculum, students represent geometric transformations of Cartesian space as a matrix times an input vector and, subsequently, as a matrix times a matrix of concatenated input vectors (Andrews-Larson, Wawro, & Zandieh, 2017; Wawro, Larson, Zandieh, & Rasmussen, 2012). This way of representing transformations begins with students' work with the "Italicizing N" task. In this unit, students complete a series of tasks to determine matrices for various transformations based on a description of the transformations' effect on specific input vectors. For instance, based on Fig. 1, students often generate the matrix equations $\begin{bmatrix} a & b \\ c & d \end{bmatrix}\begin{bmatrix} 0 \\ 3 \end{bmatrix} = \begin{bmatrix} 1 \\ 4 \end{bmatrix}$ and $\begin{bmatrix} a & b \\ c & d \end{bmatrix}\begin{bmatrix} 2 \\ 3 \end{bmatrix} = \begin{bmatrix} 3 \\ 4 \end{bmatrix}$ as they try to determine the matrix $A$ that acts on the pre-image "N" to produce an image of a larger, italicized "N." Some students represent these two equations as a product between the unknown matrix and a matrix of concatenated input vectors: $\begin{bmatrix} a & b \\ c & d \end{bmatrix}\begin{bmatrix} 0 & 2 \\ 3 & 3 \end{bmatrix} = \begin{bmatrix} 1 & 3 \\ 4 & 4 \end{bmatrix}$. Students then rewrite these as systems of equations and solve for the variables $a$, $b$, $c$, and $d$ to determine the matrix of the transformation in the standard basis.

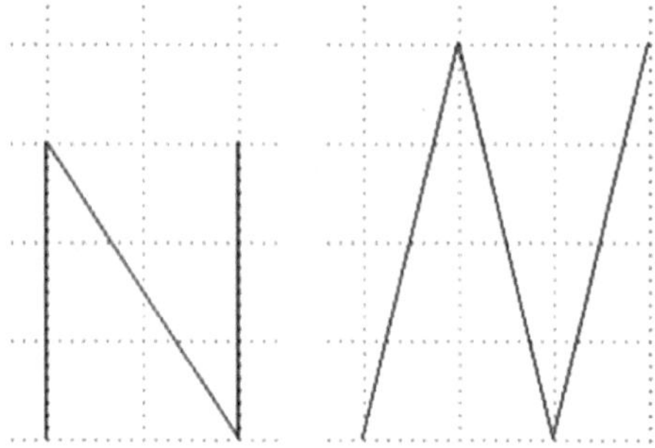

**Fig. 1** Pre-image and image in the italicizing N task

Later, in Unit 2 Task 3, students explore the composition of linear transformations by representing the same transformation as before in two steps: one matrix that stretches the "N" to make it taller and another matrix to take the taller "N" as input and "italicize" it by shearing. The teacher builds from student work to assist them in developing an understanding of the composition of linear transformations as matrix multiplication through a substitution between the two equations. Finally, Unit 2 culminates in a task that engages students in determining the matrices that "undo" the three transformation matrices developed in Tasks 1 and 3, leading to the formal definition of the inverse of a matrix and a linear transformation. Throughout Unit 2, students are continually shifting between matrix equations and systems of linear equations to solve for unknown values in a given matrix.

Unit 3 begins with a task that describes a transformation from $\mathbb{R}^2$ to $\mathbb{R}^2$ that stretches vectors along two directions (represented by the linear equations $y = x$ and $y = -3x$) by the stretch factors 1 and 2, respectively (Zandieh et al., 2017). Building upon the approaches developed during Unit 2, students often produce the matrix equations $\begin{bmatrix} q & r \\ s & t \end{bmatrix}\begin{bmatrix} 1 \\ 1 \end{bmatrix} = \begin{bmatrix} 1 \\ 1 \end{bmatrix}$ and $\begin{bmatrix} q & r \\ s & t \end{bmatrix}\begin{bmatrix} -1 \\ 3 \end{bmatrix} = \begin{bmatrix} -2 \\ 6 \end{bmatrix}$. This typically leads to the development of a system of four equations with four unknowns. Along with this activity, students are asked to sketch the result of the transformation of the plane, which helps lead to a discussion about representing the plane relative to a basis comprised of vectors in the stretch directions and considering the linear transformation relative to that basis. This in turn motivates a change of basis, which instructors can readily represent with a commutative diagram and the diagonalization equation, $A = PDP^{-1}$.

Although students typically solve Unit 3 Task 1 using the equations above, occasionally, students might represent their work using the equations $\begin{bmatrix} q & r \\ s & t \end{bmatrix}\begin{bmatrix} 1 \\ 1 \end{bmatrix} = 1\begin{bmatrix} 1 \\ 1 \end{bmatrix}$ and $\begin{bmatrix} q & r \\ s & t \end{bmatrix}\begin{bmatrix} -1 \\ 3 \end{bmatrix} = 2\begin{bmatrix} -1 \\ 3 \end{bmatrix}$, in which the stretch factor is explicitly written as a scalar on the right-hand side of the matrix equation. These equations are what we are calling in this chapter "matrix times vector equals scalar times vector" ($mtv = stv$) equations. Specifically, we use the $mtv = stv$ label to denote representations of the eigen equation that use numbers and variables in arrays of matrices and n-tuples. Although to the expert, these equations are simply a more specified version of the generalized eigen equation $A\vec{x}=\lambda\vec{x}$, we want to distinguish student use of different types of symbolizations to emphasize transitions in their reasoning. As part of making this distinction we call the equation $\begin{bmatrix} q & r \\ s & t \end{bmatrix}\begin{bmatrix} -1 \\ 3 \end{bmatrix}=2\begin{bmatrix} -1 \\ 3 \end{bmatrix}$ an $mtv = stv$ equation but call the equation $\begin{bmatrix} q & r \\ s & t \end{bmatrix}\begin{bmatrix} -1 \\ 3 \end{bmatrix}=\begin{bmatrix} -2 \\ 6 \end{bmatrix}$ an $mtv = v$ equation. This choice may seem odd because the equations are distinguished only by whether the scalar is multiplied by the entries in the vector on the right-hand side of the equation. However, the equation $\begin{bmatrix} q & r \\ s & t \end{bmatrix}\begin{bmatrix} -1 \\ 3 \end{bmatrix}=\begin{bmatrix} -2 \\ 6 \end{bmatrix}$ may initially appear to students to be simply another example of equations such as $\begin{bmatrix} a & b \\ c & d \end{bmatrix}\begin{bmatrix} 0 \\ 3 \end{bmatrix}=\begin{bmatrix} 1 \\ 4 \end{bmatrix}$ that they encountered in Unit 2. We see equations such as $\begin{bmatrix} q & r \\ s & t \end{bmatrix}\begin{bmatrix} -1 \\ 3 \end{bmatrix}=\begin{bmatrix} -2 \\ 6 \end{bmatrix}$ as a fulcrum between the ideas about linear transformations from $\mathbb{R}^2$ to $\mathbb{R}^2$ the students learned in Unit 2 and the new ways of reasoning about $mtv = stv$ and $A\vec{x}=\lambda\vec{x}$ equations that the students need to learn in Unit 3.

In particular, $mtv = v$ equations like $\begin{bmatrix} q & r \\ s & t \end{bmatrix}\begin{bmatrix} -1 \\ 3 \end{bmatrix}=\begin{bmatrix} -2 \\ 6 \end{bmatrix}$ connect to students' existing, concrete ways of thinking about linear transformations geometrically and to the matrix equation notation $A\vec{x}=\vec{b}$. Another important connection is that $mtv = stv$ equations can be converted into $mtv = v$ equations and then rewritten as a system of equations, which students use to solve for unknown variables. Finally, $mtv = stv$ equations can be used to support connecting these aspects of linear transformations more formally with the general eigen equation. Thus, the notation used in $mtv = stv$ equations allows students to engage in specific, contextualized mathematical problem solving that is leveraged to support general notions of eigenvectors and eigenvalues.

We have provided this outline highlighting prior tasks in the curriculum to emphasize the types of thinking and solution strategies students in our courses typically have available when they approach the problems in Unit 3 Task 3. Of specific importance are their ways of representing transformations from $\mathbb{R}^2$ to $\mathbb{R}^2$ as a matrix times a vector (or matrix of concatenated vectors) and translate to a system of equations to solve for unknown values in these matrices and vectors, specifically using the $mtv = stv$ and $A = PDP^{-1}$ equations.

## 3 Discussing Task 3 and Results from Students' Solutions

As stated in the introduction the overall learning goal for Task 3, which is composed of three problems, is for students to explore the relationships involved in the equation $A\vec{x} = \lambda\vec{x}$ and to develop intuitive notions of eigenvalue and eigenvector. As with earlier tasks, we cast the problems in this task geometrically, in terms of stretch factors and stretch directions, but we ask students to provide numeric solutions, giving students the impetus to create and manipulate symbolic expressions to find those numeric solutions. The three problems are ordered in increasing level of difficulty. Having already asked students (in Task 1) to find a matrix given stretch factors and stretch directions, we now recast this by switching which information is given and which is requested, as follows:

P1. The matrix and the stretch directions are given and students are asked to find the stretch factors.
P2. The matrix and the stretch factors are given and students are asked to find the stretch directions.
P3. The matrix is given, and students are asked to find both the stretch factors and the stretch directions.

In creating the Task, we have chosen to restrict the problems so that students would work in $\mathbb{R}^2$, i.e., with $2 \times 2$ transformation matrices (Fig. 2). This keeps the systems small enough so that the students can realistically solve three of them within a single 50–75 min class period and also ensures that students encounter only linear and quadratic polynomials in their work.

This sequence is intended to allow students to develop a connection between the problem statements, which are given in terms of stretch factors and stretch directions, and the general eigen equation $A\vec{x} = \lambda\vec{x}$. As discussed above, the *mtv* = *stv* equation emerges from student work on the problems in the Task. At first the equation is more of an expression by students of the fact that the transformation matrix stretches or shrinks the stretch vector by the amount of the stretch factor. As the Task progresses, students must use variables to represent unknown stretch

1. The transformation defined by the matrix $A = \begin{bmatrix} 1 & -8 \\ -4 & 5 \end{bmatrix}$ stretches images in $\mathbb{R}^2$ in the directions $y = \frac{1}{2}x$ and $y = -x$. Figure out the factor by which anything in the $y = \frac{1}{2}x$ direction is stretched and the factor by which anything in the $y = -x$ direction is stretched.
2. The transformation defined by the matrix $B = \begin{bmatrix} -8 & 2 \\ -55 & 13 \end{bmatrix}$ stretches images in $\mathbb{R}^2$ in one direction by a factor of 3 and some other direction by a factor of 2. Figure out what direction gets stretched by a factor of 3 and what direction gets stretched by a factor of 2.
3. The transformation defined by the matrix $C = \begin{bmatrix} 7 & -2 \\ 4 & 1 \end{bmatrix}$ stretches images in $\mathbb{R}^2$ in two directions. Find the directions and the factors by which it stretches in those directions.

**Fig. 2** Problem statements in Unit 3 Task 3

factors and stretch directions. From this need, variations on the $mtv = stv$ equation emerge in students' work. It is rare for students to use $\lambda$ as the symbol for stretch factors; symbols such as $c$ or $k$ are more common. It is not until their work in this task is connected by the instructor to the broader mathematical community's eigenvector and eigenvalue conventions that students switch to the more common $\lambda$. Thus, in this chapter, we use $k$ or $c$ in our generic discussions of student symbolizations to help emphasize that students are not yet familiar with the terminology or common notation associated with eigenvectors and eigenvalues.

Because of their prior work in the unit, students are typically able to connect the stretch direction and stretch factor language with matrix multiplication notation, identifying how the product of a matrix and vector can come to represent a vector being stretched under a transformation of a vector space. This is consistent with student work in Unit 3 Task 1, in which students are asked to determine the matrix of a transformation that stretches vectors along two given lines by respective factors. In the time between Unit 3 Task 1 and Unit 3 Task 3, the students will have completed two lessons involved in developing notions of change of basis matrices as a means for representing linear transformations that stretch along a basis of stretch directions. This also provides students with the ability to incorporate the equation $A = PDP^{-1}$ into their work.

In Problem 2 (Fig. 2), students are given a different matrix and stretch factors and are asked to find the corresponding stretch directions. They should notice that there are infinitely many ways to describe the stretch direction for a given stretch factor. Also, importantly, students are not able to merely calculate the product of the matrix times a vector or the stretch factor times a vector as they may have before, but instead must use a generalized stretch direction vector in their approach the problem. Because of this, we conjecture that students are more likely than before to write a matrix equation with the product of the stretch direction and stretch factor on the right-hand side. Problem 3 only provides students with the matrix and asks them to find both stretch directions and stretch factors. In this problem, students will need to recognize that they cannot solve for any of the unknowns directly, but that there are infinitely many solutions for the stretch direction. In addition, students' work (specifically, on problem 3) can be leveraged here and later in Task 4 to develop the idea of the characteristic polynomial and how finding its roots for a given matrix is equivalent to determining the stretch factors of that matrix.

In the following subsections, we provide examples of common student approaches to Problems 1–3. We have chosen the examples of student work based on how representative they are of students' approaches and also based on their usefulness for being leveraged to support more general and formal ideas of eigentheory.

### 3.1 Finding Stretch Factors

As shown in Fig. 2, Problem 1 provides students with stretch directions and a given matrix and asks them to find the stretch factor for each stretch direction. Students initially realize they will need to find at least one vector that lies on the line $y = 1/2x$ and at least one vector that lies on the line $y = -x$. Two common choices are $\begin{bmatrix} 2 \\ 1 \end{bmatrix}$ and $\begin{bmatrix} -1 \\ 1 \end{bmatrix}$, respectively. Students then determine the factor by which each of these vectors is stretched when multiplied by the given matrix.

The first example of student work that we discuss (Fig. 3) exemplifies a typical approach that we have seen after several implementations of the IOLA curriculum. This group of students began by multiplying the given matrix $A$ times the vectors $\begin{bmatrix} 2 \\ 1 \end{bmatrix}$ and $\begin{bmatrix} -1 \\ 1 \end{bmatrix}$, which yielded the vectors $\begin{bmatrix} -6 \\ -3 \end{bmatrix}$ and $\begin{bmatrix} -9 \\ 9 \end{bmatrix}$, respectively. This is a form of the $mtv = stv$ equation in which the scalar multiple is distributed into the vectors on the right-hand side. From this, the students re-wrote the vectors on the right-hand side of the equation as scalar multiples of the vectors on the left-hand side of the equation. Although not written on their board, the students indicated in class that they (correctly) interpreted their work to imply that the desired stretch factors were 3 and −9.

In our second example, students leveraged the equation $A = PDP^{-1}$ (see Fig. 4a). To do this, they relied on the knowledge that, for a given diagonalizable matrix $A$, its stretch factors are the diagonal entries of $D$ and its stretch directions, in column vector form, are the respective columns of the matrix $P$. More specifically, this group parameterized the matrix $D$ with the unknown diagonal entries $a$ and $d$, determined the matrices $P = \begin{bmatrix} 2 & 1 \\ 1 & -1 \end{bmatrix}$ and $P^{-1} = \begin{bmatrix} 1/3 & 1/3 \\ 1/3 & -2/3 \end{bmatrix}$ from the given information, and substituted these matrices (and also the given matrix $A$) into the equation $A = PDP^{-1}$. Following this, they multiplied the three matrices on the right and set the resulting matrix equal to the given matrix for the transformation. This

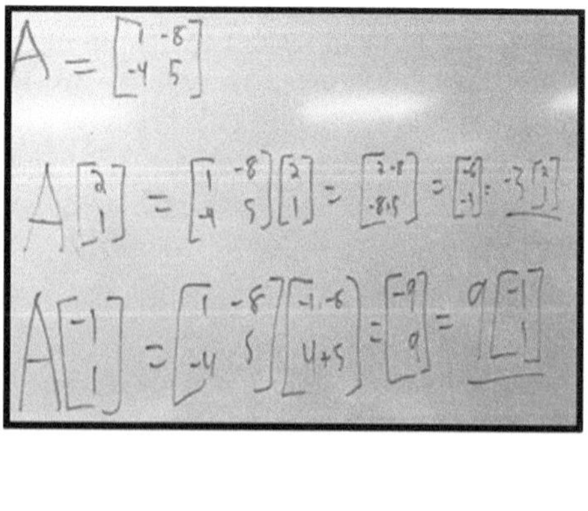

**Fig. 3** Most common approach to Task 3 Problem 1

(a) (b)

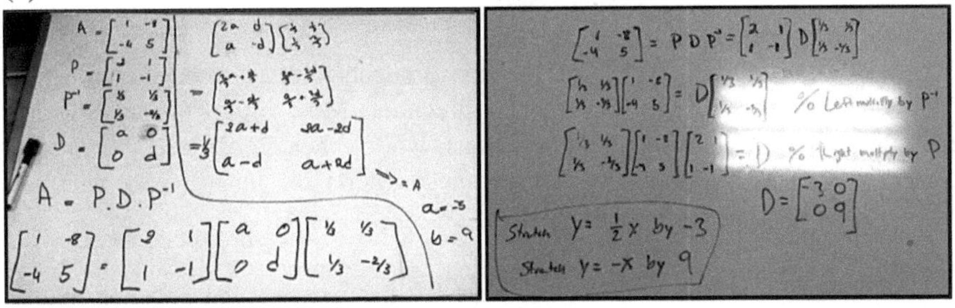


**Fig. 4** Students' work on Task 3 Problem 1 relying on $PDP^{-1}$

allowed the students to solve for $a$ and $d$ by setting corresponding components of the matrices equal to each other.

In the last approach we discuss, another student group also used the diagonalization equation $A = PDP^{-1}$ (Fig. 4b). In particular, this group determined how to use the given information in the diagonalization equation, manipulate the equation, and solve for the matrix $D$. They wrote the diagonalization equation $A = PDP^{-1}$ with the given matrix $A$. They represented the stretch direction of $y = \frac{1}{2}x$ and the stretch direction $y = -x$ as the column vectors $\begin{bmatrix} 2 \\ 1 \end{bmatrix}$ and $\begin{bmatrix} 1 \\ -1 \end{bmatrix}$, respectively, and they used that information to create matrix $P$ and substitute it and $P^{-1}$ (we are not sure how they computed the inverse) into the diagonalization equation (Fig. 4b, line 1). The students explained that they left multiplied by $P^{-1}$ and right multiplied by $P$ to solve for $D$ (Fig. 4b, lines 2–3). The product $P^{-1}AP$ yields $\begin{bmatrix} -3 & 0 \\ 0 & 9 \end{bmatrix}$ (Fig. 4b, line 4), which the students equated to $D$ and interpreted in terms of stretch factors and directions, namely that the transformation represented by $A$ stretch $y = \frac{1}{2}x$ by $-3$ and $y = -x$ by 9.

Because the stretch directions are given in equation form, the students must choose a single vector in each direction. This is consistent with and builds on the students' work in Unit 3 Task 1, which first introduced the notions of stretch direction. As we saw in the first example, students are typically able to recognize that they only need to multiply the given matrix times a vector along the stretch direction and notice that the product is a scalar multiple of the original in order to answer the question. As demonstrated, students sometimes write this as an *mtv* = *stv* equation with the scalar factored out on the right-hand side (last row in Fig. 3). We have found this to be less common in our implementation of the curriculum, with students usually determining the stretch factors without explicitly factoring the right-hand side. However, as we demonstrate in the next section, Problem 2 tends to support students' production of the *mtv* = *stv* equation with the scalar factored.

## 3.2 Finding Stretch Directions

In contrast to Problem 1, Problem 2 provides students with a matrix and two stretch factors and asks them to find the stretch directions. Most groups use an $mtv = stv$ equation to generate a system of equations, while other groups use the equation $B = PDP^{-1}$. Figure 5 shows a very detailed version of student work using the $mtv = stv$ equation. This group used $\begin{bmatrix} a \\ c \end{bmatrix}$ as a generic stretch direction vector that is multiplied by the given matrix $B$ on the left-hand side of the equation and the given scalar, 3, on the right-hand side of the equation. The vector $\begin{bmatrix} b \\ d \end{bmatrix}$ is their generic vector that is multiplied by the matrix $B$ on the left and scalar 2 on the right. The group then used each of these matrix equations to generate a system of two equations with two unknowns. The students combined like terms to convert each system into standard form for systems with the variables on the left and a constant (in this case 0) on the right-hand side of the equation. The students do not state on the board why, but in each case they use the first of the two equations to write an expression of one variable in terms of the other ($a = \frac{2}{11}c$ and $d = 5b$) and then convert these equations to a specific vector in each direction: $\begin{bmatrix} a \\ c \end{bmatrix} = \begin{bmatrix} 2 \\ 11 \end{bmatrix}$ and $\begin{bmatrix} b \\ d \end{bmatrix} = \begin{bmatrix} 1 \\ 5 \end{bmatrix}$. The students even reference a connection to the $B = PDP^{-1}$ relationship at the bottom of their work by listing a matrix, $P$, with the two vectors they found as its column vectors.

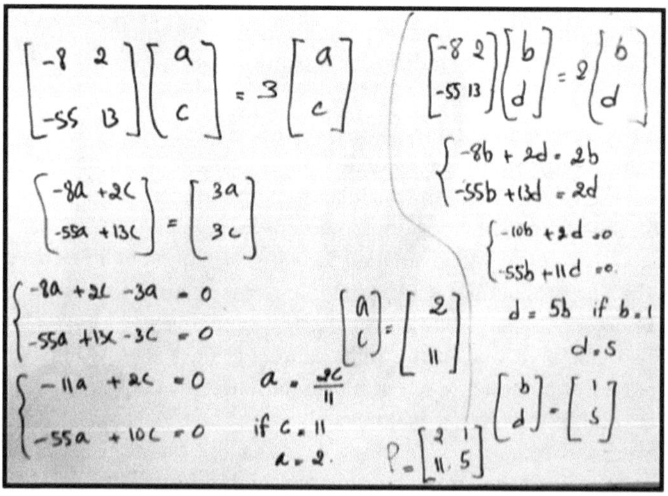

**Fig. 5** Problem 2 solved by converting $mtv = stv$ into a system of equations

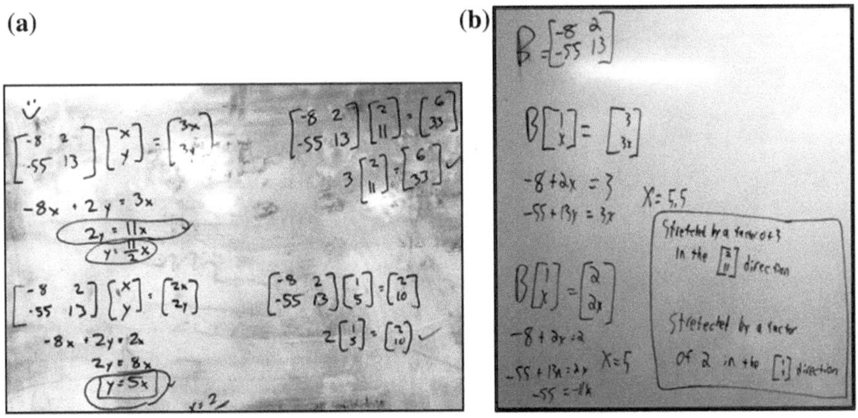


**Fig. 6** Additional student work on Problem 2 using $mtv = stv$ equations

Variations of this method include work similar to that in Fig. 6a, in which the group only wrote the first of the system's two equations on their boards. This is sufficient since the two equations describe the same line and, thus, only one needs to be considered. Figure 6a is also different than Fig. 5 in that the students used $\begin{bmatrix} x \\ y \end{bmatrix}$ as their generic vector in each case and circled the results of $y = 11/2x$ and $y = 5x$. In this way they seemed to be emphasizing the standard format for a line through the origin where $y$ is typically written in terms of $x$. This group also found a particular vector in the direction of each line and multiplied that vector times the original matrix to check that indeed was multiplied by 3 (or 2).

In Fig. 6b we see a unique variation on this strategy. These students chose a vector $\begin{bmatrix} 1 \\ x \end{bmatrix}$, with 1 in the first component and therefore only one variable, relying on the fact that any vector in the stretch direction will work. (Their strategy would fail only if the eigenvector lies along the direction $\begin{bmatrix} 0 \\ 1 \end{bmatrix}$.) Because of the choice of $\begin{bmatrix} 1 \\ x \end{bmatrix}$ each of their equations solves for a single value of $x$, e.g., $x = 5.5$ when the scalar is 3. These students then converted their answer into a stretch direction stated as a vector with integer values, e.g., $\begin{bmatrix} 2 \\ 11 \end{bmatrix}$ instead of $\begin{bmatrix} 1 \\ 5.5 \end{bmatrix}$. This method emphasizes the stretch direction as a vector direction without stating it as the equation of a line as in the circled part of Fig. 6a. We also point out here that the groups whose work appears in Fig. 6 did not write the right-hand side of the equation as a scalar times a vector, but instead distributed each stretch direction into the vector on the right. This is a nontrivial distinction from other forms of the $mtv = stv$ equation, specifically because the students' distribution of the stretch factor into the stretch vector does not lend itself to the manipulation of a more

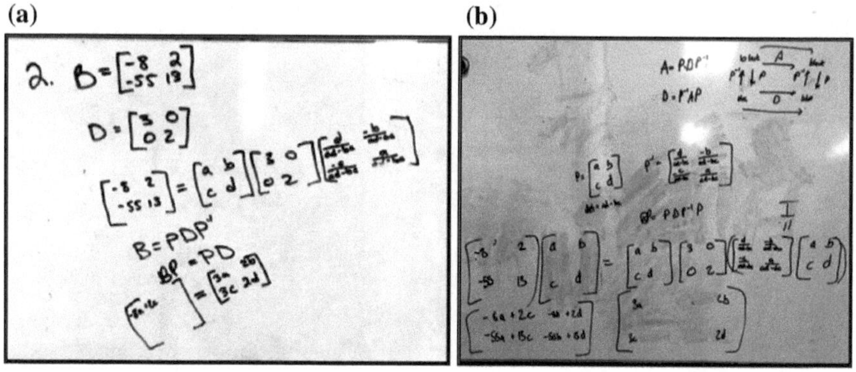

**Fig. 7** Student work on Problem 2 using $B = PDP^{-1}$

general $A\vec{x} = k\vec{x}$ equation that could be used to lead to the equation $(A - kI)\vec{x} = \vec{0}$. Accordingly, it is important for instructors to point out these distinctions and, if necessary, draw out the connections during whole class discussion.

In Fig. 7 we see student work using $B = PDP^{-1}$. Neither of these are resolved to a final solution. The method using $PDP^{-1}$ creates a more complicated matrix with fractions (Fig. 7a). Resolving this equation into $BP = PD$ creates simpler matrices (Fig. 7b). Once these matrices are multiplied and set equal, the next step would be to set the corresponding components of each of the resultant matrices equal. This would create four equations that are identical to the systems of equations created using the *mtv* = *stv* method. However, neither of these groups continued on the white board beyond creating the two resultant matrices.

## 3.3 Finding Both Stretch Factors and Stretch Directions

Students are typically able to draw on a variety of their prior approaches to solve Problem 3, which, in contrast to Problems 1 and 2, provides neither the stretch directions nor the stretch factors of the transformation. Because of this, in order to solve the problem, students must identify either the stretch factor or stretch direction and then use one to solve for the other.

There are several ways in which students can find the stretch factors first. Two of these are illustrated in Fig. 8. In each case students constructed an *mtv* = *stv* equation with variables for both components of the eigenvector and a variable for the eigenvalue. In Fig. 8a we see that one group set up proportions to generate a single equation in terms of *k*. Although this group of students did not make it explicit, the ratio is the slope of the line described by each equation. With the proportion in terms of *k*, the students developed a quadratic equation. They may have noticed in solving Problem 2 (or recognized because they are solving for a single eigendirection) that both equations in the system of equations describe the

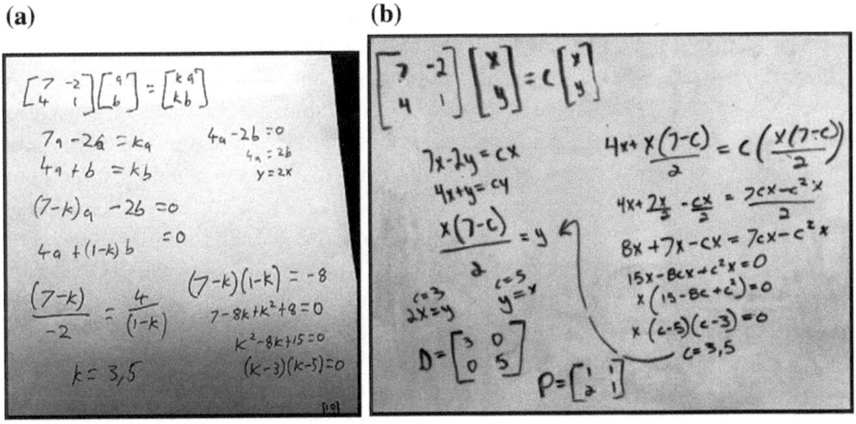


**Fig. 8** Two groups' solutions to Problem 3

same line and thus have the same slope. After solving the quadratic for the stretch factor, the students were then able to determine the corresponding stretch directions, one of which is shown on their whiteboard (Fig. 8a).

The other group opted to solve the first equation for $y$ and substituted it into the second equation (Fig. 8b). The second group then manipulated the resulting equation into the equation $x(c - 5)(c - 3) = 0$. This group did not indicate whether the x-component in the stretch direction might be zero, but focused on solutions for stretch factors. After determining the stretch factor values of 3 and 5, this group substituted these values into the original system of equations and interpreted the result of the substitution (the equations $2x = y$ and $x = y$) as stretch directions. A single vector from each direction was then chosen for the two columns of the matrix $P$.

The ways in which these two groups manipulated the system of equations can be leveraged to support a discussion of the characteristic polynomial and the standard manipulations used to calculate it. Specifically, it is helpful to juxtapose the two systems of equations in Fig. 8a with the matrix equations $\begin{bmatrix} 7 & -2 \\ 4 & 1 \end{bmatrix}\begin{bmatrix} a \\ b \end{bmatrix} = k\begin{bmatrix} a \\ b \end{bmatrix}$ and $\begin{bmatrix} 7-k & -2 \\ 4 & 1-k \end{bmatrix}\begin{bmatrix} a \\ b \end{bmatrix} = \begin{bmatrix} 0 \\ 0 \end{bmatrix}$, as well as the more generalized equations $A\vec{x} = k\vec{x}$ and $(A - kI)\vec{x} = \vec{0}$. We have found that this helps students to draw parallels between the three pairs of symbolizations so that each can be used to make sense of the other.

Furthermore, the instructor can draw on the Invertible Matrix Theorem to motivate the need to calculate $\det(A - kI)$ and, in so doing, introduce the notion of the characteristic polynomial. Such a discussion would begin with the instructor pointing out (or supporting students to identify) the need for a nonzero vector as a solution to the original eigen equation and therefore to the equation $(A - kI)\vec{x} = \vec{0}$.

Students can then use their knowledge of the equivalences in the Invertible Matrix Theorem to discuss in class the properties of $A - kI$ needed for $(A - kI)\vec{x} = \vec{0}$ to have a nonzero solution. As part of this review, students should see that one such property is $\det(A - kI) = 0$. The instructor can help students to see that the equation $\det(A - kI) = 0$ is in fact the equation (or a variation of the equation) that they have already used to calculate the stretch directions. In telling the students that the name of this equation is the "characteristic equation", the instructor serves as a broker connecting the students' mathematics to the mathematical terminology used by the larger mathematics community (Rasmussen, Zandieh, & Wawro, 2009). More generally the instructor may choose to leverage the student work to make connections to the derivation of the standard method for calculating eigenvalues and eigenvectors through the equation $(A - kI)\vec{x} = \vec{0}$.

Another method for solving this problem is to find the stretch directions first. Figure 9a, b show one group's work, which we have separated into two images. As with the other groups, this group began with the *mtv* = *stv* equation and used it to generate a system of equations. However, in each equation of the system, they solved for the stretch factor, *k*, and set the remaining algebraic statements equal to each other in an equation that reflects a proportion. The group then simplified this equation to yield a quadratic in two variables: $4a^2 - 6ab + 2b^2 = 0$. Factoring this and drawing on the zero product property, the group was able to produce the two equations $b = 2a$ and $b = a$, which they recognized as the stretch directions. Following this, the group found the corresponding stretch factors by selecting a single vector along each stretch direction and continuing in a manner similar to their approach to Problem 1.

Although the first two approaches are much more common, Fig. 10 illustrates a unique approach that also incorporates the equation $AP = PD$, derived from the equation $A = PDP^{-1}$. The students' work is difficult to parse because the students

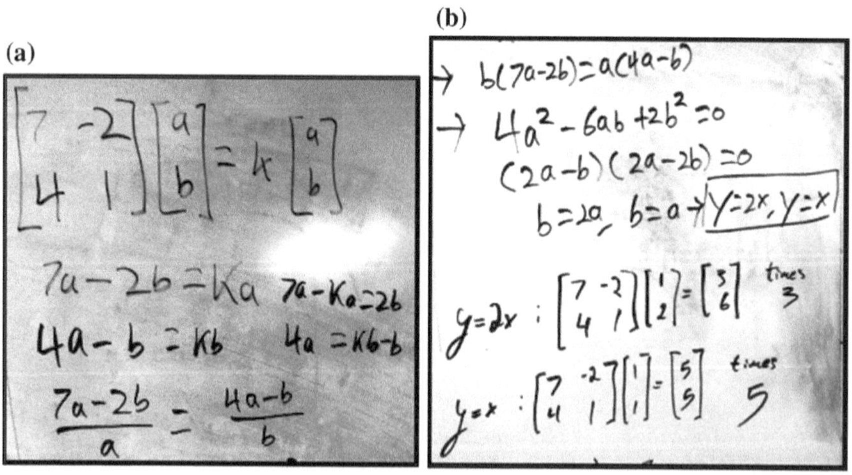


**Fig. 9** Student response to Problem 3 that uses *mtv* = *stv* approach

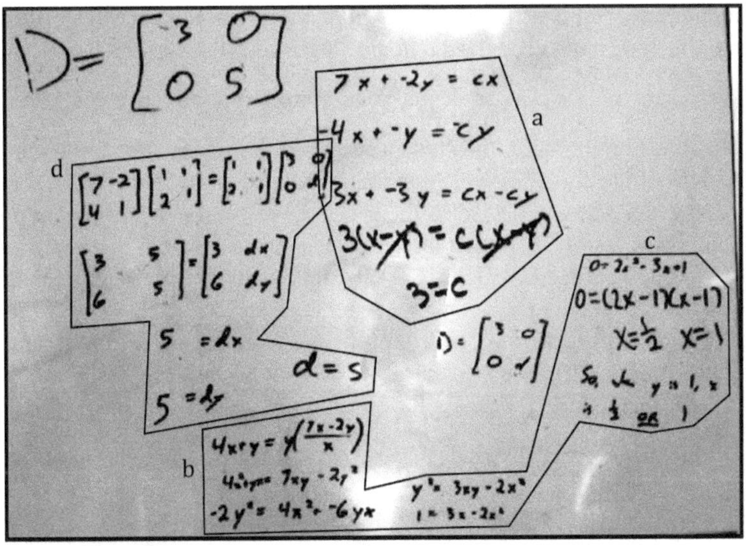


**Fig. 10** Group solution using a system of equations and $AP = PD$ to finish Problem 3

did not show all of their work or denote implications. However, we can tell that the group began by generating a generic system of equations from the matrix with unknown stretch direction vectors and stretch factors (Fig. 10a). With this system, the group was able to combine the two equations and factor the resulting equation to yield $3(x - y) = c(x - y)$. The group then canceled the binomial $(x - y)$ from each side of the equation to produce $c = 3$. Although it is not written on their whiteboard, this last step tacitly assumes that $x - y \neq 0$.

Another aspect of the work in Fig. 10a is that it can be generalized to an interesting fact about eigenvalues of $2 \times 2$ matrices. That is, if the column entries add to the same number or (as in this case) subtract to opposite numbers then this number (or one of the opposite numbers) will be an eigenvalue of the matrix. For instance, in this case, the columns of the given matrix subtract to give $7 - 4 = 3$ and $-2 - (-1) = -3$, yielding an eigenvalue of 3. Although students who develop this approach will likely not try to generalize this fact, it might be helpful for instructors to ask students to develop arguments for or against the generalizability of this pattern.

In their work, the students interpreted $c = 3$ as the first of two stretch factors, which they represented with the diagonal matrix $\begin{bmatrix} 3 & 0 \\ 0 & d \end{bmatrix}$ in the matrix equation $AP = PD$ (Fig. 10d). The students in this group also generated another equation from the system by substituting for $c$ to generate $4x + y = y\left(\frac{7x - 2y}{x}\right)$, which could then be simplified (Fig. 10b). After substituting, the students were able to generate the quadratic equation $y^2 = 3xy - 2x^2$.

Importantly, because of the students' prior work solving for stretch directions and stretch factors, they recognized that the solutions to this quadratic correspond to the components of the vector $\begin{bmatrix} x \\ y \end{bmatrix}$, representing the stretch directions of the transformation. Furthermore, the students realized that, with stretch direction vectors, the ratio of the components is important, rather than a specific value for $x$ and $y$. This understanding is reflected in the group's substitution of 1 for $y$ to produce the equation $1 = 3x - 2x^2$, which the students are able to solve more readily as a quadratic in one variable ($x = 1$ or ½). The group then interpreted the solutions of this quadratic equation as $x$ components in vectors with 1 in the $y$ component (Fig. 10c) and, more generally, as a ratio between $x$ and $y$. Although there is no written evidence that the group was aware of the implications, they chose a nonzero $y$-value in the stretch direction vector. Their interpretation of these solutions is shown in Fig. 10d where they substituted the vectors $\begin{bmatrix} 1 \\ 2 \end{bmatrix}$ and $\begin{bmatrix} 1 \\ 1 \end{bmatrix}$ into the columns of the $P$ matrix in the equation $AP = PD$. In this last step, the students used this explicit form of the $AP = PD$ equation to solve for the remaining stretch factor of 5.

Students might not recognize that this approach would not generalize to matrices with stretch directions that align with standard basis vectors—specifically eigenvectors that have zero in the component that the students set equal to 1. This being said, the approach reflects an understanding that the stretch directions are proportion-based, rather than fixed vectors. Although this group's approach is not as common as others, we find value in the types of conversations that such work can introduce into whole-class discussion. We also value the diversity in student approaches, whether they find the stretch factors first, the stretch directions first, or some combination of the two.

## 4 Concluding Remarks

In this chapter, we have delineated the usefulness of student fluidity between the eigen-equation in the various forms of matrix equations, systems of linear equations, and the equation $A\vec{x} = \lambda\vec{x}$. The tasks in Unit 3 were developed in such a way as to build and extend work that students have previously done with $A\vec{x} = \vec{b}$ equations and their various equivalent forms. The examples of student responses to the three problems in Unit 3 Task 3 that we provided in this chapter illustrate several important types of reasoning that support a robust understanding of eigentheory. Specifically, the task allows students to leverage their existing ways of representing linear transformations with matrix equations composed of numbers and variables—what we have denoted as $mtv = stv$ or $mtv = v$ equations. Students are then able to interpret these matrix equations as systems of equations in order to shift their reasoning towards developing approaches to solving equations of the form $A\vec{x} = k\vec{x}$.

As the task progresses, each subsequent problem varies which of the three components (eigenvector, eigenvalue or both) are unknown. This was designed intentionally to allow students to interpret the outcome of their activity in terms of stretch directions and stretch factors based on their work on the previous problem, as well as in Unit 3 Tasks 1–2. In this way, the students' work with Unit 3 Task 3 is meant to involve referential activity, a key component of the instructional design theory of Realistic Mathematics Education (Gravemeijer, 1999).

Unit 3 Task 3 culminates in the instructor leveraging students' solutions to Problem 3 and generalizing their use of the $mtv = stv$ and $mtv = v$ equations. In addition, the students we have worked with have begun to generalize the various relationships in the eigen-equation beyond the specific $2 \times 2$ examples of the task. This is meant to lead to an introduction and discussion of the characteristic polynomial, with its standard derivation resulting directly from generalizing activity. Furthermore, the instructor plays a crucial role as broker between the classroom and broader mathematical community by connecting students' work with stretch factors and stretch directions with the more widely known terms of eigenvalues and eigenvectors, respectively. Through this discussion, students' activity is guided toward a reinvention of eigentheory from a meaningful, problem-based approach.

## References

Andrews-Larson, C., Wawro, M., & Zandieh, M. (2017). A hypothetical learning trajectory for conceptualizing matrices as linear transformations. *International Journal of Mathematical Education in Science and Technology*, 1–21. https://doi.org//10.1080/0020739X.2016.1276225

Freudenthal, H. (1991). *Revisiting mathematics education*. Dordrecht: Kluwer Academic Publishers.

Gol Tabaghi, S., & Sinclair, N. (2013). Using dynamic geometry software to explore eigenvectors: The emergence of dynamic-synthetic-geometric thinking. *Technology, Knowledge and Learning, 18*(3), 149–164.

Gravemeijer, K. (1999). How emergent models may foster the constitution of formal mathematics. *Mathematical Thinking and Learning, 1*, 155–177.

Henderson, F., Rasmussen, C., Zandieh, M., Wawro, M., & Sweeney, G. (2010). Symbol sense in linear algebra: A start toward eigen theory. *Proceedings of the 13th Annual Conference on Research in Undergraduate Mathematics Education.* Raleigh, N.C. Retrieved from: http://sigmaa.maa.org/rume/crume2010.

Larson, C., Zandieh, M, Rasmussen, C., & Henderson, F. (2009). Student Interpretations of the Equal Sign in Matrix Equations: The Case of Ax = 2x. *Proceedings for the Twelfth Conference On Research In Undergraduate Mathematics Education.*

Larson, C., & Zandieh, M. (2013). Three interpretations of the matrix equation Ax = b. *For the Learning of Mathematics, 33*(2), 11–17.

Rasmussen, C., Zandieh, M., & Wawro, M. (2009). How do you know which way the arrows go? The emergence and brokering of a classroom mathematics practice. In W.M. Roth (Ed.), *Mathematical representation at the interface of body and culture* (pp. 171–218). Charlotte, NC: Information Age Publishing.

Salgado, H., & Trigueros, M. (2015). Teaching eigenvalues and eigenvectors using models and APOS Theory. *The Journal of Mathematical Behavior, 39*, 100–120.

Sinclair, N., & Gol Tabaghi, S. (2010). Drawing space: Mathematicians' kinetic conceptions of eigenvectors. *Educational Studies in Mathematics, 74*, 223–240.
Stewart, S. & Thomas, M. O. J. (2006). Process-object difficulties in linear algebra: Eigenvalues and eigenvectors. In Novotná, J., Moraová, H., Krátká, M. & Stehlíková, N. (Eds.), *Proceedings 30th Conference of the International Group for the Psychology of Mathematics Education* (Vol. 5, pp. 185–192). Prague: PME.
Thomas, M. O. J., & Stewart, S. (2011). Eigenvalues and eigenvectors: Embodied, symbolic and formal thinking. *Mathematics Education Research Journal, 23*(3), 275–296.
Wawro, M., Larson, C., Zandieh, M., & Rasmussen, C. (2012). A hypothetical collective progression for conceptualizing matrices as linear transformations. In S. Brown, S. Larsen, K. Marrongelle, and M. Oehrtman (Eds.), *Proceedings of the 15th Annual Conference on Research in Undergraduate Mathematics Education* (pp. 1-465–1-479), Portland, OR.
Wawro, M., Rasmussen, C., Zandieh, M., & Larson, C. (2013). Design research within undergraduate mathematics education: An example from introductory linear algebra. In T. Plomp & N. Nieveen (Eds.), *Educational design research—Part B: Illustrative cases* (pp. 905–925). Enschede: SLO.
Wawro, M., Rasmussen, C., Zandieh, M., Sweeney, G. F., & Larson, C. (2012). An inquiry-oriented approach to span and linear independence: The case of the magic carpet ride sequence. *PRIMUS: Problems, Resources, and Issues in Mathematics Undergraduate Studies, 22*(8), 577–599. https://doi.org//10.1080/10511970.2012.667516.
Zandieh, M., Wawro, M., & Rasmussen, C. (2017). An Example of Inquiry in Linear Algebra: The Roles of Symbolizing and Brokering, *PRIMUS: Problems, Resources, and Issues in Mathematics Undergraduate Studies, 27*(1), 96–124. https://doi.org//10.1080/10511970.2016.1199618.

# Examining Students' Procedural and Conceptual Understanding of Eigenvectors and Eigenvalues in the Context of Inquiry-Oriented Instruction

**Khalid Bouhjar, Christine Andrews-Larson, Muhammad Haider and Michelle Zandieh**

**Abstract** This study examines students' reasoning about eigenvalues and eigenvectors as evidenced by their written responses to two open-ended response questions. This analysis draws on data taken from 126 students whose instructors received a set of supports to implement a particular inquiry-oriented instructional approach and 129 comparable students whose instructors did not use this instructional approach. In this chapter, we offer examples of student responses that provide insight into students' reasoning and summarize broad trends observed in our quantitative analysis. In general, students in both groups performed better on the procedurally oriented question than on the conceptually oriented question. The group of students whose instructors received support to implement the inquiry-oriented approach outperformed the other group of students on the conceptually oriented question and performed equally well on the procedurally oriented question.

**Keywords** Eigenvalues • Eigenvectors • Linear algebra • Inquiry-oriented instruction • Student thinking

Linear algebra is a mandatory course for many science, technology, engineering, and mathematics (STEM) students. The theoretical nature of linear algebra makes it a difficult course for many students because it may be their first time to deal with abstract and conceptual content (Carlson, 1993). Carlson (1993) also posited that this difficulty arises from the prevalence of procedural and computational emphases

K. Bouhjar · C. Andrews-Larson (✉) · M. Haider
Florida State University, Tallahassee, FL, USA
e-mail: cjlarson@fsu.edu

M. Zandieh
Arizona State University, Mesa, AZ, USA
e-mail: zandieh@asu.edu



S. Stewart et al. (eds.), *Challenges and Strategies in Teaching Linear Algebra*,
ICME-13 Monographs, https://doi.org/10.1007/978-3-319-66811-6_9

in students' coursework prior to linear algebra, and that it might therefore be difficult for students to connect new linear algebra topics and their previous knowledge. To address this issue, researchers have developed instructional materials for Inquiry-Oriented Linear Algebra (IOLA; http://iola.math.vt.edu/) and approaches to help students develop more robust, conceptual ways of reasoning about core topics in introductory linear algebra (e.g. Andrews-Larson, Wawro, & Zandieh, 2017; Wawro, Rasmussen, Zandieh, & Larson, 2013; Zandieh, Wawro, & Rasmusen, 2017).

Instructors who were not involved in the development of these kinds of research-based, inquiry-oriented instructional materials have been shown to encounter challenges when implementing such materials (Johnson, Caughman, Fredericks, & Gibson, 2013). Under an NSF-supported project Teaching Inquiry-Oriented Mathematics: Establishing Supports (TIMES), Johnson, Keene, and Andrews-Larson (2015) designed and implemented a system of instructional supports based on research in instructional change in undergraduate mathematics education, teacher learning, and professional development in settings ranging from K-20 (e.g. Henderson, Beach, & Finkelstein, 2011). These supports included sequences of student activities with implementation notes, a three-day summer workshop, and weekly online workgroups during the semester instructors implemented the materials in their teaching. This chapter examines differences in performance and reasoning of students whose instructors received these supports through the TIMES project (TIMES students) as compared to students whose instructors did not receive these supports (Non-TIMES students). In particular, we examine assessment data to identify differences in student performance and reasoning about eigenvectors and eigenvalues.

In this work we draw on data from an assessment that was developed to align with four core introductory linear algebra topic areas addressed in the IOLA instructional materials: linear independence and span; systems of linear equations; linear transformations; and eigenvalues. and eigenvectors. In the assessment, there were two questions that addressed eigenvalues and eigenvectors: question 8 and 9. Question 8 was a procedurally oriented question related to the eigenvalue of a given matrix and question 9 focused on conceptual understanding of the eigenvectors. The research questions for this analysis are:

- How does the performance of students whose instructors received TIMES instructional supports for teaching linear algebra compare to the performance of other students?
- How did students reason about eigenvectors and eigenvalues in the context of questions designed to assess aspects of students' procedural and conceptual understanding? How did reasoning differ for students of TIMES and Non-TIMES instructors?

# 1 Literature

Many have argued that the shift from predominantly computational and procedural approaches to mathematics many students experience before college to more theoretical approaches causes a lot of difficulties for students as they transition to university mathematics. Linear algebra is a course in which students struggle to develop conceptual understanding (Carlson, 1993; Dorier & Sierpinska, 2001; Dorier, Robers, Robinet, & Rogalski, 2000; Stewart & Thomas, 2009). Across the literature on the teaching and learning of eigenvalues and eigenvectors, procedural thought processes feature prominently. For example, Stewart and Thomas (2006) pointed to ways in which students often learn about eigenvalues and eigenvectors, where a formal definition is often linked to a symbolic presentation and its manipulation. For the purpose of this paper, we will draw on the following formal definition for eigenvectors and eigenvalues:

Suppose $A$ is an $n \times n$ real-valued matrix and $x$ is a non-zero vector in $\mathbb{R}^n$. We say the vector $x$ is an eigenvector of the matrix $A$ if there is some scalar $\lambda$ such that $Ax = \lambda x$. Further, in this case, we say that $\lambda$ is the eigenvalue associated with the eigenvector $x$.

Thomas and Stewart (2011) highlighted a difficulty students find when faced with formal definitions for eigenvalues and eigenvectors: these definitions contain an embedded symbolic form ($Ax = \lambda x$), and instructors often move quickly into symbolic manipulations of algebraic and matrix representations such as transforming $Ax = \lambda x$ to $(A - \lambda I)x = 0$. Their findings that students struggle to make sense of formal definitions, struggle to make use of geometric representations of eigenvectors, and exhibit procedural orientations toward eigenvectors suggest that such treatments might not provide sufficient opportunities for students to make sense of the reasons behind these symbolic shifts (Stewart & Thomas, 2009; Thomas & Stewart, 2011).

In order to help students make sense of situations that might be modeled using eigenvectors and eigenvalues, Salgado and Trigueros (2015) developed a problem that tasked students with designing a mathematical model that describes the employment dynamics of a population and its long-term behavior. While this modeling problem does not foreground geometric interpretations, the researchers also developed other activities to subsequently establish a relationship between the algebraic and geometric interpretation of eigenvectors and eigenvalues. Drawing on analysis of data from 30 undergraduate students, Salgado and Trigueros (2015) argued that this instructional sequence supported students' learning by helping students link ideas about eigenvectors and eigenvalues to other previously learned concepts.

Schoenfeld (1995) used eigenpictures in the $2 \times 2$ case ("stroboscopic" pictures) to show $x$ and $Ax$ at the same time by using multiple line segments in the $x$–$y$-plane. He observed that graphical representations of eigenvalues and eigenvectors got little attention in the literature and that a picture may benefit more than algebraic presentations. It is also documented more generally in linear algebra that students struggle to coordinate algebraic with geometric interpretations (e.g. Larson & Zandieh, 2013; Stewart & Thomas, 2010) and the students' understanding of eigenvectors is not always well connected to concepts of other topics of linear algebra (Lapp, Nyman, & Berry, 2010).

To support students in developing a better understanding of the formal definition and associated interpretations of the eigenvalues and eigenvectors, researchers have developed a variety of instructional interventions (e.g. Gol Tabaghi & Sinclair, 2013; Salgado & Trigueros, 2015; Zandieh, Wawro & Rasmussen, 2017). This paper examines student learning outcomes associated with a geometrically motivated instructional approach (see Plaxco et al. 2018; Zandieh, Wawro & Rasmussen, 2017) when paired with TIMES instructional supports; the approach will be described in the Study Design section.

## 2 Theoretical Framing

Researchers often make reference to conceptual understanding and procedural understanding when discussing students' reasoning about mathematical concepts (Hiebert, 1986). Hiebert and Lefevre (1986) defined conceptual knowledge as a "knowledge that is rich in relationships. It can be thought of as a connected web of knowledge, a network in which the linking relatonships are as prominent as the discrete pieces of information" (pp. 3–4). According to Hiebert and Lefevre (1986) students have procedural knowledge if they can combine formal language and symbolic representation systems with algorithms or rules in order to complete mathematical tasks.

In this paper we also draw on Larson and Zandieh's (2013) framework for students' mathematical thinking about matrix equations of the form $Ax = b$. This framework details three important interpretations, relationships between geometric and symbolic representations within each interpretation, and the complexity entailed in shifting among interpretations. The three interpretations this framework includes are (1) a linear combination interpretation, in which $b$ is viewed as a linear combination of the column vectors of the matrix $A$ with $x$ functioning as the set of weights on the column vectors of $A$, (2) a system of equations interpretation in which $x$ is viewed as a solution and $A$ is seen as a set of coefficients, and (3) a linear transformation interpretation in which $x$ is viewed as an input vector, $b$ as an output vector, and $A$ as the matrix that transforms $x$ into $b$.

We argue these interpretations are helpful for making sense of students' reasoning, but that the framework may need to be modified or expanded to more fully account for student reasoning in the context of eigenvalues and eigenvectors. In the context of eigenvectors and eigenvalues, students need to coordinate a transformation interpretation with the equation $Ax = \lambda x$, where the matrix $A$ transforms the vector $x$ by stretching, shrinking, and/or reversing the direction of vector $x$. Additionally, students need to shift to a systems interpretation and consider when the equivalent system $(A - \lambda I)x = 0$ has a non-trivial solution in order to make sense of standard procedures for computing eigenvalues and eigenvectors.

## 3 Study Design

In previous work, we have developed an assessment aligned with the inquiry-oriented linear algebra (IOLA) instructional materials used in the TIMES project (Haider et al., 2016). This paper-and-pencil assessment consists of 9 items, most of which include an open-ended response component. The assessment was administered at the end of the semester, and students were allocated one hour to complete the assessment.

In this analysis we examine assessment data from 126 students across six TIMES instructors and 129 students across three Non-TIMES instructors from different institutions in the US. Non-TIMES linear algebra instructors were selected from either the same institutions as TIMES instructors or a similar institution (e.g. preferably one from a similar geographic area, with similar size of student population, with similar acceptance rate) to collect assessment data for comparison of TIMES and Non-TIMES students. In this study, we focused on an in-depth analysis of students' reasoning on the assessment questions related to eigenvalues and eigenvectors. Both items are shown in Fig. 1.

**8. Is $\lambda = 2$ an eigenvalue of $\begin{bmatrix} 3 & 2 \\ 1 & 2 \end{bmatrix}$? Why or why not?**

9. Suppose the vector $x$ is a real-valued eigenvector of the matrix M and that the entries of M are also real-valued.

a. What could be the result of the product M$x$? (Check all that apply.)

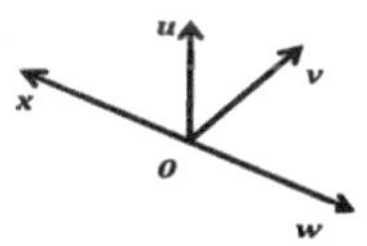


- ☐ i: M$x$ could be $u$
- ☐ ii: M$x$ could be $v$
- ☐ iii: M$x$ could be $w$
- ☐ iv: M$x$ could be $\mathbf{0}$
- ☐ v: M$x$ could be $x$
- ☐ vi: None of the above

b. Explain your reasoning for your choice(s) in part a.

**Fig. 1** Assessment items related to eigenvectors and eigenvalues (Question 9 was retrieved from http://mathquest.carroll.edu and developed as part of an NSF-supported project entitled Project MathVote: Teaching Mathematics with Classroom Voting. For related research, see Cline, Zullo, Duncan, Stewart, & Snipes, 2013)

The inquiry-oriented approach to learn eigenvalues and eigenvectors associated with this study is characterized in detail elsewhere (Plaxco et al. 2018; Zandieh, Wawro, & Rasmusen, 2017). Briefly described, this approach supports students in first learning about eigenvalues and eigenvectors as a set of "stretch" factors and directions that can be used to more easily characterize a geometric transformation. Students work through a series of tasks, first aiming to find (using standard coordinate systems) the image of a figure in a plane under a transformation that is easily described in a non-standard coordinate system. Students then work to label points in both the pre-image and the image using the standard and the more convenient coordinate systems, find matrices that rename points from one coordinate system to the other, and find matrices corresponding to the transformation described relative to both coordinate systems. The instructor works to link this work to the matrix equation $A = PDP^{-1}$ and subsequent tasks aim to leverage this conceptual basis as students learn more traditional computational methods associated with computing eigenvalues and eigenvectors.

## 4 Methods of Analysis

To answer our research questions, our analysis has two main components. The first component of our analysis is quantitative in nature, as we aim to compare learning outcomes of students whose instructors received TIMES instructional supports to those who did not. The second component of our analysis is qualitative in nature, as we work to identify students' ways of reasoning on both the more procedurally oriented assessment item (Q8) and the more conceptually oriented item (Q9). We follow Kwon, Rasmussen, and Allen's (2005) approach for distinguishing assessment items that are conceptually oriented from those that are procedurally oriented. In particular we consider Q8 to be more procedurally oriented in that there is a commonly taught procedure that students can directly invoke (with some interpretation) to produce a correct answer to the question. There is no such standard procedure for Q9, so we consider it to be more conceptually oriented. In our qualitative analysis, we also look for similarities and differences that emerge from considering the two groups.

To facilitate our quantitative analysis, we needed to score students' responses to the two assessment items. Specifically, we needed to develop a uniform system for assigning a number of points to students' responses that provide an overall assessment of the quality of their response and the understanding reflected in that response. Question 9a required students to select which subset of 6 possible options were appropriate responses, so 1 point was awarded to each of the possible options for correctly selecting or not selecting that option. Both Question 8 and Question 9b were open-ended response questions, and both of these were scored on a scale of 0 to 3 points. Three points were awarded for a fully correct response, two points were awarded for a mostly correct response (e.g. if a minor computational error was made, 2 points would be assigned), one point was awarded if the student's response provided evidence of some knowledge or understanding relevant to the question, and no points were awarded otherwise. A scoring scheme was developed to specify what kinds of responses received how many points. In order to ensure consistency among coders in how points were assigned, new examples were added to the scoring guide throughout the scoring process. A condensed version of the scoring scheme for assigning points to open-ended response questions can be found in Appendix. Additionally, the Appendix includes some explanation of how this scoring scheme aligns with our coding categories for *how* students reasoned, which are described in greater detail below. Student work exemplifying common ways of reasoning with explanation of points awarded are provided in the Findings section.

To ensure agreement regarding points assigned to each response, two researchers looked at every student's attempt and assigned a score independently before comparing with each other. If the two researchers assigned a different score to a particular student, they then discussed according to the codebook and agreed on a common score for that student. If both researchers disagreed about a particular score, then a third researcher was consulted to reach a consensus.

Once scores had been assigned to all student responses, descriptive statistics were generated to examine the overall performance of students on the eigenvalue and eigenvector questions and to compare TIMES students with Non-TIMES students for both questions. We were unable to control for factors such as students' mathematical background, major, and instructor's teaching experience, so this is an unavoidable limitation for our statistical analysis. However, we tried our best to choose TIMES and Non-TIMES students either from the same school or from similar schools. This helps us establish similarity of students in TIMES and non-TIMES classes. Then, we compared the mean scores of TIMES and Non-TIMES students using two-tailed t-tests to identify when differences of means were statistically significant.

In order to facilitate our qualitative analysis of students' reasoning, we examined student responses to the open-ended portions of question 8 and question 9. As noted before, we consider Q8 to be more procedurally oriented. After examining the data several times and refining the categories of the students' reasoning about item 8, we sorted students' responses into 5 broad categories: (1) reasoning about the determinant, (2) reasoning about $A - \lambda I$ without computing a determinant, (3) other, (4) students who explicitly indicated they did not know, and (5) students who left the item blank. A student's response was categorized as "reasoning about the determinant" if he or she solved the characteristic equation, plugged the possible given eigenvalue into the characteristic equation, or computed the determinant of the $A - 2I$ matrix and compared the result to 0. A students' response was categorized as "reasoning about $A - \lambda I$" if he or she solved the system of linear equations $(A - \lambda I)x = 0$, considered the linear independence of the columns of $A - \lambda I$, or considered whether $rref(A - \lambda I)$ had any free variables.

In examining students' responses to question 9, we found it helpful to distinguish responses that were conceptually aligned with the formal definition for eigenvectors and eigenvalues from those that were not. We were specifically interested in student reasoning that appropriately coordinated interpretations of

$A, x$, and $\lambda$ in the context of the matrix equation $Ax = \lambda x$. In particular, we say a student response "uses the eigen-concept" when there is evidence a student is coordinating $M, x$, and $\lambda$ in at least one of the following ways:

- Algebraically: The matrix $M$ is a fixed matrix that transforms the (nonzero) eigenvector $x$ in a particular way, namely such that the resulting vector $Ax$ is a scalar multiple ($\lambda$) of $x$.
- Geometrically: this can be interpreted to mean that multiplying $x$ by $A$ has the effect of
  - stretching $x$ in the same direction or opposite direction, or
  - causing the resultant vector to lie along the same line as the vector $x$.

If a student drew on a transformation interpretation to make sense of $Ax$ but did not coordinate this appropriately with $\lambda x$ in one of the ways mentioned above, we did not say that the student's response used the eigen-concept.

We grouped students' responses to question 9 into five categories: (1) responses that used the eigen-concept, (2) responses that focused on the role of the matrix $M$ in a way that did not use the eigen-concept, (3) other, (4) responses in which the student explicitly indicated he or she did not know, and (5) responses that were left blank. There were two primary kinds of responses coded as focused on the role of the matrix $M$ in a way that did not use the eigen-concept. The first one is when students focused on the role of the matrix $M$ as a transformation, but without specifying the particular way it will transform an eigenvector $x$. The second kind of response is when students suggested specific matrices $M$ that would satisfy particular equations (e.g "$Mx = x$ if $M = I$"). While this is certainly a true statement, it doesn't include evidence of understanding the special relationship between a matrix and its eigenvector(s).

After coding students' responses to Q8 and Q9, we aggregated these responses into tables, organized by the category assigned to each response and number of points awarded. We also separated TIMES from Non-TIMES students in counting the number of responses in these discrete categories. This allowed us to look for patterns in which approaches were conceptually oriented, which approaches lent themselves to arriving at correct answers, and differences in approaches taken by TIMES and Non-TIMES students.

## 5 Findings

In order to answer our research question about how TIMES students compared to Non-TIMES students, we first present our quantitative analysis of students' performance on the more procedurally-oriented question (Q8) and the more conceptually oriented question (Q9), separating students of TIMES instructors from students of Non-TIMES instructors. We then summarize findings from our coding of students' approaches to these same questions, providing examples of responses that highlight important trends in student reasoning.

## 5.1 Overview of Differences in Student Performance

We highlight three central trends from our quantitative analysis. First, TIMES students outperformed Non-TIMES on both items, with a strongly significant difference of means on the conceptual item. Second, both TIMES and Non-TIMES students did better on the procedurally oriented item than on the conceptually oriented item. Third, correlations between students' performance on both the conceptual and procedural items were weak for students in both groups, suggesting that the two items assessed relatively different aspects of student understanding. Note that the last trend is not part of answering our research questions, it is more of a side observation that emerged from our quantitative analysis.

To compare the performance of TIMES students with Non-TIMES students, we first computed the mean and standard deviation for question 8 (which was an open-ended response question with a total of 3 points possible), question 9a (which was a multiple-choice question) and 9b (which is also an open-ended response question). To make a 'cleaner' comparison, we have separately included the mean and standard deviation of part a and part b of question 9. Part a of item 9 is a multiple-choice problem with six distractors, three of which are correct choices. Per our grading scheme, students can earn a maximum of 6 points from part a, three points by selecting three correct choices and three points by not selecting incorrect choices, so chances of making a guess for correct answers are higher in 9a. We also observed that the difference in performance of TIMES and Non-TIMES students on 9a was not statistically significant with the available sample size. However, question 9b is open-ended and students can earn at most three points by providing a complete and correct explanation. Therefore, we compared question 8 with question 9b as they are naturally comparable items.

The data presented in Table 1 show that on the procedurally oriented question (Q8) the mean score of TIMES students (M = 1.98, SD = 1.24) was greater than that of Non-TIMES students (M = 1.71, SD = 1.37), but this difference of means was not statistically significant with the available sample size. Similarly, there was not a statistically significant difference in means on question 9a. However, in comparing the performance of students in both groups on question 9b (which is an open ended response style question like question 8), we noticed that TIMES students performed significantly better (M = 1.05, SD = 1.12) than the Non-TIMES students (M = 0.54, SD = 0.86). The results of the t-test indicated that this difference of means was statistically meaningful, $t(125) = 4.29$, $p < 0.001$. In this way, TIMES students outperformed Non-TIMES students on the conceptually oriented question.

Overall, students performed better on the procedurally oriented question (Q8) than the conceptually oriented question (Q9). We compared Q8 to Q9b and found

**Table 1** Summary of results of quantitative analysis

| Question | All students | TIMES students | Non-TIMES students | *p*-value (two-tailed) |
|---|---|---|---|---|
| Q8<br>3 Points | Mean: 1.85<br>SD: 1.31 | Mean: 1.98<br>SD: 1.24 | Mean: 1.71<br>SD: 1.37 | $t(125) = 1.73$<br>$p = 0.08 > 0.05$ |
| Q9 (part a only)<br>6 Points | Mean: 3.73<br>SD: 1.68 | Mean: 3.74<br>SD: 1.76 | Mean: 3.71<br>SD: 1.61 | $t(249) = 0.56$<br>$p = 0.88 > 0.05$ |
| Q9 (part b only)<br>3 Points | Mean: 0.79<br>SD: 1.03 | Mean: 1.05<br>SD: 1.12 | Mean: 0.54<br>SD: 0.86 | $t(125) = 4.29$<br>$p < 0.001$ |

that the difference of means for all students between Q8 (M = 1.85, SD = 1.31) and Q9b (M = 0.79, SD = 1.03) was also statistically meaningful with *p*-value (two-tailed) less than 0.001.

Since both problems we investigated in this study were related to eigenvectors and eigenvalues, one might think that students' performance on the two items should be correlated. However, quantitative analysis revealed a positive but weak correlation between students' performance on the two questions; the Pearson correlation coefficient was r = 0.30 for all students. Recall that a correlation coefficient measures the degree of relationship between two variables and ranges from −1 to 1, where the sign indicates the direction of the relationship and the distance from zero indicates the strength of the relationship (e.g. 1 means the two variables are highly correlated and 0 means there is very little or no correlation between the two variables). For TIMES students, the correlation between the two items was r = 0.36 as compared to the correlation for Non-TIMES which was r = 0.22. This suggests two things: first, that the two items measure *different* aspects of student understanding of eigenvalues and eigenvectors. Second, it indicates that performance on the procedurally and conceptually oriented questions was more highly correlated for TIMES students.

## 5.2 Trends in Student Reasoning on the Conceptually Oriented Question

In this section, we provide our qualitative analysis of question 8, which we consider to be the more procedurally oriented question. In particular, we highlight two common approaches to this problem: approaches that involve reasoning about the determinant, and approaches that involve reasoning about $A - \lambda I$ without computing a determinant. The majority of students who reasoned about the determinant responded correctly. Reasoning about $A - \lambda I$ was a less common approach but more frequently observed among TIMES students. Importantly, TIMES students were more often able to arrive at a correct answer by reasoning about $A - \lambda I$ than were Non-TIMES students. Further, we argue that students who reasoned about $A - \lambda I$

showed more evidence of conceptual understanding. A summary of our coding and scoring of student responses is shown in Table 2.

Reasoning about the determinant was the most common approach observed in students' responses to question 8, and students who used this kind of approach tended to do so without making conceptual errors.[1] Overall, 146 out of 255 students (57% of all students) responded to question 8 by reasoning about the determinant. We note two interesting trends within those who used this approach distinguishing TIMES from Non-TIMES students. First, more TIMES students who used determinants in their response made computational errors (usually when factoring the characteristic polynomial) than did Non-TIMES students—such errors are evidenced by 2-point responses in our coding. On the other hand, fewer TIMES students using this approach made *conceptual* errors than did Non-TIMES students —such errors are evidenced by 1-point responses in our coding. In the TIMES instructional approach (previously described under study context), the standard algorithm for finding eigenvalues and eigenvectors is intended to emerge in relation to student-invented strategies on the third of fourth day of instruction in the unit, so we conjecture Non-TIMES students may have spent more time practicing this procedure in comparison to TIMES students.

A less common approach to solve problem 8 was by reasoning about $A - \lambda I$ without computing a determinant. Overall, 48 out of 255 students (19%) used such a determinant-free approach to solve the problem. This approach was more common among TIMES students than among Non-TIMES students, and far more TIMES students successfully responded to the problem in this way without conceptual errors (evidenced by a score of 2 or 3 points in our grading scheme). Indeed, 70% (19 out of 27) of TIMES students who used this approach did so with without conceptual errors whereas only 38% (8 out of 21) Non-TIMES students who used this approach did so without conceptual errors. This indicates that more TIMES students used a determinant-free approach to solving Q8, and those who used this kind of approach did so correctly at higher rates than Non-TIMES students who used the same approach.

Students whose responses were categorized as "other" showed little or no evidence of understanding related to the definition or computation of eigenvectors and eigenvalues. We noticed that twice as many Non-TIMES students as TIMES students gave a response categorized as 'other.' However, TIMES and Non-TIMES students left the item blank at similar rates, but a larger number of Non-TIMES

---

[1]We align our conceptions of conceptual and procedural errors with our definitions for conceptual and procedural understanding. We refer to an error as conceptual when there is evidence that a student does not understand an important underlying idea or relationship. We refer to an error as procedural when a student incorrectly performs a step in a mathematical process that is not central to the idea being assessed (e.g. an error in computation or algebraic manipulation). Examples of conceptual errors include incorrectly interpreting the value of the determinant to decide if something is an eigenvalue, or computing the determinant of A rather than the determinant of $A - \lambda I$. Examples of procedural errors include incorrectly factoring the characteristic polynomial or making an error when row reducing $A - \lambda I$.

**Table 2** Summary of students' approaches and scores on Q8

| Category | | TIMES, $n = 126$ | | | | Non-TIMES, $n = 129$ | | | |
|---|---|---|---|---|---|---|---|---|---|
| | | 3 pts | 2 pts | 1 pt | 0 pts | 3 pts | 2 pts | 1 pt | 0 pts |
| Reasoning about the determinant<br><br>Total: 146 (57%)<br>TIMES: 75 (60%)<br>Non-TIMES: 71 (55%) | Solving the characteristic equation | 43 | 12 | 4 | 0 | 49 | 4 | 4 | 4 |
| | Plugging 2 into the characteristic equation | 3 | 1 | 0 | 0 | 3 | 0 | 0 | 0 |
| | Finding $\det(A-2I) \neq 0$ | 10 | 2 | 0 | 0 | 5 | 2 | 0 | 0 |
| Reasoning about $A - \lambda I$<br><br>Total: 48 (19%)<br>TIMES: 27 (21%)<br>Non-TIMES: 21 (16%) | Solving the system $(A-2I)\begin{bmatrix} x \\ y \end{bmatrix} = 0$ or $Av = 2v$ | 9 | 2 | 2 | 1 | 5 | 1 | 2 | 0 |
| | Considering linear dependence of the set of column vectors of $(A-2I) \cong \begin{bmatrix} 1 & 0 \\ 0 & -2 \end{bmatrix}$ | 2 | 0 | 1 | 0 | 0 | 2 | 1 | 1 |
| | Noting $rref(A-2I)$ does not have a free variable | 3 | 3 | 4 | 0 | 0 | 0 | 6 | 3 |
| Other | | 0 | 0 | 1 | 7 | 0 | 0 | 0 | 16 |
| Student indicated he/she did not know | | 0 | 0 | 0 | 5 | 0 | 0 | 0 | 9 |
| Item left blank | | 0 | 0 | 0 | 11 | 0 | 0 | 0 | 10 |

students explicitly mentioned that they "don't know" or "have no clue" how to solve this problem.

## 5.3 Examples of Student Reasoning on the Procedurally Oriented Question

In this section we examine examples of common approaches identified in our analysis of students' responses to question 8. We provide two example responses coded as 'reasoning about the determinant' and two example responses coded as 'reasoning about $A-\lambda I$ without using the determinant.' We highlight the use of multiple representations in these responses, as well as connections between these representations and the formal definition of eigenvectors and eigenvalues. Based on these differences, we posit that responses coded as 'reasoning about $A-\lambda I$' tend to be more conceptually rich based on flexible use of multiple representations and more explicit connections between these approaches and the formal definition of eigenvectors and eigenvalues.

The two examples shown in Fig. 2 show typical responses to question 8 coded as "reasoning about the determinant." Response 2.a. was awarded full points because the student correctly found the roots of the characteristic polynomial, presumably noted that 2 was not one of those roots, and concluded that 2 is not an eigenvalue. The response shown in 2.b. was awarded two out of three possible points because the student made computational errors in finding the roots of the characteristic polynomial that resulted in the student concluding that two was a root of this polynomial and thus an eigenvalue. It is interesting to note that response 2.b. does not explicitly set the characteristic polynomial equal to 0 in his or her written response, but the work suggests that the student is trying to factor the polynomial in a way consistent with finding the roots.

The two examples shown in Fig. 3 show typical responses to question 8 coded as "reasoning about $A-\lambda I$ without using the determinant." We note that in response 3.a., the student began with the equation $Ax=\lambda x$, rewrote this as $Ax-\lambda x=0$, and then factored this to write $(A-\lambda I)x=0$. The student then computed the entries of the matrix $A-2I$, rewrote this as a homogeneous matrix equation which he or she translated into a system of equations, correctly solved, and correctly concluded that because the solution is the zero vector that 2 **is not** an eigenvalue of the given matrix. Response 3.b. similarly considers the solution of $(A-\lambda I)x=0$ by rewriting this matrix equation as a system of equations, substituting $\lambda=2$ into this system, and finding the solution to this system to be when $x=0$ and $y=0$. However, this student incorrectly concluded from this that 2 **is** an eigenvalue. Because this is a conceptual error (thinking that finding only the trivial solution to $(A-2I)x=0$ means that 2 is an eigenvalue of $A$), this response was awarded only one point.

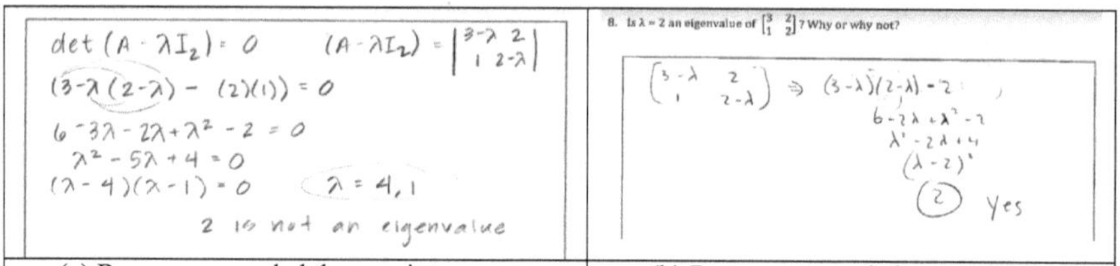

(a) Response awarded three points (b) Response awarded two points

**Fig. 2** Responses to Q8 coded as "reasoning about the determinant"

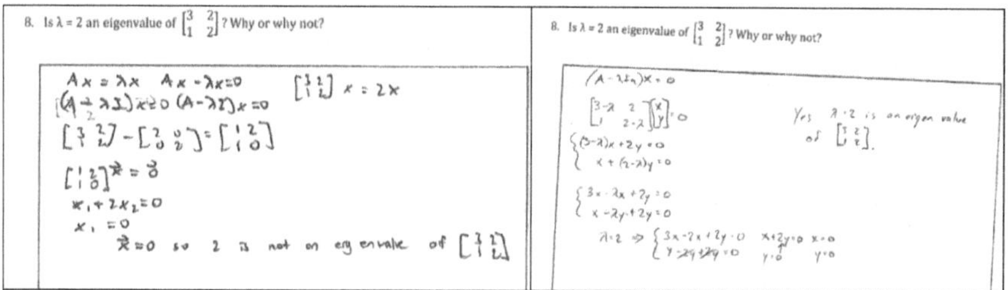

(a) Response awarded three points (b) Response awarded one points

**Fig. 3** Responses to Q8 coded as "reasoning about $A - \lambda I$"

In alignment with Hiebert and Lefevre's (1986) characterization of procedural and conceptual knowledge, we claim that responses coded as reasoning about the determinant correspond to a more procedural approach to this question. We note that those who substituted 2 in the characteristic equation and those who noted that $\det(A-2I) \neq 0$ showed some procedural flexibility indicative of conceptual aspects of their reasoning. We argue that responses coded as "reasoning about $A - \lambda I$" show more evidence of conceptual understanding of eigenvalues and eigenvectors than do responses coded as "reasoning about the determinant." Examples of responses coded as "reasoning about $A - \lambda I$ without using the determinant" included representation of the system being solved in order to determine whether or not 2 was an eigenvalue of the given matrix, whereas the examples of responses coded as "reasoning about the determinant" typically only included representation of the computation to be executed to determine whether 2 is an eigenvalue. While this doesn't mean these students didn't have a conceptual understanding of eigenvalues and eigenvectors, there is not explicit evidence of this connection in their responses. In addition, both examples of responses coded as "reasoning about $A - \lambda I$" included evidence that these students could comfortably transition between matrix equations and systems of equations, a skill that has elsewhere been documented to be both difficult for students and important for their understanding (Larson & Zandieh, 2013; Selinski, Rasmussen, Wawro, & Zandieh, 2014). This can be interpreted as evidence of connectedness of ideas and representations—which others have argued to be the very definition of conceptual understanding (Vinner, 1997; Hiebert & Lafebre, 1986).

## 5.4 *Trends in Student Reasoning on the Conceptually Oriented Question*

We now focus on responses to question 9, the conceptually oriented question. Overall, students' responses to this item were split somewhat evenly among responses that used the eigen-concept, responses that focused on the role of the matrix $M$ without using the eigen-concept, and students who wrote that they did not know or left the answer blank. However, TIMES students' responses used the eigen-concept at much higher rates than Non-TIMES students, and with greater success. Table 3 highlights trends in the approaches of TIMES and Non-TIMES students' responses.

The most commonly observed response to Q9 involved using the eigen-concept, with 99 out of 255 (39%) total responses coded in this way. This approach was more common among TIMES students than Non-TIMES students (61/126 vs. 38/129). Further, TIMES students who used this approach gave correct responses to the question at higher rate than Non-TIMES students; the ratio of TIMES students who used the eigen-concept in fully or mostly correct ways to those who used the eigen-concept in mostly incorrect ways was 44:17 whereas that ratio for Non-TIMES students is 18:20.

The second most commonly observed trend on Q9 involved responses that focused on the role of the matrix $M$ without using the eigen-concept. We noted that students using this approach tended to be mostly or completely incorrect, and that more Non-TIMES students than TIMES students used this approach (29/126 TIMES as compared to 40/129 Non-TIMES students). We noticed that 14/29 (48%) of the TIMES students used this approach did so with some conceptual understanding but not using the eigen-concept; only 12/40 (30%) Non-TIMES students also used this approach with some conceptual understanding but not using the eigen-concept. We argue these responses indicated some conceptual understanding because they drew on appropriate transformation interpretation of a matrix times a vector. However, the understanding reflected in these responses was incomplete in that the interpretation did not explicitly use the eigen-concept by coordinating that interpretation with the result of that multiplication as corresponding to a scalar times that same vector.

There was little difference between TIMES and Non-TIMES Students who used approaches classified as 'other.' In this category, we saw no evidence of using the eigen-concept. TIMES and Non-TIMES students indicated they did not know the answer at similar rates, and more Non-TIMES students left the item blank than TIMES students.

**Table 3** Summary of students' approaches and scores on item 9

| | | TIMES, $n = 126$ | | | | Non-TIMES, $n = 129$ | | | |
|---|---|---|---|---|---|---|---|---|---|
| | | 3 pts | 2 pts | 1 pts | 0 pts | 3 pts | 2 pts | 1 pts | 0 pts |
| Responses using the eigen-concept<br><br>Total: 99 (39%)<br>Times: 61 (48%)<br>Non-Times: 38 (29%) | $Mx = \lambda x$ with explanations | 9 | 14 | 4 | 0 | 5 | 3 | 2 | 0 |
| | • Stretching<br>• Same direction<br>• Scalar multiple<br>• Same line as $x$ | 10 | 11 | 12 | 1 | 2 | 8 | 16 | 2 |
| Responses focusing on the role of the matrix $M$ without using the eigen-concept<br><br>Total: 69 (27%)<br>Times: 29 (22%)<br>Non-Times: 40 (31%) | Specific entries are suggested for matrix M that would yield different outputs | 0 | 0 | 12 | 8 | 0 | 0 | 11 | 15 |
| | Descriptions are given about how the matrix M transforms the input vector to yield different outputs (without suggesting specific entries of M) | 0 | 0 | 2 | 7 | 0 | 0 | 1 | 13 |
| Others | | 0 | 0 | 0 | 11 | 0 | 0 | 0 | 15 |
| I don't know | | 0 | 0 | 0 | 10 | 0 | 0 | 0 | 11 |
| Blank | | 0 | 0 | 0 | 17 | 0 | 0 | 0 | 25 |

## 5.5 Examples of Student Reasoning on the Conceptually Oriented Question

As with Q8, we provide examples of common approaches identified in our analysis of students' responses to question 9. Specifically, we provide four examples of responses coded as "using the eigen-concept" and two examples of responses coded as "focusing on the role of the matrix $M$ without using the eigen-concept." Responses 4.a and 4.b both used the eigen-concept by writing the equation $Mx = \lambda x$ and suggesting values of $\lambda$ (e.g. 1,−1,0) that corresponded appropriately to possible outputs (Fig. 4).

Response 4.a was awarded full credit because the student linked this reasoning to all three possible outputs, whereas response 4.b was awarded just 2 out of 3 possible points due to the omission of the 0 vector as a possible output. Many students in our study who used the eigen-concept omitted the 0 vector as a possible eigenvector. We suspect this may relate to a need to distinguish the *eigenvalue* of zero from the equation $Mx = \lambda x$ having only the trivial solution when solving for the vector $x$. Responses 4.c and 4.d used the eigen-concept in a slightly different way than the previous examples. Rather than writing $Mx = \lambda x$ and suggesting appropriate values of $\lambda$, these students justified their selections of correct output vectors by describing the role of $M$ as stretching the vector $x$ by a factor or in its direction. Similar to the previous pair of examples, response 4.c was awarded 3 points for correctly identifying all three vectors (and even explaining that vectors $u$ and $v$ could not be reached by stretching $x$), whereas response 4.d was awarded just 2 points due to the omission of the 0 vector.

The next two examples presented in Fig. 5 show typical responses to question 9 that focused on the role of the matrix $M$ without using the eigen-concept. Both

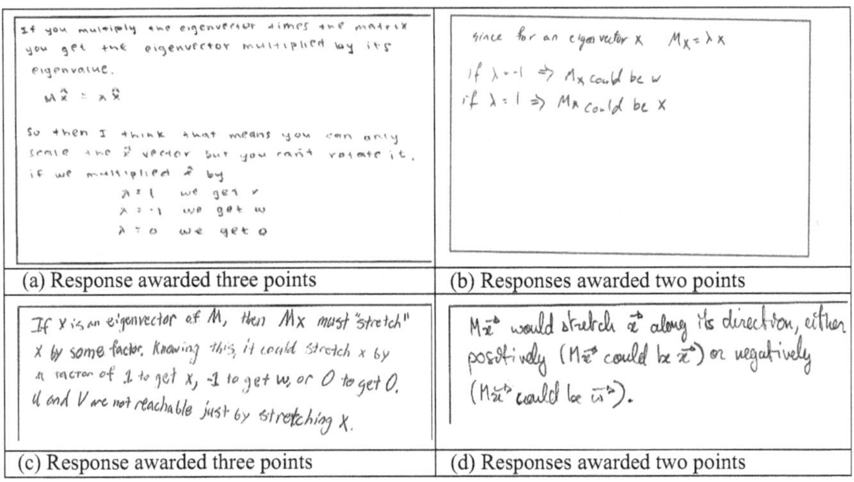


**Fig. 4** Responses to Q9 coded as using the eigen-concept

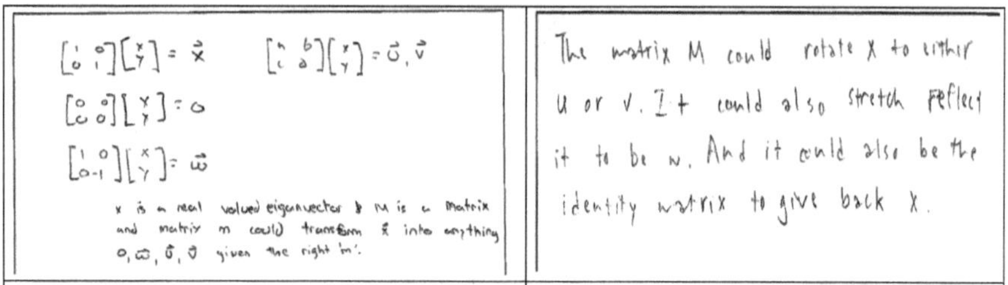


(a) Response awarded one point

(b) Response awarded one point

**Fig. 5** Responses to Q9 focused on the role of $M$ without using the eigen-concept

responses focus on the role of the matrix $M$ as a transformation that can transform the vector $x$ in many ways (not limiting to outputs that must lie along the same line as $x$). Student 5.a's response suggests that the student sees the matrix $M$ not as a fixed matrix that transforms the eigenvector in a particular way; the student suggested different matrices that correctly produced various output vectors. The student indicated $M$ could be the identity matrix $I$ to produce $x$, $-I$ (with a sign error in one entry) to produce $w$, or the zero matrix to produce the zero vector. In addition, a matrix $M$ with generic entries was suggested as a transformation that can transform $x$ into vectors $u$ and $v$.

Response 5.b. similarly focuses on the role of the matrix $M$, arguing it could rotate $x$ to produce $u$ or $v$, "stretch reflect" to produce $w$, and that it could be the identity matrix to "give back" $x$. This combination of what the student believes the matrix could be indicates that the student did not use the eigen-concept. Responses 5.a and 5.b were both awarded 1 point because both were interpreting the matrix $M$ as a transformation and making some true statements, though in ways that did not use the eigen-concept.

We argue that interpreting matrices as transformations is an important concept that students need to make sense of eigenvectors and eigenvalues, but these responses show how that alone is not enough to ensure students are using the eigen-concept. Thinking one can choose values of the matrix $M$ is in contrast with the view that a given (fixed) matrix $M$ transforms its eigenvector $x$ in a particular way such that the resulting vector $Mx$ is a scalar multiple of $x$ and thus lies along the same line as the vector $x$. Indeed, the student whose work is shown in Fig. 5b. used the term "stretch reflect," which aligns partially with the geometric interpretation of the eigenvectors and eigenvalues concept, but the student did not limit his or her interpretation of outputs to those that appropriately correspond to eigenvectors; the student saw "stretch reflect" as just one of many possible ways the matrix $M$ could transform its eigenvector(s).

## 6 Discussion

We see this chapter contributing to the literature in three primary ways. First, we document the effectiveness of a particular instructional approach that is detailed in the literature (see Plaxco et al., 2018; Zandieh, Wawro, & Rasmusen, 2017). Second, we document aspects of students' reasoning about eigenvectors and eigenvalues (including how students draw on a transformation interpretation in ways that do and do not use the eigen-concept). Finally, we consider and discuss links between conceptual and procedural understandings of eigenvectors and eigenvalues documented in our study.

Our findings showed that both TIMES and Non-TIMES students in our study performed better on the procedurally oriented assessment question than they did on the conceptually oriented question. Further, TIMES students consistently showed evidence of more robust conceptual understanding as compared to Non-TIMES students, whereas procedural performance was similar between the two groups. This is consistent with findings of previous studies examining student learning outcomes in inquiry-oriented instructional settings at the undergraduate level (e.g., Kwon et al., 2005), though we are excited that this study was conducted on a larger scale involving instructors not involved in the development of the curricular materials. These findings are consistent with a broader body of literature documenting the benefits of student-centered approaches to learning in undergraduate mathematics (Freeman et al., 2014; Laursen Hassi, Kogan, & Weston, 2014). We conclude our paper with a discussion of the kinds of conceptual understandings observed in our analysis, and the insights these offer into what is entailed in a conceptual understanding of eigenvectors and eigenvalues.

As mentioned in our theoretical framework, conceptual understanding has been broadly defined by some in terms of the richness of connections among ideas (Hiebert & Lafevre, 1986; Vinner, 1997). More recently, Star (2005) has argued that conceptions of conceptual and procedural knowledge in mathematics education are under-articulated in a way that promotes ideological rather than empirical examination, and relationships between conceptual and procedural understandings merit greater examination. With this in mind, we now reflect on the kinds of conceptual understandings observed in our analysis, and discuss three different kinds of connections we consider to be important aspects of students' conceptual understanding of eigenvectors and eigenvalues.

First, we consider the use of appropriate interpretations of a matrix times a vector to be an important aspect of students' understanding of eigenvalues and eigenvectors. On the conceptually oriented assessment question considered in this chapter, this involved drawing on a transformation interpretation of the product of a matrix $M$ and its eigenvector $x$ consistent with the characterization given by Larson and Zandieh (2013). In our data, many students showed evidence of interpreting $Mx$, the product of a matrix $M$ and its eigenvector $x$, in ways that use the eigen-concept. A smaller number of students interpreted $Mx$ with a transformation lens, but in a way that did not use the eigen-concept in that $M$ was either thought of as a matrix that could change (to yield desired outcomes) or that the product of $M$ with the vector $x$ could be anything. This is different from a transformation interpretation that uses the eigen-concept by recognizing that the vector resulting from the multiplication by a matrix with real-valued entries $Mx$ needs to yield a vector that is a scalar multiple of $x$, or that lies on the same line as $x$, or that points in the same (or opposite) direction as $x$.

This leads to our second aspect of students' understanding of eigenvalues and eigenvectors: using the eigen-concept in the context of finding eigenvalues. While many students showed evidence of using the eigen-concept in their response to the conceptually oriented assessment item, relatively few showed evidence of using the eigen-concept on the procedurally oriented question. Indeed, one could solve our procedurally oriented assessment question by applying the standard procedure for finding eigenvalues to arrive at the correct answer without explicitly using the eigen-concept; the majority of students in both groups did just this, and most did so without error. A far smaller number of students responded to the procedurally oriented question by reasoning about $A - \lambda I$ without taking the determinant. We argue this approach provided more evidence of conceptual understanding: providing and converting between multiple representations (e.g. $Ax = \lambda x$ and $(A - \lambda I)x = 0$, written as matrix equations and systems of equations), linking those representations to the eigen-concept, and offering reasons for their conclusion in terms of a matrix equation or system of equations in their response. It is possible that a student who used the standard procedure to determine if 2 is an eigenvalue on this problem also had a deep conceptual understanding of how and why that procedure works; it is also possible that a student who used the standard procedure knew this procedure only as a sequence of steps to be executed without knowing how or why the procedure worked. Further work is needed to tease out this distinction.

This leads to the final aspect of conceptual understanding of eigenvectors and eigenvalues relevant to our analysis, which includes coordinating with the Invertible Matrix Theorem (IMT). A standard procedure for finding eigenvalues and eigenvectors draws on the argument that $Ax = \lambda x$ has a non-trivial (non-zero) solution vector $x$ for some scalar $\lambda$ if and only if the equation $(A - \lambda I)x = 0$ also has a non-trivial solution; one can argue through the IMT that this happens when $\det(A - \lambda I) = 0$. As noted above, it was often unclear from the responses of students who used the standard procedure whether they understood links among the equation $Ax = \lambda x$ used in defining eigenvectors, the solution set of $(A - \lambda I)x = 0$, and the equivalencies in the invertible matrix theorem that lead to use of the determinant as a tool for determining when the solution is non-trivial. However, among students who did not use the determinant in their response to the procedurally oriented question, there was a need to draw on equivalent ideas from the invertible matrix theorem. In these responses, we observed students noting and leveraging the following relationships:

- $(A - \lambda I)$ is invertible if and only if $(A - \lambda I)x = 0$ has a trivial solution. If $(A - \lambda I)x = 0$ has only the trivial solution, then $\lambda$ is <u>not</u> an eigenvalue of the matrix $A$.

- If the columns of $A-\lambda I$ are linearly dependent or one column is a scalar multiple of the other (in the case of a 2 × 2 matrix), then $(A-\lambda I)x=0$ has nontrivial solution so $\lambda$ is an eigenvalue of the matrix $A$.
- If $rref(A-\lambda I)$ has no free variable then $(A-\lambda I)x=0$ has only the trivial solution, which means $\lambda$ is not an eigenvalue of the matrix $A$.

We argue that these kinds of responses from students who did not use the previously mentioned standard procedure offer insight into conceptual connections that are both important and potentially natural for students to make as they come to make sense of standard algorithms. Students who took a procedural approach to this question typically used the determinant to decide if 2 was an eigenvalue of the matrix, without representation of the rich set of coordinations involved in these other responses, which relate interpretations of matrix equations and systems of equations, equivalencies in the Invertible Matrix Theorem, and interpretations of the eigen-concept.

Overall, students in our study correctly solved a procedural question related to eigenvalues (as in Q8) at about twice the rate they offered an appropriate conceptual understanding of $Ax=\lambda x$ (as in Q9). This suggests there is a disconnect between students' understanding of standard procedures for finding eigenvalues and the formal definition of an eigenvector and eigenvalue, and that students are more able to execute the standard procedure than draw on conceptual understandings aligned with the formal definition. If standard instructional approaches begin by introducing students to the definition of eigenvectors and eigenvalues using the equation $Ax=\lambda x$ and its algebraic and geometric interpretations but students' work is dominated by execution of procedures such as the computation of roots of the characteristic polynomial arising from $\det(A-\lambda I)$, many students may not adequately connect their results in solving these kinds of problems with the equation $Ax=\lambda x$. This points to a need to push students to think more about core understandings as they connect to procedures rather than just assess students' ability to execute standard procedures. Indeed, many connections are needed to explain why a standard procedure for finding eigenvalues and eigenvectors works and how it connects to the formal definition of eigenvalues and eigenvectors. However, we argue that there is little value in being able to compute eigenvectors and eigenvalues without being able to appropriately interpret the meaning of the result of such computations. The inquiry-oriented approach of the IOLA instructional materials taken up by instructors who received TIMES instructional supports appears to be a promising way of beginning to address this issue, but more work is needed to better understand the ways in which students come to develop and coordinate the interpretations needed for a robust understanding of eigenvectors and eigenvalues.

**Acknowledgements** This material is based upon work supported by the National Science Foundation under Grant Numbers DRL 0634099, 0634074; DUE 1245673, 1245796, 1246083, and 1431393. Any opinions, findings, and conclusions or recommendations expressed in this material are those of the author(s) and do not necessarily reflect the views of the National Science Foundation.

## Appendix: Grading Scheme for Assigning Points to Open-Ended Response Questions 8 and 9b

| Q # | Points awarded and criteria |
|---|---|
| 8 | 3 points:<br>Method 1: Full points were awarded to students who reasoned about the determinant to arrive at the correct conclusion without making computational or conceptual errors. Examples of this kind of reasoning are shown below.<br>(i) $\det(A-\lambda I)=0$ implies $(\lambda-1)(\lambda-4)=0$ implies $\lambda=1$ or $\lambda=4$ implies $\lambda=2$ is not an eigenvalue for the matrix $A$.<br>(ii) $\det(A-2I)=-2\neq 0$ implies $\lambda=2$ is not an eigenvalue for the matrix $A$<br>(iii) $\det(A-\lambda I)=\begin{vmatrix} 3-\lambda & 2 \\ 1 & 2-\lambda \end{vmatrix}=(3-\lambda)(2-\lambda)-2=\lambda^2-5\lambda+4$. Substituting 2 in the characteristic equation gives $4-10+4=-2$ implies $\lambda=2$ is not an eigenvalue for the matrix $A$.<br>Method 2: Full points were awarded to students who reasoned about $A-\lambda I$ without using the determinant to arrive at the correct conclusion without making any computational or conceptual errors. Examples are shown below.<br>(i) $(A-2I)\begin{bmatrix} x \\ y \end{bmatrix}=0$ implies $x=0$ and $y=0$ which is the trivial solution, so $\lambda=2$ is not an eigenvector for the matrix $A$.<br>(ii) $(A-2I)\cong\begin{bmatrix} 1 & 0 \\ 0 & -2 \end{bmatrix}$, and the column vectors of this matrix are not linearly dependent, so $\lambda=2$ is not an eigenvalue.<br>(iii) $rref(A-2I)$ does not have a free variable, so $\lambda=2$ is not an eigenvalue.<br>(iv) The first column of $(A-2I)$ is not a scalar multiple of the second column so $\lambda=2$ is not an eigenvalue so $\lambda=2$ is not an eigenvalue. |
| | 2 points: Two points were awarded to students to students who take a conceptually correct approach (either by reasoning about the determinant or by reasoning about $A-\lambda I$ without using the determinant) but either<br>• made a computational error (e.g. factoring the characteristic polynomial incorrectly) or<br>• did not offer a clear conclusion about whether 2 is an eigenvalue or not, or<br>• arrived at the correct conclusion without a full explanation of why |
| | 1 point: One point was awarded to students whose response included some evidence of conceptual understanding, but who made a conceptual error (which might be accompanied by a computational error). |
| | 0 points: No points were awarded to students who left the page blank, or whose response: (i) gave no evidence of conceptual understanding, or (ii) said something like "I don't know." Example of responses we considered to include no evidence of conceptual understanding are "Yes, because $A = PDP^{-1}$" and "I say it is… because… there are 2's in the problem." |

(continued)

(continued)

| Q # | Points awarded and criteria |
|---|---|
| 9b | 3 points: Three points were awarded to students whose response appropriately coordinated with the eigen-concept, referenced (either by directly naming or by explicitly referring to their work shown in 9a) all three correct vectors, and provided a correct rationale for this selection. |
| | 2 points: Two points were awarded to students whose response provided at least two correct explanations (e.g. $Mx = \lambda x$ is written and student writes that "an eigenvector tells you the direction of stretching") but did not identify and explicitly describe what happens to all three correct vectors. |
| | 1 point: One point was awarded to students who either<br>(i) Provided one correct explanation (e.g. by either writing "$Mx = \lambda x$" or "an eigenvector tells you the direction of stretching") and explicitly connected this explanation to at most one correctly selected vector<br>(ii) Suggested components of $M$ that would transform x into one of the given choices, such as $M = I,\ -I,\ or\ 0$. |
| | 0 point: No points were awarded to responses that do not coordinate with the eigen-concept, do not suggest components of $M$ that would transform x into one of the given choices, says I don't know, or leaves the page blank. An example of student response to question 9 which was awarded 0 point was "all are the same size." |

## References


Andrews-Larson, C., Wawro, M., & Zandieh, M. (2017). A hypothetical learning trajectory for conceptualizing matrices as linear transformations. *International Journal of Mathematical Education in Science and Technology*, *48*(6), 809–829.

Carlson, D. (1993). Teaching linear algebra: Must the fog always roll in? *The College Mathematics Journal, 24*(1), 29–40.

Cline, K., Zullo, H., Duncan, J., Stewart, A., & Snipes, M. (2013). Creating discussions with classroom voting in linear algebra. *International Journal of Mathematical Education in Science and Technology*, *44*(8), 1131–1142.

Dorier, J. L., Robert, A., Robinet, J., & Rogalski, M. (2000). On a research programme concerning the teaching and learning of linear algebra in the first-year of a French science university. *International Journal of Mathematical Education in Science and Technology*, *31*(1), 27–35.

Dorier, J. L., & Sierpinska, A. (2001). Research into the teaching and learning of linear algebra. In *The teaching and learning of mathematics at university level* (pp. 255–273). Springer Netherlands.

Freeman, S., Eddy, S. L., McDonough, M., Smith, M. K., Okoroafor, N., Jordt, H., & Wenderoth, M. P. (2014). Active learning increases student performance in science, engineering, and mathematics. *Proceedings of the National Academy of Sciences*, *111*(23), 8410–8415.

Gol Tabaghi, S., & Sinclair, N. (2013). Using dynamic geometry software to explore eigenvectors: The emergence of dynamic-synthetic-geometric thinking. *Technology, Knowledge and Learning*, *18*(3), 149–164.

Haider, M., Bouhjar, K., Findley, K., Quea, R., Keegan, B., & Andrews-Larson, C. (2016). Using student reasoning to inform assessment development in linear algebra. In Tim Fukawa-Connelly, Nicole E. Infante, Megan Wawro, & Stacy Brown (Eds.), *19th Annual Conference on Research in Undergraduate Mathematics Education* (pp. 163–177). Pittsburgh, PA.

Henderson, C., Beach, A., & Finkelstein, N. (2011). Facilitating change in undergraduate STEM instructional practices: An analytic review of the literature. *Journal of research in science teaching*, *48*(8), 952–984.

Hiebert, J. (1986). Conceptual knowledge and procedural knowledge: The case of mathematics. *Hilsdale NJ: Lawrence Erlbaum Associates.*

Hiebert, J., & Lafevre, P. (1986). Conceptual and procedural knowledge in mathematics: An introductory analysis. In J. Hiebert (Ed.), *Conceptual and procedural knowledge: The case of mathematics* (pp. 1–27). Hillsdale, NJ: Lawrence Erlbaum Associates.

Johnson, E., Caughman, J., Fredericks, J., & Gibson, L. (2013). Implementing inquiry-oriented curriculum: From the mathematicians' perspective. *The Journal of Mathematical Behavior, 32* (4), 743–760.

Johnson, E., Keene, K., & Andrews-Larson, C. (2015). *Inquiry-Oriented Instruction: What It Is and How We are Trying to Help. [Web log post.]*. American Mathematical Society, Blog On Teaching and Learning Mathematics.

Kwon, O. N., Rasmussen, C., & Allen, K. (2005). Students' retention of mathematical knowledge and skills in differential equations. *School science and mathematics*, *105*(5), 227–239.

Lapp, D. A., Nyman, M. A., & Berry, J. S. (2010). Student connections of linear algebra concepts: An analysis of concept maps. *International Journal of Mathematical Education in Science and Technology*, *41*(1), 1–18.

Larson, C., & Zandieh, M. (2013). Three interpretations of the matrix equation Ax = b. *For the Learning of Mathematics, 33*(2), 11–17.

Laursen, S. L., Hassi, M. L., Kogan, M., & Weston, T. J. (2014). Benefits for women and men of inquiry-based learning in college mathematics: A multi-institution study. *Journal for Research in Mathematics Education*, *45*(4), 406–418.

Plaxco, D., Zandieh M, Wawro M. (2018) Stretch directions and stretch factors: a sequence intended to support guided reinvention of eigenvector and eigenvalue (pp. XXX)

Salgado, H., & Trigueros, M. (2015). Teaching eigenvalues and eigenvectors using models and APOS theory. *The Journal of Mathematical Behavior*, *39*, 100–120. https://doi.org//10.1016/j.jmathb.2015.06.005.

Schoenfeld, S. (1995). Eigenpictures: picturing the eigenvector problem. *The College Mathematics Journal*, *26*(4), 316–319.

Selinski, N. E., Rasmussen, C., Wawro, M., & Zandieh, M. (2014). A method for using adjacency matrices to analyze the connections students make within and between concepts: The case of linear algebra. *Journal for Research in Mathematics Education*, *45*(5), 550–583.

Star, J. R. (2005). Reconceptualizing procedural knowledge. *Journal for Research in Mathematics Education, 36*(5), 404–415.

Stewart, S., & Thomas, M. (2006). Process-object difficulties in linear algebra: Eigenvalues and eigenvectors. *International Group for the Psychology of Mathematics Education*: 185.

Stewart, S., & Thomas, M. (2009). A framework for mathematical thinking: The case of linear algebra. *International Journal of Mathematical Education in Science and Technology, 40*(7), 951–961.

Stewart, S., & Thomas, M. O. (2010). Student learning of basis, span and linear independence in linear algebra. *International Journal of Mathematical Education in Science and Technology*, *41*(2), 173–188.

Thomas, M., & Stewart, S. (2011). Eigenvalues and eigenvectors: Embodied, symbolic and formal thinking. *Mathematics Education Research Journal, 23*(3), 275–296.

Vinner, S. (1997). The pseudo-conceptual and the pseudo-analytical thought processes in mathematics learning. *Educational Studies in Mathematics, 34*(2), 97–129.

Wawro, M., Rasmussen, C., Zandieh, M., & Larson, C. (2013). Design research within undergraduate mathematics education: An example from introductory linear algebra. *Educational design research—Part B: Illustrative cases,* 905–925.

Zandieh, M., Wawro, M., & Rasmussen, C. (2017). An example of inquiry in linear algebra: The roles of symbolizing and brokering. *PRIMUS*, *27*(1), 96–124.

# Part III
# Dynamic Geometry Approaches

# Mental Schemes of: Linear Algebra Visual Constructs

**Hamide Dogan**

**Abstract** This chapter is discussing the effect of instructional dynamic visual modalities on learners' mental structures. We documented the effects by comparing the thinking modes, displayed on interview responses, of the learners who were exposed to dynamic visual representations, to those who were exposed to the traditional instructional tools. The data came from twelve first-year linear algebra students' interview responses to a set of questions on the linear independence concept. Our findings point to notable differences on the nature of the mental schemes that learners displayed in the presence and the absence of the dynamic visual modes.

**Keywords** Linear algebra · Visualization · Thinking modes
Instructional modalities

## 1 Introduction

There are both empirical and theoretical publications on linear algebra education. Some focuses on the embodied, symbolic and formal thinking both from a pedagogical perspective (Dogan, 2006; Dogan, Carrizales, & Beaven 2011; Gol Tabaghi, 2014; Salgado & Trigueros, 2015; Stewart & Thomas, 2009, 2010; Zandieh, Wawro, & Rasmussen 2017), and from a perspective of the role of instructional technologies in cognition (Dogan, 2004, 2013; Dogan-Dunlap, 2010; Mariotti, 2014). Others discuss further learning difficulties with basic linear algebra concepts (Carlson, 1997; Dorier & Robert, 2000; Dorier, Robert, Robinet, & Rogalsiu, 2000; Dorier & Sierpinska, 2001; Hillel & Sierpinska, 1994; Stewart & Thomas, 2009; Thomas & Stewart, 2011). For relevant work, see Bouhjor et al., Chap. 9, Plaxco et al., Chap. 8, and Turgut, Chap. 11 in this volume. Many of these publications state that students have problems with the abstraction level of linear algebra material. The high level of formalism in linear algebra seems to make learners have

H. Dogan (✉)
Mathematical Sciences, University of Texas at El Paso, El Paso, TX 79968, USA
e-mail: hdogan@utep.edu



S. Stewart et al. (eds.), *Challenges and Strategies in Teaching Linear Algebra*,
ICME-13 Monographs, https://doi.org/10.1007/978-3-319-66811-6_10

the feeling of lack of connection to what they already know in mathematics. Additionally, axiomatic approach to linear algebra appears to give many students the feeling of learning a topic that is unnecessary for their majors. Another area of difficulty appears to be with the multiple representational approaches used in linear algebra. Apparently, students have difficulty in recognizing different representations of the same concept. Many lack logic and set theory knowledge (Carlson, 1997; Dorier et al., 2000; Dorier & Robert, 2000; Dorier & Sierpinska, 2001; Hillel & Sierpinska, 1994). Specifically, students' lack of skills in elementary Cartesian Geometry (Dorier & Sierpinska, 2001), and their inadequate set theory knowledge (Dogan-Dunlap, 2010) seems to cause the majority of learning difficulties in linear algebra courses. Dorier and Sierpinska (2001) argue that an understanding of linear algebra requires a fair amount of cognitive flexibility. Dubinsky (1997) further adds that:

> It seems that mathematics becomes difficult for students when it concerns topics for which there do not exist simple physical or visual representations. One way in which the use of computers can be helpful is to provide concrete representations for many important mathematical objects and processes (Dubinsky, 1997, p. 104).

Some researchers consider providing initial experiences via easily accessible tools (concrete or visual) as the necessary means for a successful uncovering of meanings behind abstract ideas (Dogan, 2004, 2006, 2013, 2014b; Dogan-Dunlap, 2010; Gol Tabaghi, 2014; Harel, 1987, 1989, 1997, 2000; Mariotti, 2014; Stewart & Thomas, 2004). In fact, this very idea motivated us for our investigation. By collecting data from three groups of differing instructional tools, we were able to investigate the differences and similarities in their cognitive schemes. In this chapter, in an attempt to shed light on the effect of visual representations, we compare the modes of three groups of matrix algebra students. One group was fully exposed to dynamic visual representations, both in lectures and as part of homework assignments. The second group was given traditional instructional means, coupled with take-home assignments integrating dynamic visual representations. The third group lacked any exposure to dynamic visual representations.

Besides the difficulties reported above, the changing demographics of the linear algebra courses add challenges. In fact, due to advances in technologies, such as digital computers, used widely in engineering schools linear and matrix algebra are among the advanced mathematics courses attracting more and more students from other disciplines (Torres & Dogan-Dunlap, 2006; Tucker, 1993). These students are usually not prepared or at best ill-prepared for the high abstraction level of matrix algebra courses. Since students are lost in much of the abstraction, even the simplest ideas become difficult to comprehend. Thus it creates high stress, in turn "burn out", and as a result, high failure rates (Dogan, 2012, 2013, 2014b; Dogan-Dunlap, 2010; Dorier et al., 2000; Sierpinska, Trgalova, Hillel, & Drayfus, 1999).

According to many researchers (Dubinsky, 1997; Harel, 1987, 1989, 1997, 2000) students can cope with abstraction if flexibility between representations of the same concept is established. Abstraction might be successfully dealt with if concept images (defined as all mental pictures, properties, and processes associated with the concept), and concept definitions [defined as a form of symbols used to specify the concept (Hiebert & Lefevre, 1986)] are not contradicting one another. On the other hand, others argue that multiple representations without inquiry may not provide the cognitive support students need in coping with abstraction (Dogan, 2006, 2013; Dogan-Dunlap, 2003; Dorier et al., 2000; Gardenfors & Johansson, 2005; Harel, 1997, 2000; NRC, 2000). Indeed, it has been reported that technology with inquiry may provide the first-hand knowledge learners need to make better sense of the second-hand knowledge (Wawro, Rasmussen, Zandieh, Sweeney, & Larson, 2012). First-hand knowledge is defined as the knowledge obtained through direct experience, while the second-hand knowledge is defined as the knowledge obtained from formal descriptions (Schwarz, Martin, & Nasir, 2005).

In addition to the wide range of computer activities implemented in many linear algebra classrooms (Leon, Herman, & Faulkenberry, 1996; Roberts, 1996; Wicks, 1996), there has also been studies documenting student performances in the presence of technology-based activities (Dogan, 2004, 2013; Dogan-Dunlap, 2010, 2003; Gol Tabaghi, 2014). While some of these studies reporting learner's difficulties with geometric representations (Stewart & Thomas, 2009; Thomas & Stewart, 2011), others documented the effect of dynamic geometry software on one's cognition of linear algebra topics such as linear transformations (Gol Tabaghi & Sinclair, 2013) and eigenvealue and egenvector tasks (Gol Tabaghi, 2014; Stewart & Thomas, 2009; Zandieh et al., 2017).

Leron and Dubinsky (1995), as an example, reported that as a result of writing programs in ISETL [a programming language (Dautermann, 1992)] as solutions for abstract algebra questions, a substantial increase was observed in students' comprehension of abstract algebra concepts. ISETL allowed students, through the inquiry process, to construct their initial understanding of the concepts, thus facilitating the formation of the connections between the existing knowledge, and newly introduced concepts. Another is a study by Sierpinska et al. (1999). They, in their paper, shortly discussed the effect of Cabri (a dynamic geometry software) on their students' mental images of both linear combination and linear independence ideas. Contrary to the findings of Leron and Dubinsky (1995), in abstract algebra cognition, Sierpinska et al. (1999) reported no significant findings.

Moreover, Turgut (see Chap. 11 in this volume) documented their participants' verbal signs used in their conversation while working with a dynamic geometry software to discover properties of linear transformations. Gol Tabaghi (2014) talked about how dragging in a dynamic geometry environment changes student's awareness, thus developing meanings for eigenvalues and eigenvectors. Also, Bouhjar et al. (Chap. 9 in this volume) discussed the effect of two instructional approaches with two separate groups (one traditional and the other, a computerized program called TIMES). Their findings showed no significant difference in the two groups' performance on procedural questions. They stated however that the group who went through the TIMES program outperformed the traditional group on conceptual questions. For a description of TIMES, see Bouhjar et al., Chap. 9 in this volume. Our work reported here is likewise comparing the effect of differing instructional tools. Unlike the previous study's focus (performance on procedural vs. conceptual questions), our focus is on the type of modes students displayed in the presence and absence of dynamic geometric representations.

We investigated the particular area utilizing a set of structured clinical interviews with 12 volunteers from three sections, with differing instructional means, of a linear algebra course. In this chapter, we discuss the findings of our investigation from this particular work.

Our research question was to document the effect of visual representations in learners' mental images. We investigated the potential effects by addressing the following two tasks:

(1) The frequency of visual representations that occurred on student responses to interview questions.
(2) The nature of these visual constructs (similarities and differences in the three differing instructional groups).

## 2 Framework

Sierpinska's (2000) framework on student thinking modes provided the analysis tools for our investigation. Sierpinska (2000) reports three kinds of thinking modes occurring in linear algebra courses. These are Synthetic-Geometric, Analytic-Arithmetic, and Analytic-Structural. According to Sierpinska (2000), the three thinking modes differ mainly in the representational types. Precisely, Synthetic-Geometric category uses geometrical representations, and in this group, objects are given readily through visual modes, but not defined (Sierpinska, 2000). For instance, in this classification, a line or a plane may be considered as a pre-given object with a recognizable shape located in space (Sierpinska, 2000). As another example, consider the visual representations of a set of vectors in geometric environments, in this scenario, the linear independence of these vectors can be determined using the location of the vectors within the geometric systems.

Analytic modes, on the other hand, use numerical and algebraic representations. In these modes, objects are defined. For instance, the formal definition of linear independence uses an analytic mode (Sierpinska, 2000). Within the analytic category, Sierpinska (2000) classifies the modes further into two separate groups. One is the Analytic-Arithmetic, and the other is the Analytic-Structural mode. In contrast to the abstract, symbolic, and formal nature of the Analytic-Structural category, Analytic-Arithmetic category considers objects on their computational processes. For instance, with the Analytic-Structural classification, learners may consider a set of vectors in connection with the vector space they are a member of, and determine the set's linear independence using the dimension of the vector space. In the Analytic-Arithmetic modes, learners may proceed to the row reduction processes, solely computational approaches, to address the linear independence tasks. See Table 1 for more descriptions and sample indicators of each category.

**Table 1** Thinking modes modified from Sierpinska (2000)

| Mode of thinking | | Representations | Indicators |
|---|---|---|---|
| ***Synthetic-geometric*** | | Graphical representation provide properties of objects readily it describes an object but not define it | Student is able to determine whether vectors whose graphs are provided in $R^2$ or $R^3$ are linearly independent or dependent |
| ***Analytic modes*** | ***Analytic-arithmetic*** | Numerical representation linear combination defines the object | Student is able to construct matrix from vectors, compute row-reduced echelon form and relate reduced matrix to linear dependence and independence |
| | | Linear Combination | Student is able to find/ use linear combination of vectors and determine linear independence |
| | ***Analytic-structural*** | Objects are considered in a system | Use of the dimension of vector space in determining linear independence of vectors |

We should also add that we use the term "*algebraic*" interchangeably with the term "structural" as in the analytic-structural mode since Sierpinska's framework appears to have been focusing mainly on the algebraic/symbolic representations in her categorization of the Analytic-Structural types (2000).

## 3 Methodology

Data discussed in this chapter came from our work with twelve students enrolled in the three sections of a first-year matrix algebra course at a Southwest University in the US; one with minimum exposure to static geometric modes namely group C, and the other two implementing, in varying intensity, interactive web-modules, namely groups A and B. We used alphanumeric names to refer to the students enrolled in the three sections. For instance "A12" is used for a student enrolled in the section, A. We also used the terms "*he*" and "*she*" sporadically and interchangeably with no direct association to the gender of our participants.

## 3.1 Section Characteristics

In the three groups, topics were presented via lectures mainly through formal definitions, theorems, and arithmetic computations. Section A however differed from B markedly in that even though both sections assigned bi-weekly investigative take-home tasks integrating online modules, section A integrated these modules fully into its lectures as well. See Table 2 for a summary of the section characteristics.

In the group C, the classroom lectures of the topics were similar for the most part to that of section B. Topics were covered mainly via formal definitions, theorems, and arithmetic computations. Even though it was not the intended approach for section C, the instructor of this section regularly provided the static versions of the graphical representations in the lectures. What set this section apart from the groups, A and B, though, is the absence of the guided investigative interactive web-based take-home assignments.

To summarize, looking at Table 2, one can observe that group B was exposed to the dynamic visual representations solely via take-home assignments, and group C was introduced to the static geometric modes only in its lectures. Group A, on the other hand, is the only section that made use of the dynamic pictorial representations in both lectures and take-home assignments.

## 3.2 Web-Module and Companion Investigation

Throughout the semester, we administered seven guided investigative assignments using a set of interactive online modules. The theme of the chapter, however, came from data gathered right after an investigation on the linear independence relevant ideas. This investigation was coupled with a web-module. Three separate snapshots of the particular module can be seen in Fig. 1. In this figure, frame (a) shows a view of the input boxes for vectors and scalars. Frames (b) and (c) depict, from two separate angles, the geometric representations of various vectors (as line segments

**Table 2** Characteristics of the three matrix algebra sections

| Characteristics | Groups | | |
|---|---|---|---|
| | A | B | C |
| Same instructor | | √ | √ |
| Investigative assignments via web-modules | √ | √ | |
| Lectures integrating interactive web-modules | √ | | |
| Lectures integrating static geometric representations | | | √ |
| Lectures integrating formal definitions, theorems, and arithmetic computations | √ | √ | √ |
| Same textbook | √ | √ | √ |

with arrows) along with a few linear combinations (as dots). For instance, using the module, one can consider a plane in connection with the 3-dimensional space. Moreover, one can study vectors on their relative positions in the 1–3-dimensional spaces formed by the linear combinations of the vectors. The current location of the module can be found at Dogan (2014a).

As mentioned earlier, the online-based investigative homework was assigned only to the groups A and B. This investigation required learners to use the particular online module in Fig. 1. This was to address the mathematical tasks, on the solutions of two separate vector equations, namely $w = du + ev + lt$ for real numbers $d$, $e$, $l$, and $Au + Bv + Cw + Dt = 0$ for vectors, u, v, w, and t. While the first equation was to address the linear combination ideas, the second was to bring up the linear independence tasks. This investigation moreover asked students to compare and contrast the solutions of the two equations for twelve different sets of vectors. This task furthermore provided a detailed description of the web-module and an informative example. It guided learners via tasks that gradually became more abstract and thought-provoking. That is, earlier vectors were numerical and easier to observe their relative positions in visual environments. Latter vectors, however, became increasingly symbolic. Thus, identifying linear relations among them required not only the observation of the visual objects but also the use of more formal and abstract though processes.

For example, on the frame (b) in Fig. 1, one can observe the module providing a geometric representation of a linear combination of two vectors, $u = (1, 2, 3)$ (upper left) and $v = (1, 0, 6)$ (lower right) resulting in the vector $w = (3, 4, 12)$ (middle) (i.e. $w = 2u + v$). The dots seen in the frames (b) and (c) stand for the visual

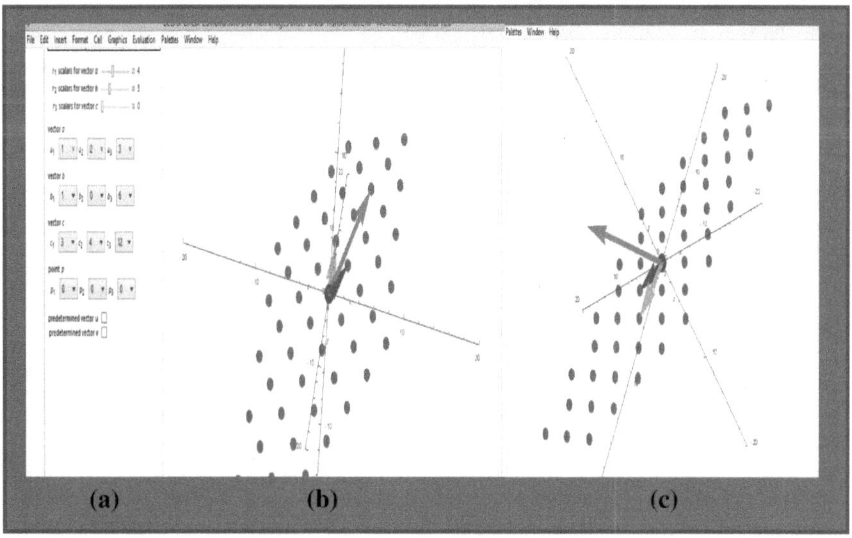


**Fig. 1** Module view

forms of some of the linear combinations. Indeed, by tracing the dots, one can easily verify the linear combination, $w = 2u + v$, visually.

To elicit more abstract thoughts, later in the investigation, the following strictly symbolic vectors ($u = a_1$, $v = a_5$, $t = a_0$, and $w = a_{10}$) were given. As seen in Fig. 1, this module includes only three input-boxes for the visual construction of linear combinations. Thus, the fourth vector, w, can only be sketched as a line segment. Hence, w's potential linear connection to other vectors has to be observed via its relative positions within the module's geometrically constructed system.

To summarize, in this investigation, having learners work with two separate equations involving vectors with solely numerical and symbolic representations, the goal was to facilitate the formation of first-hand knowledge, which may become an anchor for the mental construction of the more formal external ideas introduced mainly via lectures (second-hand knowledge).

## 3.3 Interview Questions

Students volunteered for a set of one-on-one interviews. Interviews were conducted right after the class coverage and the completion of the take-home assignment discussed above. Each interview lasted about an hour, and each began with a set of pre-determined questions, on the basic vector space concepts such as linear independence, span and spanning set. We added new questions on an as-needed basis for the clarification of the responses as well as to solicit further information on one's mental images. Pre-set questions were determined based on the most common learning difficulties reported in the literature (Dogan, 2004, 2006, 2013, 2014b; Dogan-Dunlap, 2010; Dorier et al., 2000; Gardenfors & Johansson, 2005; Harel, 1997, 2000; NRC, 2000; Sierpinska, 2000).

Each participant began the interview with a slightly modified version of the same set of questions, provided on a sheet of paper. The modifications entailed negligible changes mainly on the numerical values or the order in which the questions were administered. These questions covered a range of topics from linear independence to dimension ideas. Given on one such question sheet, questions 1 and 2, for example, were general enough to get a glimpse of the modes students brought out of their existing mental schemes of linear independence concept.

Question 1: Define the linear independence of a set of vectors.
Question 2: Give an example of a linearly dependent set of vectors.

Taken from the same question sheet, the question number 4, on the other hand, was given to elicit primarily the mental structures displayed in the absence of the numerical representations.

Question 4: Given a linearly independent set, $\{u_1, u_2, u_3, u_4\}$, in $R^n$.
Determine the linear (in)dependence of the set $\{u_1, u_2, u_3, u_4\}$. Explain.

Also, there were questions to elicit the thinking modes used to make connections between the linear independence ideas and the other relevant topics. Question 8 is an example of this kind. Indeed, this question was given to document the modes used to connect the nonzero solutions to the linear dependence ideas.

Question 8: Given the vector equation $a_1u_1 + a_2u_2 + a_3u_3 = 0$ with a solution, $a_1 = 1$, $a_2 = -2$, and $a_3 = 0$. Determine the linear (in)dependence of the set, $\{u_1, u_2, u_3\}$. Explain your answer.

## 3.4 Analysis

A qualitative approach, namely the constant comparison method (Glaser, 1992) was used to analyze the interview responses. The interviews were first videotaped, and next transcribed. The analysis of the transcripts was conducted independently by three raters following the Sierpinska's framework (2000). The average rater correlation is calculated to be about 75%. Each analysis focused on the identification and classification of the modes as displayed on the interview responses. At this point, we should note that some responses contained multiple modes. Thus these responses were included in multiple categories. Furthermore, as a result of a consensus among the three raters, we obtained a final list of nine categories. In fact, in this chapter, we are utilizing the particular list to support our arguments on the effect of visual representations in one's mental structures (see Table 3).

The final list contains nine categories of thinking modes. See Table 3 for all nine categories with their titles, and the frequency of responses in each class. The category, "*One vector comes out of a plane*," with the abbreviation "O" is one of them. This category includes responses that make use of the geometric features of linear objects such as planes, lines, and the relative positions of vectors on these linear environments. The excerpt below is a representative of the types of responses included in the particular category. It is easy to see that this excerpt belongs to the O category. In fact, the phrases; "*vectors either on the plane*", "*go up*", and "*another dimension*" are referring to a geometrically formed plane and a set of vectors located on this plane.

> Because I can't, I can't, I don't think I can form, uh… I can do some like state a scalar, like multiply this one times a scalar add it to this one to get it to go up like this…just 'cause these are on the same plane I want to think that they are gonna stay… when we look at the collection they're all just… since they are on a plane, they are just gonna form like this, or somehow, other spots on these dots… but I don't think they'll jump up to the next… like another dimension…

The particular response moreover contains phrases hinting to the mental structures that are little more computational. Thus, it was also considered for the category, LCS. This category included responses where students refer to the linear combination ideas with specific justifications. We indeed can observe this behavior in the participant's explicit mention of the arithmetic operations. The phrase, for

**Table 3** Number of responses in each category

| Label | Category | Group A | Group B | Group C | Total |
|---|---|---|---|---|---|
| O*+ | One vector comes out of a plane<br>Vectors lie on the same plane<br>Scalar multiples of vectors on plane | 62 | 34 | 45 | 141 |
| LC*+ | Linear combination just stated, no work<br>Linear independence definition<br>Matrix column dependence<br>Scalar multiple<br>Geometric and algebraic notions | 152 | 23 | 37 | 212 |
| LCS* +^ | Linear combination of vectors given<br>Algebraically<br>Geometrically<br>Operations on vectors | 231 | 109 | 265 | 605 |
| D* | Vectors go in the same or different direction<br>Different vectors<br>Angle of vectors<br>Connected vectors | 8 | 9 | 8 | 25 |
| Z^* | Zero vector<br>Algebraically adding vectors gives zero vector<br>Geometrically tracing vectors reaches zero vector<br>Set has a zero vector | 28 | 0 | 13 | 41 |
| (ZS)+^ | Solution type<br>Unique/infinite<br>Matrix forms<br>Independent variable<br>Zero row/column | 169 | 64 | 131 | 364 |
| V*+ | Vector space dimension<br>Number of vectors versus dimension of space<br>Plane versus $R^n$<br>Number of component of vectors<br>Basis vectors<br>Number of equations versus unknowns | 133 | 89 | 124 | 346 |
| E^ | Row reduced echelon form | 35 | 41 | 90 | 166 |
| L* | Overlapping<br>Lines collinear<br>Lines parallel | 4 | 0 | 28 | 32 |

*Geometric categories, + algebraic, and ^ arithmetic categories

instance, "*I can do some like state a scalar, like multiply this one times a scalar add it to this one to get it to go up like this*" clearly indicates that, at the time, there were thought processes at work, applying the vector operations of scalar multiplication and addition. Looking at Table 3, one can see that the category with the most number of responses is the LCS category (with 605 responses), followed by the category, ZS, (with 364 responses).

Based on the framework of Sierpinska (2000), we argue that some or all responses in the categories, O, LC, LCS, D, Z, V, and L can further be considered as responses using Synthetic-Geometric modes. The categories, O, LC, LCS, Z, ZS, V, and E, on the other hand, contain responses that display arithmetic or/and algebraic modes. As mentioned earlier, many responses displayed multiple mental constructs, and naturally, they were included in the multiple categories. For example, the mode classes, O, LC, LCS, and V include responses referencing both geometric and algebraic thought processes. The categories, D and L, however, are the ones including responses with strictly geometric content, and the E category contains strictly arithmetic responses.

# 4 Results

## *4.1 Comparison of Groups*

Once categorizing the modes into the three main categories, we furthermore computed the frequency of responses containing mixed and solo modes. Figure 2 provides a chart displaying the percent responses of these thinking modes. Looking at the bars in the figure, one can easily verify that all three groups produced the highest percentage of responses in the mix-use of the *Geometric/Algebraic*, and the

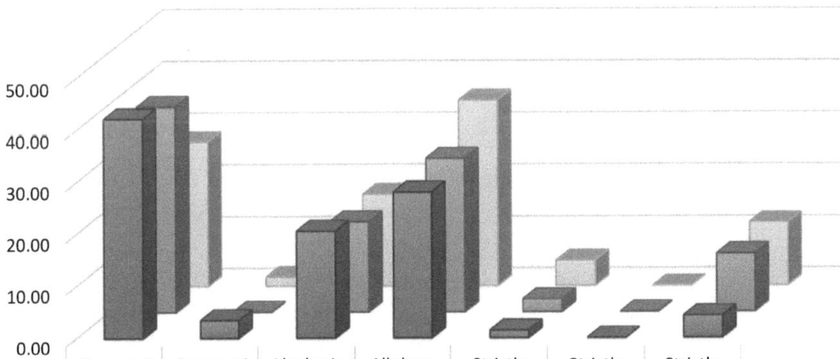


| | Geometric/Algebraic | Geometric/Arithmetic | Algebraic/Arithmetic | All three | Strictly Geometric | Strictly Algebraic | Strictly Arithmetic |
|---|---|---|---|---|---|---|---|
| A | 42.21 | 3.41 | 20.56 | 28.10 | 1.46 | 0.00 | 4.26 |
| B | 39.57 | 0.00 | 17.34 | 29.54 | 2.44 | 0.00 | 11.11 |
| C | 27.80 | 1.75 | 17.68 | 35.76 | 4.86 | 0.00 | 12.15 |

**Fig. 2** Percent of mixed/solo mode usage in groups

mix-use of *All Three* modes followed by the mix-use of the *Algebraic/Arithmetic* classifications.

An increasing behavior from the group A (28.10%) to the group C (35.76%) is observed among the responses containing all three modes as well as the responses containing strictly arithmetic modes (4.26% in group A; 11.11% in group B; 12.15% in group C). This predisposition, however, is reversed for the number of responses displaying both Geometric (Synthetic-Geometric) and Algebraic (Analytic-Structural) representations. That is, this time, the incline is from the group C (27.80%) to the group A (42.21%). Figure 2 furthermore shows that the percent responses in the mix-modes of geometric/arithmetic, strictly algebraic, and strictly geometric categories are all very small, hence negligible.

One can then conjecture that the algebraic and geometric modes were not used standalone but rather interweaved with other modes. Another word, all three groups produced responses containing algebraic modes mixed with the arithmetic representations at about same percentage. The mixing of the Analytic-Structural (algebraic) with the geometric modes, on the other hand, was at the highest frequency, and more prominent in both groups A and B with a slightly higher quantity in the group, A.

In short, comparing the percent responses from Fig. 2, group A followed by the group B shows higher tendencies to incorporate the geometric entities with the algebraic modes. Group C, on the other hand, displays tendencies to use arithmetic modes unaccompanied, or all three modes simultaneously. Looking closely at the manner in which these modes are used, we also identified drastically different mental structures at work in each group. These differences seem to be more prominent among the geometric modes. To be exact, in the groups A, and B, the initial knowledge appeared to have been shaped mainly by the geometric objects. Group C, on the other hand, revealed an initial knowledge, influenced mainly by the numerical entities.

### 4.1.1 Nature of Modes in Group A

Group A's geometric representations, for the most part, revealed dynamic features. That is, during the interview, many participants in this group used external means such as hand gestures, pencils, tables, and the interview room. Additionally, group A provided descriptions closely paralleling the module features, seen in Fig. 1.

Excerpts below, taken from A21's interview responses, for example, are revealing the overall behavior of all the participants in the group. One can see that in this excerpt, SA21 is providing a vivid geometric description of linear operations as dots. SA21 furthermore is using hand gestures to describe the shape formed by these dots, and furthermore using the formation to make arguments for the dimension cases. Thus, the excerpts are verifying the close resemblance of his mental forms to the web-module seen in Fig. 1. His use of hand gestures and external, objects such as pencils and tables, further reinforces our observations. Additionally, the phrase "*... I can do some like state a scalar, like multiply this one* *times a scalar add it to this one to get it to go up like this…just 'cause these are on the same plane I want to think that they are gonna stay…*" is indicating that in order to make sense of the arithmetic features of vector operations, SA21 is using his view of the geometric entities.

> SA21: Let's see… what, what I think is jumping at me is, this one is up here, just because we can express these as a combination does not mean that we can get this one…[student is talking about the vector $u_4$ described in the question as sticking out of the plane] *'cause this is in a different location, so maybe it is, and I would say linearly independent because we cannot…Because I can't, I can't, I don't think I can form, uh… I can do some like state a scalar, like multiply this one times a scalar add it to this one to get it to go up like this…just 'cause these are on the same plane I want to think that they are gonna stay… when we look at the collection they're all just…since they are on a plane, they are just gonna form like this, or somehow, other spots on these dots… but I don't think they'll jump up to the next… like another dimension?…* [student is moving the pencils/vectors on the table standing for a plane to imply that linear combinations will stay on the same plane].

Among the responses classified as the Analytic-Algebraic modes, in group A, we documented responses that were mostly referring to the linear combination ideas. These modes were primarily intertwined with visual structures. That is, almost all participants in group A interpreted the algebraic features of the linear dependence ideas in the context of the geometric notion of "*vectors being on a plane*". The excerpts below are examples of such perspectives. Take the participant SA12's response (below) given to the question number 3 from his interview question sheet:

Question 3: Given the set $\{u_1, u_2, u_3, u_4\}$ where the vectors $u_1$, $u_2$, $u_3$ are on the same plane, and u4 is not. Determine if the set $\{u_1, u_2, u_3, u_4\}$ is linearly independent. Explain your answer.

The phrase "*$R^2$ so then these vectors automatically become combination*" clearly indicates that SA12 understands the linear combination ideas based on his/her experiences with the visually constructed spaces, in the interactive web-module (Fig. 1). In another participant's excerpt, namely SA22, one can see similar mental structures being applied. Indeed, this participant's notion of "*vectors on a plane*" and "*linear dependency*" appears to be intertwined. As a matter of fact, this student regularly used the two ideas interchangeably, throughout her interview. In the particular excerpt, SA22 is undeniably incorporating a conditional statement similar to: "*vectors being on a plane implies that they are* [linearly] *dependent on each other.*"

> SA12: Because in $R^2$ you only need two vectors to form, two linearly independent vectors to form any vector inside of $R^2$ so then these vectors automatically become combination of the first two, and then the same thing for $R^3$ but with 3 vectors and the fourth one would have to be a linear combination of the first three, if they are linearly independent…

> SA22: well this three are on the same plane so this will be like if I gave the example from the previous one. So these three are dependent of each other or they can be. No, they are because they are on the same plane…

In the Analytic-Arithmetic category, there were no notable differences in the type of and the manner in which the modes were applied by all three groups. The top two frequencies in the Analytic-Arithmetic category came from the modes referring to the solution types (ZS), and the row reduced echelon matrix forms (E). Moreover, we observed that many responses incorporated the two modes, E and ZS, simultaneously.

The excerpts below are the sample representations of the line of reasoning applied by the participants of the group, A. In his response, SA21, for instance, is clearly focusing on the structure of the reduced echelon matrix forms (rref) and connecting this to the linear independence ideas. Indeed, his thought process is evident in the phrase: "*if this* [matrix] *ends up in, in the identity, this* [set] *would be linearly independent...*" Another participant from the same group, SA24, is using an analogous reasoning.

Distinctively, though, this participant first connects the matrix form to a solution type, and next, the solution type ideas to the linear independence concept.

> SA21: Well I know, like I would like try to get the RREF, row reduced. ...and if this ends up in, in the identity, this would be linearly independent...

> SA24: True, because when I input the matrix in the calculator, it only showed that the only solution was $a = 0$, $b = 0$ and $c = 0$, this meaning that it is linearly independent...

To summarize some of the notable characteristics of the mode use, this group revealed an initial understanding primarily structured by the geometric entities. Furthermore, many in group A tended to use this notion to make sense of the algebraic and more abstract tasks. Thus, group A naturally produced the highest percent responses mixing algebraic and geometric modes. This group, on the other hand, showed the least preference for the arithmetic means.

### 4.1.2 Nature of Modes in Group B

As mentioned earlier, the groups, A and B, showed similar tendencies in their use of geometric modes. Both groups provided descriptions paralleling the module views (see Fig. 1). Contrary to group A, group B, however, revealed disconnects between the intended aspects of the module features and their interpretations. Thus, they differed in the nature of the geometry-based mental structures.

The following excerpt from SB15's interview, for instance, shows that the geometric notions formed by this student diverge notably from the intended features of the web-module (Fig. 1). We should note that some portion of this excerpt came from his responses to a question of whether a set of 4 vectors,

$$\left\{ \begin{bmatrix} 1 \\ 2 \\ 3 \\ 4 \end{bmatrix}, \begin{bmatrix} 0 \\ 1 \\ 1 \\ 1 \end{bmatrix}, \begin{bmatrix} 2 \\ 3 \\ 1 \\ 0 \end{bmatrix}, \begin{bmatrix} 0 \\ 1 \\ 1 \\ 1 \end{bmatrix} \right\}$$

is linearly independent. Even though the numerical values were given in this task, this question appears to have evoked geometric modes in his/her mental processes.

This behavior supports our hypothesis of Group B holding an initial knowledge shaped by geometric means. To further elaborate our observations, in an attempt to address the question, this student consistently referred to various geometric shapes such as hexagons and spheres. These analogies do not closely resemble the linear objects (lines and planes) provided by the web-module. His descriptions are furthermore revealing the dynamic use of external tools such as pencils and the interview room. In this excerpt, B15 talks about dots as well but, contrary to their use in the group, A, this participant's notion of dots appears to entail only the notion of dimension ideas. In fact, with the dots, B15 focuses largely on how things from higher dimensions "*look like*" in lower dimensions.

> SB15: Then, as I go it would start getting hexagonal, so you're just getting the slices of this. So you're getting 2-dimensional objects out of something 3-dimensional. Now, in $R^4$, I think of like… let's say, I don't know bubble gum or something, or like a sphere but in that sphere on the inside it's like a very thin sphere… where it's hollow on the inside and on the outside I mean there's space. But in that little thin layer is 3-dimensional space. The fourth dimension is either is like on the outside doing the same… that's about it.... imagine, okay… like right now we are in $R^3$ [meaning the meeting room], right? You see me, I see you, and everything. But… imagine... okay, wait… uh… I guess the way I would think of it is let's just call this flatland [referring to either the table or the floor of the room] right here, and it's from… I read a… I read something where it said, okay, imagine we're trying to pass a 3 dimensional shape into space, what would it look like?... So, I'm trying to pass this pencil through here. The first thing, let's say, it's a perfect tip, it would be a little dot…

In the Analytic-Structural (Algebraic) category, most responses in this group contained the terms similar to "*one depends on the other*" and "*one is a linear combination of the other.*" Unlike the other two groups, this group used them strictly verbally without providing any justifications. The excerpts below are the cases of such instances. In the first phrase, for instance, SB6 is simply indicating that "*…this vector was a combination of another two vectors…then …it was dependent.*" In his response, B6 falls short on providing any concrete linear combination to justify his answer. Overall, the responses of SB6 indicate that, at that point, SB6's initial understanding of linear dependency may have been formed mainly in the context of superficial (with no substantial insights) linear combination ideas. As a matter of fact, this is plainly laid out in her second excerpt. In this excerpt, SB6 provides an argument for the following interview question sheet:

Interview Question: Given an nxm matrix, A, where $a_{i2} = a_{i4} + 3a_{i5}$. Determine if the set, $\{A_1, A_2, A_3, \ldots, A_m\}$ is linearly independent (Here, $A_j$ is the $j$th column of A). Explain your answer.

Even though in this question, an explicit linear combination among the columns of a matrix is given, it can be seen that SB6 is struggling to apply the explicitly given linearity between the columns of the matrix to the corresponding set of vectors. This may furthermore be interpreted as, at the time, SB6 lacking any connection between matrices and vector sets.

SB6: …yeah, I said something like this…this were…what I thought about it is, this is a set of vectors and these two…I mean this vector was a combination of another two vectors… then I think since it was a combination I said it was dependent…

SB6: …well, I don't know how to…like…like if this is…because what I am saying here is like in a matrix…one of the columns is equal to these two…so…but here it's saying that…a matrix, which is this one, this…this doesn't mean that this is equal to…another column within the same matrix…

Similar to the other two groups, many participants of the group, B, also did not reveal any differences in their use of arithmetic modes. That is, this group's responses included ideas from the solution types (ZS) and the row reduced echelon matrix forms (E). For example, the student, SB6, in the excerpt below, is revealing one such behavior. Here, SB6 is focusing on the existence of free variables as the indicators of non-trivial solution cases. Thus, this student is connecting the matrix forms, he obtained from a row reduction processes, to the solution types, in turn later in the interview, using this idea to infer about the linear independence of vector sets.

SB6: …when you have a…a matrix and it has a bunch of numbers and you can reduce it, and if at the end you have any free variable…that means it is not a unique solution because that free variable can be whatever…

To summarize the overall behavior of the group B's participants, even though the group's initial knowledge was clearly shaped by the geometric entities, unlike the group A, this group's knowledge contained irrelevant features. Nevertheless, many participants in this group, similar to group A, coupled algebraic and geometric modes in their responses. Group B, in contrast, displayed notably higher tendencies to apply arithmetic processes, and they did this in isolation.

### 4.1.3 Nature of Modes in Group C

In comparison to the groups, A and B, group C showed a drastically different behavior. The responses in this group uncovered an initial knowledge that is shaped mainly by the numerical entities. In fact, many in this group focused on the x, y, or z components of vectors to make sense of the location of vectors or the dimension of spaces. Thus, even though these responses were initially categorized as geometric, they were revealing mental structures that are mostly numerical at its core. Therefore, this group's high percent on the use of all three modes (see Fig. 2) may have been contaminated with their use of the numerical implications for the geometric entities. In light of this fact, one may choose to re-label most of the responses in this category under the category of the mix-use of Algebraic and Arithmetic.

Throughout the interview, a participant in the group C, SC6, for instance, showed tendencies to interpret the various characteristics of vector spaces using the components of vectors, rather than the geometric structures of span ideas, such as lines and dots as seen in Fig. 1. Take the excerpt below. In this excerpt, SC6 interprets the vectors as having/not having a z-component (meaning nonzero/zero values). The student gave his response, on this excerpt, to a question similar to the question number 3 stated above, for SA12, in the Sect. 4.1.1. The phrase "*vectors you always have to break them into components. You can't add them when they're in terms of their length and their angle...*" further reveals that SC6, at that point, may not have held mental structures that could facilitate any connections between arithmetic operations and their geometric counterparts. Indeed, at the time, SC6's initial knowledge appeared to have been, mostly, shaped by the numerical entities. Throughout the interview, this participant and many in this group regularly used similar knowledge to make sense of the geometric tasks.

> SC6: Right, uh… with these two, because these three vectors, the one, this one, and this one [forth vector described in the question as being off of a plane] are not co-planar, and these two have no z component [referring to vectors described as being part of a plane] then I can take the z component in this one [forth vector] which is the same as the z component on this one, and the x components in the three of them, and the y components of three of them…

Group C, furthermore, applied its "*vector component*" ideas in their responses, categorized as Analytic-Structural. In fact, this group used "*vector component*" ideas, frequently, to justify their answers involving linear (in)/dependence ideas. In the excerpt below, for instance, SC6 is applying his "*vector components*" view to support his understanding of the linear operations and their geometric interpretations. Without a doubt, at the time, C6's initial knowledge was heavily influenced by the "*vector component*" features. In his interview responses, as a matter of fact, C6 repeatedly justified the absence of the linear combinations of vectors, resulting in the desired vector, by pointing to the differences in the component values of the vectors. In her excerpt, this student, for example, provides an explanation to what would occur as a result of the scalar multiple of a vector. This behavior is clearly visible in the phrase "*…the problem is the components between the two are not the same...*" Furthermore, with the statement, "*I can't change the orientation...*" we believe SC6 is considering geometric directions strictly in the context of the vector component values.

> SC6: …I can express all of my other vectors in terms of, of a separate one; a linear combination of my other 2…but the problem is the components between the two are not the same, multiplying it by a scalar would only… I can only make it longer or shorter, but I can't change the orientation….

Group C is, however, no different than the groups, A and B, when it comes to its use of arithmetic processes. Likewise, in the Analytic-Arithmetic category, many in this group used, repeatedly, the top two frequency producers. These are the solution types (ZS), and the row reduced echelon matrix forms (E).

In the excerpt below, for instance, a participant from group C, SC3, is interpreting the absence of the identity form as the indicator of the nontrivial solution types, in turn, using this line of thinking to determine the linear dependence of vector sets.

> SC3: okay let's just go (student is writing new matrix thinking out loud) well this can be because it can reduce completely if this goes, if this row what I understand, if this row is 0 and this goes I guess if turns out to be the answer is one zero zero (student writes new identity matrix) this last row zero zero It can be linearly independent…

## *4.2 Role of Visual Constructs*

Table 4 summarizes the noteworthy similarities and differences, among the three groups, in the manner in which modes were used. As seen on the table, only groups A and B displayed initial mental structures that are shaped predominantly by the geometric entities. Considering that these two groups were the only two who had initial exposures to the dynamic geometric instructional representations, we conjecture that the particular visual instructional tools may have been the primary influencing external factors in shaping their initial knowledge. In fact, group C's initial knowledge is a further testimony to that effect. That is, this group's participants displayed highly numerically structured initial mental forms. Also, recall that the instructional dynamic web-based tasks were omitted from this group.

Dynamic visual instructional tasks, moreover, appeared to have encouraged the two groups, A and B (highest percent being in the group A), to integrate their geometric modes often with the abstract structures. Group C's low percent responses in the Geometry/Algebra mix-mode category further reinforces this observation.

Last but not least, the effect of visual instructional modalities may have led to the use of the external tools and gestures. As a matter of fact, throughout the interviews, it was only the groups, A and B, who exhibited the various kinds of tools and gestures. Once again, this behavior was lacking among the participants of group C.

Recall that group C was introduced to the static geometric sketches via its lectures. These sketches appeared to have very little influence on the group's initial knowledge. Recall that group C displayed an initial knowledge dominated, for the most part, by the numerical entities. Even though this group did include geometric ideas in their responses, this behavior emerged only for the questions that were inherently geometric. Even then, group C attempted to use their numerically shaped initial knowledge.

Another distinctive behavior is observed among the participants of the groups, B and C. The two groups exhibited a higher tendency to use arithmetic processes. In light of the group A's lower percent responses in this category, we, consequently, conjecture that the absence of the dynamic visual representational tools, from its lectures, may have resulted in the computational cognitive tendencies.

**Table 4** Mode-use characteristics of the three groups

| Characteristics | Groups | | |
|---|---|---|---|
| | A | B | C |
| Initial knowledge highly geometric | √ | √- | |
| Initial knowledge highly arithmetic | | | √ |
| Gesture use | √ | √ | |
| Higher use of geometric features mixed with algebraic modes | √ | √ | |
| High arithmetic tendency | | √ | √ |

√- stands for modes containing irrelevant characteristics

## 5 Conclusion

We discussed the aspects of twelve matrix algebra students' thinking modes from their interview responses to the questions on linear independence. Interviews were conducted right after the completion of an investigative assignment, using an interactive online module. This module provided, mainly, the dynamic visual representations. Interview responses were qualitatively analyzed using a framework by Sierpinska (2000). We, furthermore, provided the comparisons of three separate groups of a first-year matrix algebra course in the manners in which the modes were applied.

To summarize our findings, notable differences were observed in the role of visual instructional representations. The geometric instructional modalities appeared to have shaped the initial knowledge of the groups, A and B. On the contrary, the initial knowledge of group C revealed predominantly numerical characteristics. Moreover, both groups A and B appeared to have used their initial geometry-based knowledge to make sense of more abstract algebraic ideas. Group C, on the other hand, showed tendencies to apply their numeric-based vector component ideas to make sense of not only the abstract topics but also the geometric concepts.

In conclusion, the prior introduction of dynamic geometric tasks, provided that they are integrated both in-class and outside-class activities concurrently, may have noteworthy benefits in structuring a strong supportive initial knowledge. Though, if the geometric modes are introduced only as take home assignments (absent from lectures), the effect may be tainted with the undesired outcomes such as the formation of the irrelevant ideas as in the case of the group B.

**Acknowledgements** Work reported in this chapter is made possible partially by a grant from NSF (CCLI-0737485).

## References

Carlson, D. (1997). Teaching linear algebra: must the fog always roll in? *Resources for Teaching Linear Algebra*. MAA Notes, Vol. 42. pp. 39–51.

Dautermann, J. (1992). *ISETL: A Language for Learning Mathematics*. St. Paul: West Educational Publishing.

Dogan, H. (2004). Visual instruction of abstract concepts for non-major students. *International Journal of Engineering Education (IJEE)*. Vol. 2, pp 671–676.

Dogan, H. (2006). Lack of set theory-relevant prerequisite knowledge. *International Journal of Mathematics Education in Science and Technology (IJMEST)*. 37(4). pp. 401–410. June.

Dogan, H. (2012). Emotion, confidence, perception and expectation: case of mathematics. *International Journal of Science and Mathematics Education (IJSME)*. Vol. 10. pp. 49–69.

Dogan, H. (2013). Cognitive traits of instructional technology: linear algebra. *Proceedings of 25th International Conference on Technology in* Collegiate Mathematics (ICTCM). Boston, Massachusetts, March 21–24. http://archives.math.utk.edu/ICTCM/i/25/S077.html.

Dogan, H. (2014a). *Web-module on Linear Independence*. Retrieved from http://demonstrations.wolfram.com/SetsOfLinearCombinationsAndTheirImagesUnderLinearTransformat/. Accessed July 2 2017.

Dogan, H. (2014b). Multiple tasks derived from an interactive module: linear transformations and eigenspace. *Proceedings of 26th anniversary International Conference on Technology in Collegiate Mathematics (ICTCM)*. Paper S123. San Antonio, Texas, March 20–23. http://archives.math.utk.edu.

Dogan-Dunlap, H. (2003). Technology-supported inquiry based learning in collegiate mathematics. *The Electronic Proceedings of the 16th annual International Conference on Technology in Collegiate Mathematics (ICTCM),* Chicago, November 2003. http://archives.math.utk.edu/.

Dogan-Dunlap, H. (2010). Linear algebra students' modes of reasoning: geometric representations. *Linear Algebra and Its Applications (LAA)*. Vol. 432. pp. 2141–2159.

Dogan, H., Carrizales, R. & Beaven, P. (2011). Metonymy and object formation: vector space theory. In Ubuz, B. (Ed.) *Proceedings of the 35th Conference of the International Group for the Psychology of Mathematics Education*, (Research Reports) Vol. 2, pp. 265–272. Ankara, Turkey: PME.

Dorier, J. & Robert, A. (2000). On a research programme concerning the teaching and learning of linear algebra in the first year of a French science university. *International Journal of Mathematics Education in Science & Technology (IJMEST),* 31(1), 27–35.

Dorier, J. L, Robert, A, Robinet, J., & Rogalsiu, M. (2000). The obstacle of formalism in linear algebra. In JL. Dorier (Ed.) *On the Teaching of Linear Algebra* (85–124). Mathematics Education Library, Vol. 23. Springer, Dordrecht.

Dorier, J., & Sierpinska, A. (2001). Research into the teaching and learning of linear algebra. In Derek Holton (Ed.) *The Teaching and Learning of Mathematics of University level* (255–273) Kluwer Academic Publishers, DorDrecht.

Dubinsky, E. (1997). Some thoughts on a first course in linear algebra at the college level. *Resources for Teaching Linear Algebra*. MAA Notes. Vol. 42. pp. 85–106.

Gardenfors, P., & Johansson, P. (2005). *Cognition, Education, and Communication Technology.* Lawrence Erlbaum Associates, Inc. New, Jersey.

Glaser, B. (1992). *Emergence vs. Forcing: Basics of Grounded Theory Analysis*. Sociology Press. Mill Valley, CA.

Gol Tabaghi, S. (2014). How does dragging changes student's awareness: developing meanings for eigenvector and eigenvalue. *Canadian Journal Science, Mathematics and Technology Education*. 14(3). pp. 223–237.

Gol Tabaghi, S. & Sinclair, N. (2013). Using dynamic geometry software to explore eigenvectors: the emergence of dynamic synthetic-geometric thinking. *Technology, Knowledge and Learning*. 18(3). pp. 149–164.

Harel, G. (2000). Three principles of learning and teaching mathematics. In JL. Dorier (Ed.) *On the Teaching of Linear Algebra* (177–1890). Mathematics Education Library, Vol. 23. Springer, Dordrecht.

Harel, G. (1997). The linear algebra curriculum study group recommendations: moving beyond concept definition. *Resources for Teaching Linear Algebra.* MAA notes. Vol. 42. pp. 107–126.

Harel, G. (1989). Learning and teaching linear algebra: difficulties and an alternative approach to visualizing concepts and processes. *Focus on Learning Problems in mathematics*. 11(2). Spring Edition.

Harel, G. (1987). Variations in linear algebra content presentations. *For the Learning of Mathematics*. 7 (3). pp. 29–34.

Hiebert, J., & Lefevre, P. (1986). Conceptual and procedural knowledge in mathematics: an introductory analysis. In J. Hiebert (Ed.) *Conceptual and Procedural Knowledge: The Case of Mathematics* (1–3) Lawrence Erlbaum Associates, London, Hillsdale, NJ.

Hillel, J., & Sierpinska, A. (1994). On one persistent mistake in linear algebra. *18th PME Proceedings*. Vol. 3. Research Papers.

Leon, S., Herman, E., & Faulkenberry, R. (1996). *ATLAST Computer Exercises for Linear Algebra*. Upper Saddle River, NJ: Prentice Hall.

Leron, U., & Dubinsky, E. (1995). An abstract algebra story. *American Mathematical Monthly*. 102(3). pp. 227–42.

Mariotti, M. A. (2014). Transforming images in a DGS: the semiotic potential of the dragging tool for introducing the notion of conditional statement. In S. Rezat, M. Hattermann &A. Peter-Koop (Eds.) *Transforming –A Fundamental Idea of Mathematics Education* (155–172) New York: Springer.

National Research Council (NRC). (2000). *How people learn: Brain, Mind, Experience and School*. National Academy Press. Washington, D.C.

Roberts, L. (1996). Software activities for linear algebra: concepts and caveats. *PRIMUS*. Vol. 6. No. 2. June.

Salgado, H. & Trigueros, M. (2015). Teaching eigenvalues and eigenvectors using models and APOS theory. *The Journal of Mathematical Behavior*. Vol. 39. pp. 100–120.

Schwarz, D., Martin, T. & Nasir, N. (2005). Design for knowledge evolution: towards a perspective theory for integrating first- and second-hand knowledge. In Gardenfors, P., and Johansson, P., (Eds.). *Cognition, Education, and Communication Technology* (22–55). Lawrence Erlbaum Associates, Inc. New, Jersey.

Sierpinska, A. (2000). On some aspects of students' thinking in linear algebra. In JL. Dorier JL. (Ed.) *On the Teaching of Linear Algebra* (pp. 209–246). Mathematics Education Library, vol 23. Dordrecht: Springer.

Sierpinska, A., Trgalova, J., Hillel, J., & Drayfus, T. (1999). Teaching and learning linear algebra with cabri. Research forum paper. *The Procedıngs of PME 23*, Haifa University, Israel. Vol 1, pp. 119–134.

Stewart, S., & Thomas, M. O. J. (2010). Student learning of basis, span and linear independence in linear algebra. *International Journal of Mathematical Education in Science and Technology*. 41(2). pp. 173–188.

Stewart, S., & Thomas, M. O. J. (2009). A framework for mathematical thinking: the case of linear algebra. *International Journal of Mathematical Education in Science and Technology*. 40(7). pp. 951–961.

Stewart, S., & Thomas, M. O. J. (2004). The learning of linear algebra concepts: instrumentation of CAS calculators. *Proceedings of the 9th Asian Technology Conference in Mathematics (ATCM)*, Singapore. Pp. 377–386.

Thomas, M. O. J., & Stewart, S. (2011). Eigenvalues and eigenvectors: embodied, symbolic, and formal thinking. *Mathematics Education Research Journal*. Vol. 23. pp. 275–296.

Tucker, A. (1993). The growing importance of linear algebra in undergraduate mathematics. *College Mathematics Journal*. p. 24.

Torres, C. & Dogan-Dunlap, H. (2006) Technology use in ability-grouped high school mathematics classrooms. *E-Journal of Instructional Science and Technology (e-JIST)*. Vol. 9, No. 1.

Wawro, M., Rasmussen, C., Zandieh, M., Sweeney, G. F., & Larson, C. (2012). An inquiry-oriented approach to span and linear independence: the case of the magic carpet ride sequence. *PRIMUS: Problems, Resources, and Issues in Mathematics Undergraduate Studies*. Vol. 22. No. 8. pp. 577–599.

Wicks, J. R. (1996). *Linear Algebra: An Introductory Laboratory Approach with Mathematica*. Addison-Wesley Publishing Company, Inc. Reading, Massachusetts.

Zandieh, M., Wawro, M., & Rasmussen, C. (2017). An example of inquiry in linear algebra: the roles of symbolizing and brokering. *PRIMUS: Problems, Resources, and Issues in Mathematics Undergraduate Studies*. Vol. 27. No. 1. Pp. 96–124.

# How Does a Dynamic Geometry System Mediate Students' Reasoning on 3D Linear Transformations?

**Melih Turgut**

**Abstract** In this chapter, I discuss the integration of the Dynamic Geometry System (DGS) with the teaching and learning of two (2D) and three-dimensional (3D) linear transformations. To do this, certain tools and functions of the DGS, in particular, the dragging, slider and grid functions, and the Rotate and move 3D Graphics and ApplyMatrix construction tools of GeoGebra are focused on. Through semiotic potential analysis, a task is designed for students' construction of mathematical relationships among the characterization of the transformation matrix, determinant of the transformation matrix, and area and volume of given and transformed figures. A task-based clinical interview was conducted with a pair of undergraduate linear algebra students. Data from video records, student production and field notes was analysed within a semiotic lens with reference to the theory of semiotic mediation. The results appear to confirm that the DGS can be considered as an effective tool of semiotic mediation for characterizing 3D linear transformations. Such an approach to the data also provides a detailed understanding for students' reasoning steps from the use of artifact to creating mathematical meaning.

**Keywords** Semiotic mediation • Teaching-learning linear algebra
DGS • 3D linear transformations

## 1 Introduction

The notion of *linear transformation* is a core concept in linear algebra and is a unique topic associated with the notion function. Even though undergraduate linear algebra students have meaning[1] for the notion of functions from high school,

[1]Hereafter, the word *meaning* refers to 'systems of practices related to the object' (Godino, Batanero, & Font, 2007) in a socio-cultural constructivist perspective.

M. Turgut (✉)
Faculty of Education, Eskisehir Osmangazi University, Eskisehir, Turkey
e-mail: mturgut@ogu.edu.tr

 
S. Stewart et al. (eds.), *Challenges and Strategies in Teaching Linear Algebra*,
ICME-13 Monographs, https://doi.org/10.1007/978-3-319-66811-6_11

they cannot easily establish a connection among functions, transformations and linear transformation (Bagley, Rasmussen, & Zandieh, 2015; Zandieh, Ellis, & Rasmussen, 2012, 2013, 2017). One reason for this could be the introduction to students of the notion of linear transformation through fundamental rules or with matrix representation as researchers indicate (Turgut, 2017), and this could also trigger students' failure to construct mathematical meaning of non-linear transformations (Dreyfus, Hillel, & Sierpinska, 1998). Making connection between functions, transformations and linear transformations can be considered as a type of *theoretical thinking* (Sierpinska, 2005). However, by giving particular attention to the instructor's role, it is possible to provide an environment where students establish a bridge between a *matrix* and the notion of *transformation* (Andrews-Larson, Wawro, & Zandieh, 2017).

The integration of Digital Technologies (DT) to mathematics teaching has received great attention by educators for the construction of mathematical meanings within the socio-cultural theory perspective (Falcade, Laborde, & Mariotti, 2007; Leung, Baccaglini-Frank, & Mariotti, 2013; Mariotti, 2013, 2014), where the central role of DT is based on its *mediator* role. Researchers have exploited the potential of DT to design a teaching-learning environment, as well as looking at students' learning in a *mediation* process. In order to talk about the interactions among DT, mediator and mediation, a specific model comes across; the theory of semiotic mediation (Bartolini Bussi & Mariotti, 2008). In my recent studies (Turgut, 2015, 2017; Turgut & Drijvers, 2016), which are parts of an extensive postdoctoral (design-based) research project, I refer to this theory to analyse the semiotic potential of certain tools and the functions of the Dynamic Geometry System (DGS) for students' construction of key notions for learning linear algebra, as well as to analyse the emergence of mathematical thinking.

The postdoctoral project was designed with two major steps; *local instruction theories* (Gravemeijer, 2004) within two different contexts. The notion of linear transformations is first acknowledged in the use of a specific DGS (GeoGebra) to create an environment for students' construction of mathematical connections between the notion of function, transformation and linear transformation, as well as for characterizing matrices for specific (geometric) transformations in $\mathbb{R}^2$ and $\mathbb{R}^3$. For the first step, nine tasks (Didactic Cycle I) were designed to provide students with an environment for transition to formal mathematics, where in the first eight tasks of Didactic Cycle I, the students characterized relationships between functions, matrices, linear transformations and lengths and areas of the figures in the DGS. The second step of the project was to design a DGS environment (Didactic Cycle II, including seven tasks) for students' meaning making on linear combination, span, linear independency-dependency and basis notions. All sixteen tasks were piloted within the same group of linear algebra students.

In this chapter, a case from the project will be presented, after the eight tasks in Didactic Cycle I. The present case, which is the ninth task of Didactic Cycle I, concerns the mediator role of the DGS in the learning and characterizing of 2D and 3D linear transformations, and also the making of connection between them.

First, certain tools and functions of the DGS for students' construction of a mathematical link among a $3 \times 3$ transformation matrix, the determinant of the matrix, and concepts of area and volume of given and transformed figures will be analysed. Next, students' reasoning within a semiotic lens will be considered. Therefore, in this chapter, the focus is on the following research question: *How does the DGS mediate students' reasoning on 3D linear transformations?*

For this purpose, the next section contains the study's theoretical framework, the theory of semiotic mediation (Bartolini Bussi & Mariotti, 2008), while the third section provides the methods employed and the mathematical context of the study. The fourth section presents an analysis of the data triangulation coming from interviews. The chapter ends with a conclusion section where certain limitations and a number of doubts are addressed.

## 2 Theoretical Framework

The Theory of Semiotic Mediation (TSM) was adapted into the mathematics classroom by Bartolini Bussi and Mariotti (2008), based on Vygotsky's notion of semiotic mediation embedded in social constructivism (Ernest, 2010). Following Vygotksy, the TSM postulates an idea that mathematical meanings can be constructed when the teacher, as a *mediator*, intentionally uses an artifact[2] for students to accomplish a carefully-designed task in a social-communicative environment. In other words, in the TSM, constructing knowledge can be considered as a kind of 'instrumented activity' in a social context (Mariotti, 2009). The TSM aims to transform students' personal meanings on a proposed topic into culturally-accepted mathematical meanings through the employment of a task. Consequently, the interpretation of a task is a kind of *mediation* process, and this process mainly focuses on the emergence of *signs* that are attached to students' learning. Along this direction, the TSM is described within two main notions; *the semiotic potential of an artifact* and *the design of didactic cycles* by the teacher.

The semiotic potential of an artifact means the evocative power of the proposed artifact for students' construction of mathematical meanings (Mariotti, 2009, 2013). Therefore, the search for an artifact's evocative power needs careful analysis, both didactical and epistemological, and also a phenomenological way to elaborate students' (possible) utilization schemes when they use the artifact. The teacher then needs a learning route considering the students' pre-knowledge on key concepts, the goal of the task, the personal senses and mathematical meanings emerging in the use of the artifact, and the teacher's role. In light of this route, the teacher designs task sequences for future didactic intervention, where the teacher exploits the

[2]Here, the notion of an artifact has a general sense. Arzarello (2013) points out that an artifact could be considered as 'a material with its own physical and structural characteristics made for specific tasks' (p. 8).

artifact as a tool of semiotic mediation. Application of didactic cycles could form a story beginning from human-computer interaction to formal mathematics, and within a semiotic perspective, from signs that show a relationship between the artifact and the task to signs that show a relationship between the artifact and progressive mathematics (Mariotti, 2012). The teacher's role in the application of didactic cycles is to orchestrate student learning. Therefore, the teacher should ask carefully constructed questions to make students focus on the task, and he or she should be aware of the production of signs and, specifically, the emergence of shared mathematical meanings.

The teacher could search for three types of signs to understand students' learning, where a complex semiotic source occurs while the students encounter the artifact; (i) artifact signs (aS), (ii) mathematical signs (mS), and (iii) pivot signs (pS). Artifact signs could appear in the students' immediate use of the artifact, which are commonly implicit and foster the students' initial perceptions on the artifact's tools and/or their functions. Mathematical signs refer to signs that show students' mathematization, such as expressing a conjecture, a hypothesis or a definition, which may be indicators of culturally-accepted mathematical meanings within the task context. Pivot signs can be considered as potential polysemy for the construction of an interpretative link between personal senses and mathematical signs. Of course, a sign has broad meaning including different kinds of resources; for example, words, mimics, sketches or gestures. For the sake of page constraint, the focus will be on the (verbal) signs that appear in the discourse. For example, as described in Bartolini Bussi and Mariotti (2008), when high school students interact with a specific DGS (in that case Cabri®) to explore the effects of a unknown *macro,*[3] certain specific signs appear through a haptic sense; 'it moves', 'it does not move', 'point', 'point on the object' and so on, can be considered as aS. Next, students produce mathematical expressions for a function; 'a relationship that links two points', which appears as mS. In addition, in the discourse composed by the students and the teacher, 'independent' and 'dependent' appear, and these can be classified as pS because they indicate a double relationship between the artifact context (e.g., points are independent) and the mathematical context (e.g., independent variable). However, it should be noted that no linear transition from aS to mS exists. In some cases, no pS may appear in the context of the discussion.

## 3 Methodology

This chapter is part of an extensive design-based research (DBR) (Bakker & van Eerde, 2015) project, which includes two sets of didactic cycles as local instruction theories (Gravemeijer, 2004). As briefly mentioned in the introduction, there are

[3]A specific tool belongs to the Cabri environment. For an analysis of its semiotic potential, see (Mariotti, 2009, 2013; Falcade et al., 2007).

two different contexts; Didactic Cycle I (9 tasks) for learning linear transformations, and Didactic Cycle II for learning linear combinations, span, linear independency-dependency and basis. In Didactic Cycle I, Task 1 and 2 are designed for the students' invention of a transformation as a special function and transition to the idea of linear transformation. Task 3 and 4 are for characterizing the transformation matrix (if it is invertible, zero, unit, and so on), but also for geometric applications, such as stretching, shrinking or reflecting the figures. Task 5 is for the reinvention of fundamental rules of linear transformations, and Task 6, 7 and 8 are for characterizing isometric transformations in $\mathbb{R}^2$. The task presented in this paper is the final one (the ninth task) for students' making connection with geometric properties in $\mathbb{R}^2$ and $\mathbb{R}^3$ and matrix (linear) transformations.

A task-based clinical interview was conducted together with two students (A and B, both females and 20 years old) who had exhibited average performance in previous mathematical courses and, with respect to their GPAs, A had 76.30 and B had 77.10 (in the range of 0–100). The students had received certain courses, such as (2D & 3D) geometry, abstract mathematics and fundamental calculus with general physics. In addition to regular class lectures, A and B had attended implementation of eight previous tasks in Didactic Cycle I. Therefore, they had experience in the geometric applications of linear transformations in the dynamic geometry environment. In other words, they knew the role of the determinant (e.g., as area) of matrix transformation and how the determinant, and thereby matrix entries, could affect lengths and areas of geometric figures in a specific DGS. In the application of the tasks, they experienced the artifact GeoGebra; in particular, certain specific tools and functions, such as dragging, slider, 3D rotate, measure and ApplyMatrix tools and grid function on Algebra, 2D and 3D Graphics window, as will later be discussed.

Data was collected through a task-based video recorded interview with a laptop facing the students, with a separate camera focusing on the students' working environment. The experiment lasted about an hour and screen recorder software was also used to capture the students' employed techniques with the DGS. The data (discourse was translated into English) was analysed through a semiotic lens considering categories of signs in the TSM.

Because the TSM particularly shows students' mathematical thinking process with digital technologies, along the data analysis, I selected moments that were mathematically rich based on students' discussion and interaction with the DGS. Next, specifically, I explored verbal signs that show students' initial explorations and observations reflecting their personal meanings on the situation, where I categorised them as an aS. Through exploring the situation, when they expressed a proposition, a remark or conjectures showing their personal meanings returning to (new) mathematical meanings, I categorised those as mS. When specific signs appeared that indicate a corridor from aS to mS (but not exactly meaning aS or mS), I categorised them as pS.

### 3.1 Mathematical Context and Semiotic Potential of DGS

Generally speaking, a (geometric) transformation in Euclidean space can be defined as mapping, from $\mathbb{R}^n$ to $\mathbb{R}^m$ by $T: \mathbb{R}^n \to \mathbb{R}^m$ for $(x_1, x_2, x_3, \ldots, x_n) \to T(x_1, x_2, x_3, \ldots, x_n)$ (Lay, 2006), but if the output of this specific function are linear expressions, then the representation can be expressed through matrices by the following:

$$T: \mathbb{R}^n \to \mathbb{R}^m, T(x_1, x_2, x_3, \ldots, x_n) = \begin{bmatrix} a_{11} & a_{12} & a_{13} & \cdots & a_{1n} \\ a_{21} & a_{22} & a_{23} & \cdots & a_{2n} \\ \vdots & \vdots & \vdots & \ddots & \vdots \\ a_{m1} & a_{m2} & a_{m3} & \cdots & a_{mn} \end{bmatrix} \begin{bmatrix} x_1 \\ x_2 \\ \vdots \\ x_n \end{bmatrix}.$$

In short, for $\forall w \in \mathbb{R}^n$, $T(w) = Aw$, where $A$ (i.e., coefficients) is a $m \times n$ and $w$ is a $n \times 1$ matrix. Consequently, because of matrix algebra, $T$ satisfies two properties for $\forall u, v \in \mathbb{R}^n$ and $k \in \mathbb{R}$

$$\begin{cases} T(u+v) = T(u) + T(v), \\ T(ku) = kT(u). \end{cases}$$

The case $m = n = 3$ is focused on (because of the DGS availability that will be explained), since there is a real vector space $\mathbb{R}^3$. In many textbooks, geometric applications of linear transformations in $\mathbb{R}^2$ are well-presented. However, an emphasis on relationships between the determinant of matrix (and therefore linear) transformations and associated geometric meanings in $\mathbb{R}^3$ or $\mathbb{R}^n$ seem to be missing. Generally, during a course, students learn the role of the determinant of the matrix transformation in $\mathbb{R}^2$, and they may make over-generalizations on such a view to higher dimensions. With respect to the linear transformation of figures and 3D objects in $\mathbb{R}^3$, there are two interrelated situations that could arise; (a) transformation of planar figures in $\mathbb{R}^3$, and (b) entire transformation of 3D objects. In the first case, the relationship between lengths and areas of figures is dependent on matrix *entries*, but is different to that in $\mathbb{R}^2$ because the determinant of a $3 \times 3$ matrix could be zero when areas of the initial figure and construction[4] could be non-zero. In the second case, the determinant of the transformation matrix is a factor between *volumes* of the initial 3D object and the construction. The main objective of this chapter is to provide a DGS environment for students to construct the epistemological views expressed above, where it is considered that they would also make generalizations to higher dimensional spaces. For the emergence of such views, it is postulated that the DGS, in our case GeoGebra, has semiotic potential with the following tools and functions for my specific purpose:

[4]Hereafter, the notion of *construction* referring to initial drawing's (figure or a 3D object) matrix transformation is used.

- The *Dragging tool* in 2D and 3D windows has a core function in any DGS that allows the user to draw and manipulate objects freely. At the same time, if the user constructs an object through commands and tools of the DGS, initial figures can be dragged. However, the constructed objects cannot be dragged; for details see (Jones, 2000; Mariotti, 2000, 2014). Consequently, the dragging tool could evoke for variation as well as co-variation between independent and dependent objects, and also invites the user to establish conjectures and theoretical thinking (Gol Tabaghi, 2014; Leung et al., 2013; Lopez-Real & Leung, 2006; Mariotti, 2014). In the context of this paper, the function of dragging is to explore the relationship between the initial drawing and construction of objects by manipulating free points or figures and to create an environment for students to make conjecture on the relationships between the transformation matrix, its determinant, lengths, areas and volumes.
- The *Slider tool* works with the function of dragging, but first it has to be constructed. The user can assign a real number to the slider and can define such a real number as a parameter connected to an equation, a figure or a computation. By dragging it, the movement of the slider would change the parameter, and therefore, the equation, the position of the figure or even the computation. In our case, if the sliders are defined as entries of a matrix, then because of the movement of the sliders, not only is the transformation matrix affected, but also the lengths, areas, and even the volumes. To summarize, consider the slider as providing the user a sense of dynamic variation (Turgut & Drijvers, 2016) for students' conjecturing on the effects of matrix entries (and determinant value) on the geometric properties of figures and 3D objects.
- The *ApplyMatrix construction tool* can be used through an 'Input' line. However, in order to use this tool, the user needs a square matrix ($2 \times 2$ or $3 \times 3$) and a figure or an object on which to apply the matrix transformation. Consequently, this provides an environment for students to observe different manipulations of transformed figures or objects in 2D or 3D Graphics windows under specific linear transformations, if the user assigns sliders as matrix entries. Therefore, for our case, the dragging of free points and objects, and the movement of sliders (connected to matrix entries), by the students would evoke a meaning for constructing relationships among the concepts of the determinant of the matrix, the transformation matrix and its entries, length, area and volume of the objects.
- The *Grid function* can be activated through right-clicking, and provides the user a Cartesian view by representing the coordinates of points. In our case, it enables the user to compute lengths and areas of initial and construction objects quickly.
- The *Rotate and move 3D Graphics tools* enable the user to move a Cartesian system on the 3D Graphics window and provides the user with a sense of *spatial orientation* (Turgut & Drijvers, 2016), where the user can explore the manipulation of initial and transformed figures and objects from different perspectives. Consequently, these tools could enable students to establish conjectures on different views of objects under certain linear transformations.

## 3.2 Description of the Task

The steps of the task are divided into two parts with each part being formulated as follows:

***Part I***

*Step I*: Open GeoGebra and define three sliders ($a$, $b$, $c$) on the (2D) Graphics window.

*Step II*: Using the sliders' values, construct a $3 \times 3$ matrix, naming it matrix 1 $= \begin{bmatrix} a & 0 & 0 \\ 0 & b & 0 \\ 0 & 0 & c \end{bmatrix}$.

*Step III*: Construct a triangle on the 3D Graphics window and apply matrix transformation onto it. Calculate the lengths and areas of the initial and constructed triangles. Drag the free points and move the sliders. Answer the following with your partner:

1. *What are the roles of sliders on the 3D Graphics window?*
2. *What is the relationship between the determinant of the matrix and the lengths and areas of the different triangles?*
3. *Drag the sliders to make one slider equal to zero and the others non-zero. Explore each possibility and explain why the construction triangle has a non-zero area, although the determinant of the transformation matrix is zero.*

***Part II***

*Step IV*: *Open a new GeoGebra sheet and repeat the first and second steps. Moreover, activate the grid function on the Graphics window.*

*Step V*: *Draw a pyramid defining the points on the x, y and z-axes on the 3D Graphics window, and calculate the lengths, areas and volume of the pyramid.*

*Step VI*: *Apply matrix transformation to each surface of the pyramid and calculate the construction's lengths, areas and volume.*

*Step VII*: *Explore the lengths, areas of surfaces and volumes of the initial object and construction. Discuss your observations with your partner.*

*Step VIII*: *Explain, what would happen if you have a $4 \times 4$ or an $n \times n$ matrix?*

Of course, a synergy exists between the phenomenological analyses expressed above, the students' pre-knowledge and the goals of the task. Here, in Steps I, II and III, it is assumed that the students first comprehend the sliders' role on the length and areas of the figures and, because they will use ApplyMatrix command, they will relate the case with plane matrix transformations. Exploring the different situations through the dragging function and slider tool, they will notice that although the determinant (of the transformation matrix) is zero, the construction has a non-zero area. They will explore the situation and comprehend the relationship between the determinant of the partitioned matrix in the transformation matrix, and lengths and areas of the figures. Next, moving the case to three and higher dimensional contexts

(Steps IV–VIII), the students will make reasoning on the dimensions of the length, area and volume and rows of the transformation matrix. After this, they will make connection between the determinant of the given matrix and volume. Exploring the different cases, they will generate their results to higher dimensional spaces of $\mathbb{R}^n$, where the teacher's main role here is to orchestrate the students' making connection between characterization of the transformation matrix, area and volume notions, and generating their results.

## 4 Emergence of Signs in the Students' Reasoning on 3D Linear Transformations

At the start of the interview, the teacher introduces the protocol of the task to the students. The students then begin to follow the steps as described in Sect. 3.2, where they first construct sliders and thereafter use the slider's values as parameters, forming the matrix (as 'matrix1' at the Algebra window in Fig. 1). They open the 3D Graphics window and draw a triangle defining one point on the $z$-axis, and two points on the $x$-axis. Next, they compute the lengths and area of the triangle using software tools and apply matrix transformation to the triangle using their experience coming from eight previous tasks in Didactic Cycle I. By following all details in the task steps, they finally obtain the following GeoGebra interface (Fig. 1), before they explore the questions given in the third step of the task.

After this, the students together begin to explore questions in the third step of the task. Interestingly, before exploring the role of the sliders, the students briefly look

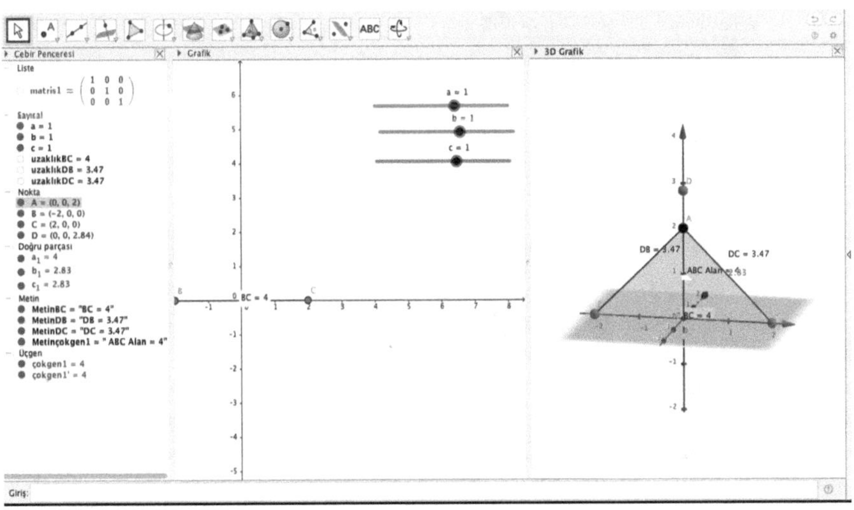


Fig. 1 Students' initial DGS interface

for the construction (triangle) because, initially, the software assigned $a = b = c = 1$ (as in Fig. 1). At that time, the following discussion took place:

38 T: … what happens when they all [*meaning the sliders*] equal 1?
39 A: [*dragging sliders*] Aha … [*looking for the construction*] …
40 T: What about the construction here?
41 A: Ha. They are overlapping, when they [*meaning the sliders*] are all 1.
42 B: Also, because the areas of the triangles are the same.
43 A: It is due to the determinant. Actually, it is due to the determinant of the matrix transformation.
44 T: We will discuss that. Now, let's focus on the questions in the task.

In this part of the discussion (#39, #41–43) specific signs appear. For example, 'overlapping when they are all 1' and 'the areas of the triangles are the same' can be considered as aS. Such expressions by the students are not only signs that their focus is on the sliders' roles, but also an indicator that the semiotic potential of the amalgam of the dragging, slider and ApplyMatrix construction tools evoke the emergence of characterization regarding the students' personal meanings of the transformation of a triangle; the determinant of the transformation matrix gives a relationship between the areas of the triangles (#42–43), due to their experience coming from the previous task settings in Didactic Cycle I. Following this, the teacher intervenes to focus on the questions. Thereafter, the students' views on the effects of the sliders emerge with interlacement of specific aS, pS, and mS.

82 A: Their roles are changing the determinant, so that the relationship between the areas [*meaning the areas of initial triangle and construction*] changes.
83 B: It also changes the lengths, since the figure changes.
84 T: Ok, could you explore other sliders? For example, slider *a* …
85 A: I think the sliders are affected according to the axes. For example, here, because it [*slider a*] is the first component [*meaning the transformation matrix's first component*], it affects the figure [*meaning the construction*] along the *x*-axis. Similarly, *b* affects the figure along the *y*-axis, and *c* does with respect to the *z*-axis.
86 T: How can you confirm this? Explore your assertions.
87 A: [*B is dragging the slider b*] … Why does it [the *construction*] not move?
88 B: … Let's change the viewpoint [*using Move and Rotate tool*].
89 A: Others have effect, but *b* does not.

Initially, (#82–83), the students' focus is still with the sliders, because they still could not characterize the sliders' roles explicitly, which is apparent from the emergence of an aS, such as 'changing the determinant', 'areas', 'lengths', 'figure changes' and 'along *x*-axis'. However, the word 'effect' transforms to a pS (#85), which indicates an interpretative link between the students' personal meanings and the case. In other words, 'effect' has a double relationship with artifact and emerging mathematical notion, where finally, an mS appears when student A conjectures for changes and movements through the semiotic potential of tools and functions of the DGS, where she relates those 'changes' with the matrix's entries.

The teacher is aware that student A's conjecture is correct, but is not valid for the transformation of two-dimensional figures in $\mathbb{R}^3$ and therefore asks the reason and an explanation of the situation (#86). However, the students finally observe that the construction does not move along the *y*-axis (#87–89).

For the next step, they explore the effect of dragging sliders with different *a*, *b* and *c* values, and are surprised that slider *b* does not change (or move) the construction, which contradicts their initial experience. The discussion on the issue took a while:

103 B: … the construction triangle should move when I drag *b*.
104 A: We searched for *a*, it moves along the *x*-axis, also for *c*. Therefore, I also expected the same. But it does not!
105 B: Why is it not moving? …
106 T: Let's go back to the task's steps. [*Repeating the steps*] …
107 A: [*they are together checking the areas*] … They are the same. Why?
108 B: … determinant … Aha.
109 A: But, it should be possible when the determinant equals 1.
110 B: Here, the determinant is not 1 …
… …
116 B: … Also, the lengths should change, but they remain. How is the length independent from the transformation matrix?

Specific signs, such as aS, 'move', '*x*-axis', 'checking the areas' and 'lengths should change' still appear (#104–105, #107–110, #116). This also means that they still did not express an explicit mathematical reason for slider *b*'s effect on the construction. Another fact is that student B still thinks that the 'transformation matrix' could affect the 'lengths' (#116). However, after reading the instructions to be followed in the task, they immediately relate the situation with the determinant, where the 'determinant' can now be acceptable as a pS (#107–110).

Through the students' personal senses to the notion of determinant, they observe a situation that contradicts their conjecture as they already expressed in #109 and #116, while they explore the relationship between the determinant value and the ratio of the areas:

133 A: … [*B drags sliders*] … Does the area [*meaning the construction's area*] change?
134 B: It depends on the determinant's value. Let's calculate it now …
135 A: We can find it like this [*showing the paper*] … Aha. The determinant is different, but the areas are the same …
136 T: Could you follow the steps of the task? For instance, what do you think about the third question in Step III?
137 A: But now the determinant is zero! Oops.
… …
146 A: Let's make $b = 0$. … Actually, if any of *a*, *b* or *c* is equal to zero, then it [*meaning the determinant*] will be zero again. …

Some of the aS 'change' and 'the areas' show their personal meanings regarding the situation (#135–146). However, they finally notice that the determinant does not give the factor between the areas of drawing and construction (#135, #137). The teacher reiterates the instructions in Step III to move beyond considering the pivot role of the 'determinant'. At this point, an mS appears, and student A expresses the role of entries of the matrix where her explanation is independent from the dragging of the sliders (#146).

Next, the students continue their exploration. However, the teacher asks the students in which plane they initially draw the triangle to make them relate why the construction does not move along the $y$-axis with the matrix entries and the determinant. This evokes emergence of specific mathematical meanings:

170 B: [*thinking*] … let me explain my view. We drew the triangle on the plane generated by the $x$ and $z$-axes; consequently we don't have any $y$. Similarly, the change in $b$ does not affect anything, other values affect. But, although we drag the sliders, the areas are always the same. I still cannot find the reason …

171 T: How do you find the relationship between the areas of the triangle?

172 A: Normally, the determinant gives the factor. … But it is like … For example, we explored a two-dimensional object, but in a three-dimensional environment. Is it because of this?

177 T: Then explain, for our triangles here [*pointing the screen*], what kind of matrix determinant gives the relationship between the areas?

178 A: Two-dimensional …

179 B: A $2 \times 2$ matrix.

180 T: Can you find such a$2 \times 2$ matrix … here?

185 A: Yes [*pointing to the second row and the second column of the matrix*], if we can cancel like this.

Student B expresses her feelings about the situation, saying that no relationship exists between slider $b$ and movement on the $y$-axis, because of the plane where the initial triangle was drawn (#170). However, this is a complete aS that interlaces with 'effect', 'move' and so on. Such signs indicate the student's immature explanation, because B is manipulated by the steps of the task. Therefore, the teacher exploits the pivotal role of the 'determinant' and again asks how they analyse the relationship between the areas of drawing and construction (#171). Student A immediately relates the situation with the dimensions of the objects and notices that the zero-determinant issue could be due to drawing a 'two-dimensional' object in a 'three-dimensional' environment, which can be accepted as an mS (#172). The teacher then guides the students to make a connection between the situation with a planar figure's transformation, the transformation matrix and the areas (#177, #180). Student A easily finds that the *partitioned matrix* creates a relationship between the areas (#178–180, #185), which is a pS, because it opens a door to exploration and emergence of mathematical meanings.

For the next step, the students calculate the partitioned matrix's determinant for different sliders' values and to check their understanding. Finally, they express their conclusions for the first part of the task, which explicitly show the evolution whereby personal meanings transform into mathematical meanings:

192 A: … we never considered a three-dimensional matrix transformation, I mean in $\mathbb{R}^3$, but the figure was two-dimensional … We never experienced anything and, because of this, it was surprising that dragging *b* [*meaning dragging slider*] did not affect anything.

195 T: Can you generalize such a conclusion?

196 B: We should consider our working environment's dimension [*meaning the plane or space*] and the dimension of the constructed object. Then we can observe variations and movements of the construction.

197 A: Yes. If we consider an $n \times n$ matrix, then manipulations of the transformation can be observed in $\mathbb{R}^n$.

Student A expresses that they first worked on matrix transformations in $\mathbb{R}^3$, but at the same time, she constructs a geometric meaning of the transformation of planar figures in $\mathbb{R}^3$ (#192). While student A's explanation is independent from the transformation of triangles and can be acceptable as an mS, student B points out 'variations and the movements of the construction' (#196), which is an aS. This is a sign that also indicates that student B's focus is still on the 'movements'. However, student A generalizes the situation through conjecturing on the '$n \times n$ matrix' and '$\mathbb{R}^n$', which are mS (#197) and ends the first part of the task.

At the beginning of the second part of the task, the students build sliders and associated matrix. Thereafter, they activate the grid function on the 2D Graphics window, and draw a pyramid on the 3D Graphics window. To follow the steps, they next begin to calculate the lengths of the edges and areas of the surfaces on the pyramid using Measure and Move and Rotate in the 3D tools. They apply matrix transformation to each surface of the pyramid and they not only calculate the lengths and areas of the surfaces, but also the volumes of the drawing and construction. They then begin to drag the sliders to explore the steps of the task. Figure 2 shows their working environment in the DGS.

As a next step, they explore the effects of the sliders on the DGS interface. In this part, some aS also appear, such as 'along *x*-axis' and 'move'. In this way, they realize the effects of the sliders in three-dimensional space, where they also explore negative values for the sliders. At this moment, an important aS 'volume' appears which reflects the students' observations from the artifact, coming from previous experiences:

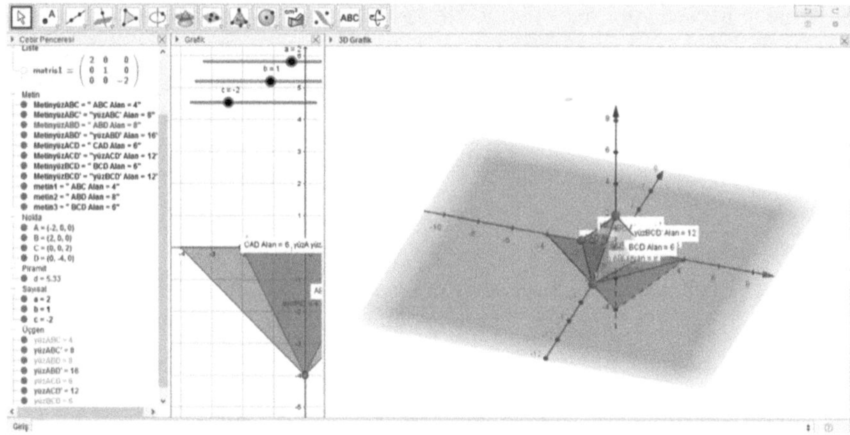


**Fig. 2** Students' DGS interface in the second part of the task

261 A: … when we find the determinant, we will find the volumes, right?
262 T: How did you arrive at this conclusion?
263 A: It is a factor between them [*meaning volumes*]… We are asked to explore the relationship between volumes… The determinant of the matrix could give the relationship between volumes.
264 T: Why?
265 A: Because, there will be a transformation … I think, the determinant will give the relationship between volumes … but … I don't know how …

Using spatial perception and her view expressed before (#172), student A thinks that there will be a relationship between the 'determinant' and 'volume' of the construction, because it is asked for (#261). However, these are aS, because she cannot express a mathematical reason (#261–265) of her view. However, this situation opens a door to explore different sliders' values on the determinant and the relationship between volumes. After this, they validate the conjecture of A for the relationship between the determinant and volume employing and integrating their experience from the first part, in which the mS appear as a collection of constructions of mathematical meanings:

285 B: Now the absolute value of the determinant is 4, the initial object's volume is 12, and also the transformation object's volume is 48. It is again validated … How did this happen? This is because, the object is three-dimensional and all sliders effect along all the axes, since the determinant value affects the volumes … If we consider a two-dimensional thing in a three-dimensional environment, then the transformation affects the areas.
… …

290 A: It was different there because the space was $\mathbb{R}^3$, but exact transformation was as in $\mathbb{R}^2$, just like the planar figures. This would be different, for example, if we worked in $\mathbb{R}^4$. I mean if we had a $4 \times 4$ matrix, and again if we worked with pyramids, then we would cancel one row and column to relate the volumes, similar to the previous case.

… …

299 B: If you search for the direct effects of the determinant value on higher-level geometric relationships between objects, you should note that the dimension of the matrix and the dimensions of the objects need to be the same.

… …

302 A: If we work in $\mathbb{R}^3$, we use a $3 \times 3$ matrix… because $a_{11}, a_{22}$ and $a_{33}$ [*meaning transformation matrix's entries*] are slider values and other entries are zero, the sliders directly affected the movements along the $x$, $y$, $z$ axes. Of course, all the zeros in the entries helped us to see such relationships regarding movements and to calculate the determinant of the matrix.

First, through the pivot roles of 'volume' and 'determinant', B explains the relationship between the pyramid's dimension and the sliders' effects (#285). Contrary to her initial views (e.g. #116, #170), an mS appeared where she finally constructs meaning. This is because the sliders have a 'three-dimensional' effect, and the 'determinant' gives the relationship between the volumes of the objects. Student B also explains why, in the previous part, the transformation only 'affects the area' of construction (#285), which is an mS. However, a progressive emergence of an mS is observed (#290, #299, #302): (i) Student A generalizes the situation to the space $\mathbb{R}^4$, making connection between the volume of the 3D object and the dimension (rows and columns) of the matrix; (ii) Student B points out 'higher-level geometric relationships' (meaning hyper-volume or hyper-area and so on), and relates this with the dimension of the matrix and the dimension of the object to observe 'direct effects'; (iii) Student A constructs the meaning of the role of given zeros in the entries of the transformation matrix, i.e., observed movements are due to selecting the sliders' values as the diagonal of the matrix, and consequently, the sliders affect the objects along the axes.

In our case, the discussion orchestrated by the teacher yields a complex *semiotic chain* (Bartolini Bussi & Mariotti, 2008), which shows synergy between specific signs. The following figure (Fig. 3) is presented to represent the evolution of signs and also to picture how students' personal meanings transform into mathematical meanings.

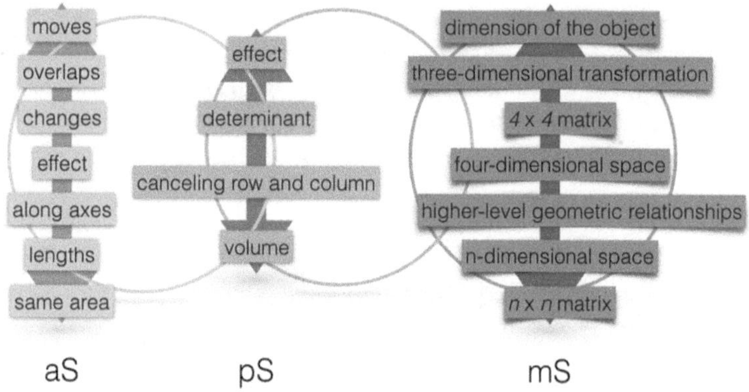


**Fig. 3** A semiotic chain describes the emergence of signs in reasoning on 3D linear transformations

## 5 Conclusions

This chapter focuses on the following research question: How does DGS mediate students' reasoning on 3D linear transformations? A task was designed in the context of $\mathbb{R}^3$ through analyzing the semiotic potential of certain tools and functions of the DGS; in particular the dragging, slider, ApplyMatrix construction tool, rotate and move 3D Graphics tools and grid function of GeoGebra. A task-based clinical interview conducted with a pair of linear algebra students was analyzed within a semiotic mediation perspective and, approaching the data in such a way provided evolution of how the students' personal meanings coming from phenomenological experience, transformed into (new) mathematical meanings. The students' initial experiences were based on the matrix transformations in $\mathbb{R}^2$ concerning the relationships between the lengths, areas of figures and determinant of the transformation matrix. In addition, within our case, they finally not only re-invented the role of the determinant notion in linear transformations in $\mathbb{R}^3$, but also generated their results to $\mathbb{R}^n$.

Specifically, in the above process, the TSM perspective describes certain tools and functions' mediating role in the reasoning on 3D linear transformations and, therefore, how the semiotic potential of certain tools and functions evoke the emergence of aS, pS and mS that foster the students' characterization and reasoning on 3D linear transformations. Analysis of the data provides specific verbal signs that attach to students' meaning-making steps, first with the use of artifact, from tools and functions of DGS use to higher-dimensional spaces, i.e., de-contextualization from the proposed artifact. Therefore, the present study confirms how certain tools and functions of a DGS might be used as effective tools of semiotic mediation for teaching learning 3D linear transformations.

Moreover, the TSM perspective also shows the relationships between the task and employed techniques in the students' reasoning. Even though the following are not linear, it can be expressed that the students first analysed the effects of the sliders on the construction, and secondly they found that the sliders affect length, but do not volume. They checked the matrix and related the determinant value by comparing the areas of drawing and construction and also dragging initial points of the figure. They finally related the dimension of the object and the transformation matrix. In the following steps of the task, they established a relationship between volume and determinant value. They conjectured on the case and generalized to higher dimensional space $\mathbb{R}^n$. In such processes, the ApplyMatrix construction tool command seems to have had a *vague role*. However, all actions on the interface were due to it and such a role contributed to the students' invention of a number of key points, which are also discussed in a relevant paper (Turgut, 2017). However, in my analysis, *instrumentation* (i.e., instrumental genesis) perspective (Trouche, 2004; Vérillon & Rabardel, 1995) of the ApplyMatrix construction tool was in shadow, which can be discussed in a further setting to elaborate students' utilization schemes for designing new tasks.

Interrelation of the dragging and slider tools allowed students to explore different (independent and dependent) cases and to establish conjectures, which were in line with the dragging tool's epistemic role as described in the literature (Falcade et al., 2007; Gol Tabaghi, 2014; Leung, 2008; Lopez-Real & Leung, 2006). The slider's main role was in providing a dynamic variation that was similarly discussed in (Sierpinska, 2000; Turgut & Drijvers, 2016). The students used the Rotate and Move 3D Graphics tool a number of times to check the *robustness* (Jones, 2000) of the figures and to change the viewpoint while they dragged the objects and sliders. The grid function was used while the students calculated the volumes of the objects. The last two were also indissoluble effects in the learning process that also appear in (Turgut, 2017), but they did not have much potential comparison to others.

Although tools and functions of the DGS above worked well, one limitation is that the students never mentioned the notion of *similarity*. Selecting integer values on the sliders could form a similarity effect between the triangles on the 3D Graphics window. The reason for this may be the students' concentration on finding the sliders' effects on the construction. Reasoning to the similarity notion could open a door to a notion of the three-dimensional similarity of objects and, in this way, it would be possible for students to invent further characterizations. A second limitation concerns the task setting, which is limited to a pair of students. The TSM perspective emphasizes the role of interpersonal exchanges during mathematical discussions. A classroom teaching experiment could provide a meaningful environment for students to conceptualize shared mathematical meanings and to reflect on students' reasoning steps in a broader sense. These could be areas of interest for future research.

**Acknowledgements** I would like to thank my supervisor Prof. Dr. P.H.M. (Paul) Drijvers for his kind and countless contributions to the project and insightful comments on this paper. I also thank the Scientific and Technological Research Council of Turkey (TUBITAK), under the 2219-International Post-Doctoral Research Fellowship Programme (grant no: 1059B191401098), which supported this research. Finally, thanks go to anonymous reviewers and Christine Andrews-Larson for making constructive suggestions, which substantially improved the presentation of the paper.

## References

Andrews-Larson, C., Wawro, M., & Zandieh, M. (2017). A hypothetical learning trajectory for conceptualizing matrices as linear transformations. *International Journal of Mathematical Education in Science and Technology, 48*(6), 809–829. https://doi.org//10.1080/0020739X.2016.1276225.

Arzarello, F. (2013). *In search of the cognitive and cultural roots of mathematical concepts.* Paper presented at the VI. Colloquium on History and Technology in the Teaching of Mathematics (HTEM), The Federal University of São Carlos (UFSCar), Brasil, 15–19 July 2013.

Bagley, S., Rasmussen, C., & Zandieh, M. (2015). Inverse, composition, and identity: The case of function and linear transformation. *The Journal of Mathematical Behavior, 37*, 36–47. https://doi.org//10.1016/j.jmathb.2014.11.003.

Bakker, A., & van Eerde, D. (2015). An Introduction to Design–Based Research with an Example From Statistics Education. In A. Bikner-Ahsbahs, C. Knipping, & N. Presmeg (Eds.), *Approaches to Qualitative Research in Mathematics Education: Examples of Methodology and Methods* (pp. 429–466). Dordrecht: Springer Netherlands.

Bartolini Bussi, M. G., & Mariotti, M. A. (2008). Semiotic mediation in the mathematics classroom: Artifacts and signs after a Vygotskian perspective. In L. English, M. Bartolini Bussi, G. Jones, R. Lesh, & D. Tirosh (Eds.), *Handbook of International Research in Mathematics Education* (2nd ed., pp. 746–783). Mahwah, NJ: Erlbaum.

Dreyfus, T., Hillel, J., & Sierpinska, A. (1998). Cabri based linear algebra: Transformations. In I. Schwank (Ed.), *Proceedings of the First Conference of the Europen Society for Research in Mathematics Education* (pp. 209–221). Osnabrück: Forschunsinstitut für Mathematikdidaktik.

Ernest, P. (2010). Reflections on Theories of Learning. In B. Sriraman, & L. English (Eds.), *Theories of Mathematics Education: Seeking New Frontiers* (pp. 39–47). Berlin, Heidelberg: Springer Berlin Heidelberg.

Falcade, R., Laborde, C., & Mariotti, M. A. (2007). Approaching functions: Cabri tools as instruments of semiotic mediation. *Educational Studies in Mathematics, 66*(3), 317–333. https://doi.org//10.1007/s10649-006-9072-y.

Godino, J. D., Batanero, C., & Font, V. (2007). The onto-semiotic approach to research in mathematics education. *ZDM Mathematics Education, 39*(1), 127–135. https://doi.org//10.1007/s11858-006-0004-1.

Gol Tabaghi, S. (2014). How dragging changes students' awareness: Developing meanings for Eigenvector and Eigenvalue. *Canadian Journal of Science, Mathematics and Technology Education, 14*(3), 223–237. https://doi.org//10.1080/14926156.2014.935528.

Gravemeijer, K. (2004). Local instruction theories as means of support for teachers in reform mathematics education. *Mathematical Thinking and Learning, 6*(2), 105–128.

Jones, K. (2000). Providing a foundation for deductive reasoning: Students' interpretations when using Dynamic Geometry Software and their evolving mathematical explanations. *Educational Studies in Mathematics, 44*(1), 55–85. https://doi.org//10.1023/a:1012789201736.

Lay, D. C. (2006). *Linear Algebra and Its Applications* (3 ed.). Boston: Pearson Addison-Wesley.

Leung, A. (2008). Dragging in a dynamic geometry environment through the lens of variation. *International Journal of Computers for Mathematical Learning, 13*(2), 135–157. https://doi.org//10.1007/s10758-008-9130-x.

Leung, A., Baccaglini-Frank, A., & Mariotti, M. A. (2013). Discernment of invariants in dynamic geometry environments. *Educational Studies in Mathematics, 84*(3), 439–460.

Lopez-Real, F., & Leung, A. (2006). Dragging as a conceptual tool in dynamic geometry environments. *International Journal of Mathematical Education in Science and Technology, 37*(6), 665–679. https://doi.org//10.1080/00207390600712539.

Mariotti, M. A. (2000). Introduction to proof: The mediation of a Dynamic Software Environment. *Educational Studies in Mathematics, 44*(1), 25–53. https://doi.org//10.1023/a:1012733122556.

Mariotti, M. A. (2009). Artifacts and signs after a Vygotskian perspective: the role of the teacher. *ZDM Mathematics Education, 41*(4), 427–440. https://doi.org//10.1007/s11858-009-0199-z.

Mariotti, M. A. (2012). ICT as opportunities for teaching–learning in a mathematics classroom: The semiotic potential of artefacts. In T. Y. Tso (Ed.), *Proceedings of the 36th Conference of the International Group for the Psychology of Mathematics Education* (Vol. 1, pp. 25–35). Taipei, Taiwan: PME.

Mariotti, M. A. (2013). Introducing students to geometric theorems: how the teacher can exploit the semiotic potential of a DGS. *ZDM Mathematics Education, 45*(3), 441–452.

Mariotti, M. A. (2014). Transforming images in a DGS: The semiotic potential of the dragging tool for introducing the notion of conditional statement. In S. Rezat, M. Hattermann, & A. Peter-Koop (Eds.), *Transformation–A Fundamental Idea of Mathematics Education* (pp. 155–172). New York: Springer.

Sierpinska, A. (2000). On some aspects of students' thinking in linear algebra. In J.-L. Dorier (Ed.), *On the teaching of linear algebra* (pp. 209–246). The Netherlands: Kluwer Academic Publishers.

Sierpinska, A. (2005). On practical and theoretical thinking and other false dichotomies in mathematics education. In M. G. Hoffmann, J. Lenhard, & F. Seeger (Eds.), *Activity and Sign* (pp. 117–135). US: Springer.

Trouche, L. (2004). Managing the complexity of human/machine interactions in computerized learning environments: Guiding students' command process through instrumental orchestrations. *International Journal of Computers for Mathematical Learning, 9*(3), 281–307.

Turgut, M. (2015). Theory of semiotic mediation in teaching-learning linear algebra: In search of a viewpoint in the use of ICT. In K. Krainer, & N. Vondrová (Eds.), *Proceedings of the 9th Congress of European Society for Research in Mathematics Education* (pp. 2418–2424). Prague, Czech Republic: Charles University in Prague, Faculty of Education and ERME.

Turgut, M. (2017). Students' reasoning on linear transformations in a DGS: a semiotic perspective. In T. Dooley & G. Gueudet (Eds.), *Proceedings of the 10th Congress of the European Society for Research in Mathematics Education*. Dublin, Ireland: DCU Institute of Education and ERME.

Turgut, M., & Drijvers, P. (2016). *Students' thinking modes and the emergence of signs in learning linear algebra*. Paper presented at the 13th Congress on Mathematical Education, Hamburg, Germany, 24–31 July 2016.

Vérillon, P., & Rabardel, P. (1995). Cognition and artifacts: A contribution to the study of though in relation to instrumented activity. *European Journal of Psychology of Education, 10*(1), 77–101. https://doi.org//10.1007/BF03172796.

Zandieh, M., Ellis, J., & Rasmussen, C. (2012). Student concept images of function and linear transformation. In S. Brown, S. Larsen, K. Marrongelle, & M. Oehrtman (Eds.), *Proceedings of the 15th Annual Conference on Research in Undergraduate Mathematics Education* (pp. 320–328). Portland, Oregon: SIGMAA.

Zandieh, M., Ellis, J., & Rasmussen, C. (2013). Students reconciling notions of one-to-one across two contexts. In S. Brown, G. Karakok, K. Hah Roh, & M. Oehrtman (Eds.), *Proceedings of the 16th Annual Conference on Research in Undergraduate Mathematics Education* (Vol. 2, pp. 297–306). Denver, Colorado: SIGMAA.

Zandieh, M., Ellis, J., & Rasmussen, C. (2017). A characterization of a unified notion of mathematical function: the case of high school function and linear transformation. *Educational Studies in Mathematics, 95*(1), 21–38. https://doi.org//10.1007/s10649-016-9737-0.

# Fostering Students' Competencies in Linear Algebra with Digital Resources

**Ana Donevska-Todorova**

**Abstract** This chapter discusses current research regarding the teaching and learning of concepts in linear algebra with the aid of (digital) resources. In particular, it looks into potential of digital resources to foster *students' competencies* in linear algebra. The aim of the chapter is to explain how technology-enhanced teaching and learning environments may contribute to developing competencies in multiple representations, visualization as well as procedural and conceptual understanding. The chapter culminates with a suggested *nested model of three modes of thinking* of concepts in linear algebra, which is suitable for designing teaching and learning environments.

**Keywords** Linear algebra · Competencies · Nested model of three modes of thinking · Technology

## 1 Introduction

It seems that the question whether technologies could be used in mathematics education is long behind us. This undoubtedly includes the teaching and learning of linear algebra content. While historically some questions regarding the role of digital and non-digital resources in linear algebra instruction have been addressed, many remain unanswered. To give a sense of the scope of the remaining questions, consider the following. What makes the use of a particular digital resource efficient? What are the ideal qualities of technology-based materials for the teaching and learning linear algebra and how can we measure these qualities? What are the advantages of one type of software over another, for example, a *Computer Algebra System* (CAS) versus a *Dynamic Geometry System* (DGS)? When and how should each be applied? How can we best disseminate research-based materials and sustain investigations about their values? Which new forms of digital support may increase

A. Donevska-Todorova (✉)
Humboldt-Universität zu Berlin, Berlin, Germany
e-mail: todorova@math.hu-berlin.de



S. Stewart et al. (eds.), *Challenges and Strategies in Teaching Linear Algebra*,
ICME-13 Monographs, https://doi.org/10.1007/978-3-319-66811-6_12

the motivation, communication and collaboration in a linear algebra course, e.g. flipped or inverted linear algebra classroom? These questions are still present and of significance in the current debates. Rather than try to answer any one of these questions in detail, this chapter presents a way to frame an examination of these types of questions.

This chapter builds on the discussions on the teaching and learning linear algebra in two relatively different groups at the 13th International Congress on Mathematical Education (ICME13). Firstly, the topic study group (TSG43) about the uses of technology in upper secondary mathematics education focused on the implementation of technologies from cognitive and epistemological perspectives, as well as accessibility to and the roles of emerging technologies. It also studied interrelations between technology and specific mathematical contents. Secondly, one of the key issues proposed within the discussion group (DG) for Teaching Linear Algebra at the ICME13 was the incorporation of technology specifically in the teaching of this subject. This chapter aspires to establish connections between the perspectives deliberated within these groups. It describes a diversity of technologies that can be used as a supplement to traditional educational media. The chapter begins by considering how the development of particular students' *competencies* for linear algebra may be fostered by appropriate technology-based environments such as CAS or DGS. The aim of the chapter is to suggest a *model* for multiple representations of concepts that are important when designing efficient (digitally based) environments in order to support the development of particular students' competencies in linear algebra.

## 2 Theoretical Background

Discussions about technology utilization in university linear algebra courses started with considerations of how "super calculators" or "commercial systems for both numeric and symbolic computation" (Carlson, Johnson, Lay, & Porter, 1993, p. 45) may be relevant for the content related knowledge of mathematics. In the last twenty years, the discussions have continued by also considering the role of technology for didactical purposes. For example, Day and Kalman (1999) point out that computers could be efficient not only for "eliminating computational drudgery", but also for providing interactive "environments for actively exploring properties of mathematical structures and objects" (Day & Kalman, 1999, p. 12). Rapid intensification of digitalization in general also parallels new educational trends. Curricula have been re-oriented towards learning outcomes and *competencies*. Similar to the principles and standards (NCTM, 2000) in the USA, there are six general competencies for tertiary level of mathematics in Germany (the numeration is used only for easier reading):

(K1) bringing forward arguments and proofs,
(K2) problem solving,
(K3) mathematical modeling,
(K4) representing mathematical concepts,
(K5) interacting with symbols, formal and technical elements of mathematics and
(K6) communicating mathematically (Kultusministerkonferenz, 2012, pp. 14–17).

Each of them is relevant for, and meets the goals of the teaching of linear algebra. Though there are certainly no firm boundaries between them, the development of one or more of them may be supported by meaningful implementation of digital resources deployed during the teaching and learning processes. This could be done through interactive explorations in modeling and problem solving, by promoting understanding through the use of multiple representations or by reducing systematic procedures when handling large data sets. Such students' competencies and the possible effects of technology on their development are the focus of this chapter. An overview of the theoretical background precedes this description.

Investigating complex phenomena like mathematics education in the presence of technology is challenging because of the fast pace of technological change and the lack of specific theories for studying the teaching and learning a particular mathematical content, e.g. linear algebra, with digital aids. This has been explored in recent literature (e.g. Donevska-Todorova & Trgalova, 2017; Turgut & Drijvers, 2017). This also appears as a new issue in the call for topic-working group 17 at the 10th conference of the European Society for Research in Mathematics Education (CERME 10). A recent review (Drijvers et al., 2016) considered whether digital technology improves students' learning of a particular mathematical content (e.g. geometry) through quantitative studies, and why it may be the case through qualitative studies focusing on the teacher as an important factor. Another survey paper (Sinclair et al., 2016) stated seven 'threads' of contributions which affect the teaching and learning geometry with technologies at different levels of education including pre- and post compulsory. A question that comes out of this research is if these '*threads*' might also refer to the teaching and learning other mathematical domains including linear algebra. I focus on two of the min particular.

*The first 'thread'*, "developments and trends in the use of theories" (Sinclair et al., 2016, p. 1) relates to whether the use of general *theories* about the teaching and learning mathematics with digital equipment is adequate for a specific mathematical domain such as geometry, or linear algebra. In the absence of a particular theoretical framework or apparatus for investigating the teaching and learning linear algebra with digital resources, this chapter suggests connecting suitable theoretical frameworks. In order to give the reader a sense of what is meant by a later categorization, a possible network that may consider three groups of theoretical frameworks is:

(1) *general theories on mathematics education*,
(2) *theories on technologies in mathematics education and*

(3) *specific theories for the teaching and learning linear algebra* (*with or without technological support*).

The suggested groups of theories are certainly global and do not exclude other. Nonetheless, they refer to the works of the TSG43 and the DG at the ICME13 and are here meant to serve as examples. Exploited theoretical frameworks from each of the groups have to be adjusted and linked for achieving the goals of the research of teaching and learning linear algebra. General theoretical frameworks (1) which may be useful in the sense of *competencies* may be the construct *concept image-concept definition* (Tall & Vinner, 1981), regarding the competency K1. Further, the three worlds of mathematical thinking (Tall, 2004) or the action-process-object-schema (APOS) theory (Dubinsky & McDonald, 2001) is relevant for studying the development of the competency K4. These frames directly relate to the above consideration ('*thread*') about the underrepresentation of defining (K1)—versus overemphasis of representing (K4) issues in technology-enhanced environments. This already shows a natural bridge between the general theories (1) and the theories associated to digital media (2).

Further on, the theory of *semiotic mediation* (Bartolini & Mariotti, 2008), also used within the ICME13-TSG43, may be exploited for investigating exact effects of particular tools (drag/drop, touch/move, hide/show, slide, zoom in/zoom out) on the learning linear algebra (e.g. distinguishing scalar from vector operations, or referring geometric meaning to algebraic concepts, etc.). Studying historical and epistemological developments of the concepts in linear algebra is relevant for designing technology-based teaching/learning environments and increasing their semiotic potential for didactical purposes. Taking the multifaceted nature of concepts in linear algebra such as analytic-arithmetic or synthetic-geometric, into consideration may contribute the creation of environments to foster development of multiple representations (K4). I come to this point in Sect. 3.3. Another theoretical framework specific for technology-rich settings is instrumental genesis (Trouche, 2005), which has a potential to facilitate arrangement of instruction of linear algebra, among other, and to provide relevant data at interpersonal, classroom, resource or institutional level of a multiple-level data analysis of communication and collaboration competencies (K6).

Finally, specific theories for the teaching and learning linear algebra (3) as the one referring to students' difficulties with the unifying and general theory of linear algebra and the obstacle of formalism (Dorier, 2000), multiple *modes of description* (Hillel, 2000) and multiple *modes of thinking* (Sierpinska, 2000) may also be valuable in the teaching of technology supported instruction. I elaborate these issues more in detail also in Sect. 3.3 in relation to the competencies K1, K4 and K5.

Looking back to the survey the '*thread*' "advances in the understanding of the teaching and learning of *definitions*" (Sinclair et al., 2016, p. 2) in geometry supported with technology, may as well refer the insufficient number of studies directly addressing key mathematical issues as *defining* concepts in linear algebra. This issue about the defining mathematical concepts is in particular relevant for developing a competency of formal proving (K1). Further, depending on the way concepts in linear algebra are defined, e.g. analytically or geometrically, their representations may also vary, which affects the development of the competency K4. Finally, each definition of concepts in linear algebra uses a particular symbolic and formal mathematical language that directly influences the growth of the competency K5. I investigate the possibilities to strengthen mutual development of these competencies and, based on chosen theories from (1) to (3), suggest a *model* that I consider important when teaching or creating teaching/learning trajectories or environments for concepts in linear algebra (in Sect. 3.3).

*Research question*

Drawing upon the theoretical concerns above, including (1)–(3), the main research question that arises is: how could the development of students' competencies in linear algebra be facilitated by technology usage in instruction and learning?

By collecting, comparing, contrasting and synthesizing data for digital environments suitable for gaining competencies in linear algebra, I provide insights to each of the competencies briefly (in Sects. 3.1, 3.2 and 3.4), however set my focus on the competencies K1, K4 and K5 (in Sect. 3.3).

## 3 Content Specific and Process Oriented Competencies in Linear Algebra

This section offers some insights in some of the previously mentioned content specific (or subject matter) and process oriented competencies for a tertiary level mathematics K1–K6 with reference to linear algebra.

### *3.1 Defining, Proving and Understanding*

The inverse treatment of axiomatic properties for defining, at tertiary level of linear algebra, on the one hand and describing concepts, at upper secondary level of linear algebra, on the other hand, signalizes possible obstacle for learning (Donevska-Todorova, 2017b, p. 2). Sometimes concepts and their properties remain to occur as separate mathematical objects in the students' minds (Donevska-Todorova, 2017b, p. 6). For example, while associativity is a defining property of vector spaces at university level linear algebra, it is perceived as a characteristic of the operation addition of vectors that are previously defined as classes of parallel arrows with same length and orientation in upper high school. Although, there exist some studies, which have considered descriptive (a posteriori) defining of concepts after exploring properties with DGS (in addition to other media), in author's knowledge there are no studies on students' deeper understanding of the need for axioms and definitions for avoiding infinite regress and circularity (de Villiers,

1998). The dependence of the development of deep conceptual understanding on the definitions of concepts has seldom been in focus (e.g. Donevska-Todorova, 2015; Hannah, Stewart, & Thomas, 2016). Based on semi-structured clinical interviews with participating students in a linear algebra class, Bagley, Rasmussen, and Zandieh (2015, p. 36) have found that all participating students think that “the result of composition of a function and its inverse should be 1”. In a linear algebra context, functions appear as linear transformations from one vector space into another, preserving axiomatic properties as addition and scalar multiplication, but such function conceptions have also rarely been directly examined. An exemplary study about transformations in a Cabri-based environment has been undertaken by Dreyfus, Hillel & Sierpinska, 1998. Another exemplary study (Donevska-Todorova, 2016), points out students’ difficulties about the introduction and understanding of linear, bi-linear and multi-linear transformations on a real vector space which have been discovered in pre-service teachers when working on discussing questions and multiple-solution tasks. Many abstract mathematical concepts, function (in this context linear transformation) among them, can be understood either operationally, as processes, or structurally as objects but the operational and the structural aspects do not replace, rather complement each other (Bagley, Rasmussen, & Zandieh, 2015, p. 37). Yet, there are studies, which have discovered students’ predominant possession of *procedural* versus *conceptual* understanding for example for concepts as determinants and suggest that this discrepancy may be overcome with a possible technology-based environment for a semiotic mediation (Donevska-Todorova, 2016, p. 10). The problem of defining concepts in linear algebra is certainly further related to argumentation and proving. Students’ abilities for proving have been examined, e.g. that a set is a subspace of a vector space (Britton & Henderson, 2009, p. 963) however insufficiently from the aspect of technology facilitation.

The competency about defining concepts, possibly with technological assistance in visualizing and representing axiomatic properties, is connected to two other of the above-mentioned competencies (K4 translations between multiple representations and K5 symbolic language and formal nature of linear algebra). I will revisit this point in Sect. 3.3.

## *3.2 Computation, Symbol Manipulation and Programming*

The historical evolution of technological devices starting with hand-held calculators through graphical calculators to powerful CAS shows a quick ongoing expansion. This growth has many possibilities and challenges for the teaching and learning of linear algebra. A common agreement among researchers is that the usage of CAS should be an integral part of mathematical instructions (e.g. Janetzko, 2016). Development of competencies for programming besides those for computing or symbol manipulation (K5) may be supported by CAS because of their embedded powerful apparatus (e.g. Díaz et al., 2011). Manual solutions of systems of linear equations (SLE), e.g., by the Gaussian elimination method, are meaningful both at school and university level. In particular, understanding the meaning of the solutions sets, the structure and the algorithm are among the most important learning goals. On the one hand, exercise and assessment tasks are usually suited to lead towards achieving these goals without a technological support. On the other hand, engagement in algorithm development and computer programming may contribute in the development of spatial reasoning and coding capabilities (e.g. Francis, Khan, & Davis, 2016) or development of undergraduate students' proficiency, as a combination of fluency and confidence, in a pillar of scientific inquiry having form of simulation, optimization and modeling (Buteau, Muller, Marshall, Sacristán, & Mgombelo, 2016). When solving SLE with large number of equations and unknowns e.g. by students in economic studies, interpretations, rather than calculations of results is worth more. However, by implementing CAS calculators for checking answers or performing single step direct calculations to compute, for example a determinant, or an inverse of a matrix, Stewart and Thomas (2004) have found that even enthusiastic students require sustained attention in a technology-based learning environment. The efficiency in calculating inverse or exponential matrices of large dimension (over a hundred), eigenvectors and eigenvalues of matrices with real and complex entries have made CAS become an integral part of contemporary mathematics (Caridade, Encinas, Martín-Vaquero, & Queiruga-Dios, 2015). Digital image processing by the use of CAS in a Mathematic Virtual Laboratory (MVL) developed on a Moodle platform has been suggested for making some linear algebra concepts, as matrix operations and their properties, more concrete and clearer to geological and industrial engineering students by Caridade et al. (2015). The authors also envision possibility for similar resourceful usage in high school mathematics. MATLAB,[1] for example, is often perceived and recommended as one of the most natural CAS for linear algebra as it has been developed purposely for matrix operations (Dios, Martínez, Encinas, & Encinas, 2012; Jin & Bi, 2011). Its usage in instruction is however not straightforward for the reason that, students have to be familiar with the programming language in advance. A linguistic perception of mathematics includes syntactical, semantical and pragmatic aspects of the algebraic language and it can easily be handled by a CAS, nevertheless, students have difficulty to understand what a CAS does and how its output is to be interpreted (Oldenburg, 2016). Some difficulties of engineering students when CAS are sporadically used may be overcome by a user interface, called CATO, for different CASs written in Java as Mathematica,[2] Maxima[3] or the mathematical toolbox of MATLAB (Janetzko, 2016).

The roles of CAS may also be observed as "multiply-linked graphical, numerical, and symbolic manipulation utilities" (Heid & Edwards, 2001, p. 128). Powerful technology-based mathematics packages as Mathematica or Maple[4]

[1]https://www.mathworks.com/products/matlab/.

[2]https://www.wolfram.com/mathematica/.

[3]maxima.sourceforge.net/.

[4]https://www.maplesoft.com/products/maple/.

enable rich approaches for teaching and learning undergraduate linear algebra. Their capabilities for performing exhausting calculations and symbolic manipulations, e.g. solutions of (large) SLE or matrix multiplication, may transform or even replace some classroom activities. In particular, their relevance for graphical, in addition to the numerical and symbolic representations is connected to the competencies (K4).

## 3.3 *Visualization, Representation, Exploration and Generalization*

Students' difficulties with a priory *visualizations* have already been noticed and reported in research and an overemphasis of the visual potential of technologies in improving conceptual understanding per se, is considered as a naïve attitude (Lagrange, Artigue, Laborde, & Trouche, 2001, p. 7). Yet, it seems that a careful implementation of an appropriate DGS with an integrated algebra in it, rather than a CAS environment, may be helpful for visualizing and multiple-representing concepts in linear algebra. While a pure synthetic-geometric approach may be quite challenging the students to apprehend, building linear algebra on a coherent multi-domain base, e.g. geometry, functional calculus and modern axiomatic may be more beneficial. In this sense, appropriate DGE for concepts in linear algebra may be of help and I try to elaborate how.

DGS have the potential for dynamic and simultaneous changes of multiple representations. In continuation, I prefer a usage of rigorous terminology that is specific to the research field of teaching and learning linear algebra. Namely, instead of considering the "algebraic, geometrical and abstract presentation" (e.g. Fig. 1 in Konyalioglu, Isik, Kaplan, Hizarci, & Durkaya, 2011, p. 4042), I use the vocabulary different *modes of description* (Hillel, 2000) and *modes of thought* (Sierpinska, 2000). Besides, a triple of *distant components* of linear algebra concepts, the relations between which are not identified (Fig. 1) seems a bit problematic.

The authors also suggest a teaching approach according to these components by considering the "geometric presentation" for dimensions less than or equal to three, while the "algebraic presentation" for dimensions greater that three (Konyalioglu et al., 2011, p. 4042). In my opinion, there seems to be no reason why not considering the algebraic one also for dimensions less than or equal to three. Moreover, the order: first, algebraic definition, second geometric meaning and third, abstract representation (Konyalioglu et al., 2011, p. 4043) does not necessarily need to take place in the teaching of linear algebra even in high school. On the contrary, beginning with a geometrical context may foster students to deep intuitive thinking, motivate *explorations* and therefore contribute to the development of alternative *competencies*. Furthermore, the teaching of linear algebra at the university level goes along with the nature of mathematics as a science, so the concepts are

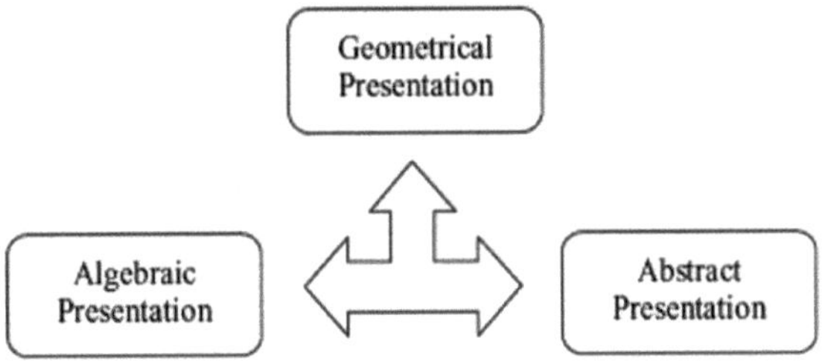


**Fig. 1** Abstract, algebraic, and geometric presentations of concepts in linear algebra on Fig. 1 in Konyalioglu et al. (2011, p. 4042), Copyright (2018), with permission from Elsevier

introduced by definitions through the abstract mode of description. Abstract concepts gain their meaning in contexts and both the algebraic and the geometric modes allow such concretizations. Therefore, a severe order in the introduction of the concepts, as suggested by the authors above, is not a necessity. Similarly, to this view, Dray and Manogue (2008) have suggested a geometric introduction for an exemplary concept, the dot product of vectors, which may continue with arithmetic-algebraic and culminate with analytic-structural aspects. This seems to be a more natural sequence due to the primarily geometric introduction of vectors in physics, lower-secondary mathematics and upper-secondary linear algebra and because of the vector-input and scalar-output of the dot product. For empirical results with this sequencing for the introduction of the dot product in a dynamic geometry environment, see Donevska-Todorova (2015).

### 3.3.1 A Nested Model of the Three Modes of Thinking of Concepts in Linear Algebra

In contrast to the triple presentation of distant constituents of concepts in linear algebra given above (Fig. 1) and because of a doctoral study (Donevska-Todorova, 2017a), I would rather suggest a *nested model* for presenting the modes of thinking. I actually situate the algebraic and the geometric modes of description (Hillel, 2000) and corresponding analytic-arithmetic and synthetic-geometric modes of thinking (Sierpinska, 2000) as nested constituents of the analytic-structural mode (Fig. 2). The prior two allow concretization of the abstractness of the concepts.

How does the *nested model* (Fig. 2) refer to the theoretical considerations (1)–(3) and the competencies *K1* to *K6* in Sect. 2 exactly? In other words, how does the nested model help in the analysis towards answering the posed research question? I try to explain this through its nodes and arrows.

Starting form (1) *general theories about mathematics education*, the model allows a development of a wider and deeper *concept image* for the *concept definition* (Tall & Vinner, 1981) of a particular concept in linear algebra. For example, the node *analytic-arithmetic mode thinking* in the model refers to thinking of vectors in as ordered n-tuples, while the node *geometric mode* refers to vectors as equivalent classes of arrows that are equal in length, orientation and direction. These two nodes show how vectors as elements of vector spaces in an *analytic-structural mode* of thinking

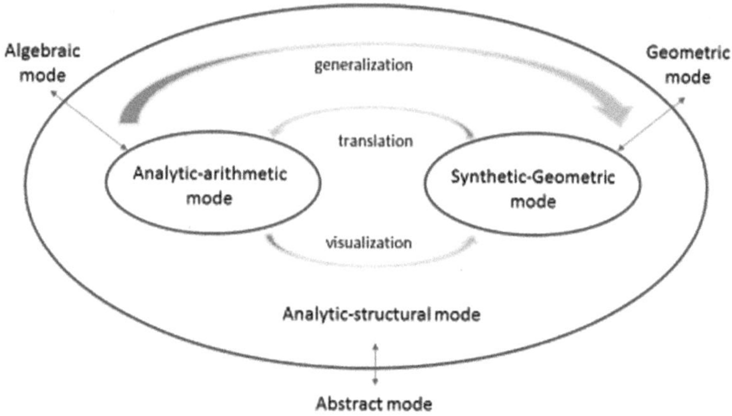


**Fig. 2** Nested model of the three modes of thinking and description of concepts in linear algebra

perceive their concretization in a context (e.g. in $\mathbf{R}^2$ or $\mathbf{R}^3$). Perceiving these three different *concept definitions* of a single concept (e.g. vectors) is enabled through different modes of description (Hillel, 2000) which, likewise the corresponding modes of thinking (Sierpinska, 2000) belong to (3) *specific theories for the teaching and learning linear algebra*.

Due to the importance of precise definitions of concepts in linear algebra, based on which argumentations and formal proofs develop (competency K1), it may be worthy to further utilize the *nested model* when aiming to also advance the competency for multiple representing and symbol manipulating (*modes of description and language*) of concepts (K4 and K5).

Looking at the *nested model* again, the arrows represent the relations and interplays between the nodes. For example, *translations* from the geometric into the arithmetic-analytic mode of thinking or the other way around, like *visualization* from the arithmetic-analytic to the geometric mode of thinking and even *generalization* of concepts from 2D and 3D to *n*D may take place simultaneously (Fig. 2). In particularly, this may come into focus in a DGE, which brings us to the use of (2) *theories about technologies in mathematics education.*

Switching from one mode of thinking into another and vice versa may significantly be stimulated, e.g. by the *dragging tool* in a DGE. Such devices could serve as instruments of semiotic mediation (Bussi & Mariotti, 2008) in exploring multiple modes of description. A digital simultaneous multiple-dynamic manipulation, in contrast to single-static paper-pencil control could be achieved e.g.by means of sliders which facilitate numerical variations. The numerical dependences represented by sliders allow transparency of the difference between scalars and vectors which is typical for the content of linear algebra, e.g. for the teaching and learning of linear combinations of vectors, linear (in)dependence of vectors, vector spaces, linear transformations, bi- and multi-linear forms, etc. What makes a DGS toll a specific instrument of semiotic mediation could deeply be observed by the

Vigotskian perspective for the transformation form inter- to intra-personal mental processes (Falcade, Laborde, & Mariotti, 2007). Further on, geometric, and simultaneously arithmetic-algebraic dependences could also be examined by the dragging facilities of points or vectors and there already exist DGEs for such purposes. For example, a recursive exploration space (Hegedus, Dalton, & Moreno-Armella, 2007) can mediate a mental concept formation and therefore participate in development of mathematical cognition (e.g. the concept of dot product of vectors in Donevska-Todorova, 2017b). Students co-act with the environment by exchanging their roles in switching from one into another mode of description and thinking which modifies (though not negates) the paper, as a *frozen* (Hegedus, Dalton, & Moreno-Armella, 2007) into the DGE as a *fluid medium* for thinking of mathematical concepts (Donevska-Todorova, 2017b). This study shows that the challenge of supporting the learning of abstract concepts or even completely abstract structures, e.g. vector space or subspace, by the DGS has by now been approached to a certain degree by interactive dynamic artifacts for one or more of their defining axiomatic properties.

## 3.4 Communication and Collaboration

A Spanish group of authors has been looking at generic, content-independent competencies like team-working, self-learning, critical thinking, problem solving and technical communication with the use of CAS Derive (García López, García Mazario, & Villa Cuenca, 2011) and CAS Maxima in a later study (García, García, Del Rey, Rodríguez, & De La Villa, 2014). They have concluded that both CAS have allowed not only improvement of students' academic performance but also increase of students' motivation, satisfaction, self-confidence and team working. Bulgarian scholars have examined a combined, traditional and CAS-based environment for an acquisition of competencies in higher education and have concluded that it is helpful for action competencies related to emotional, social and value-related components (Varbanova & Durcheva, 2016, p. 54).

Although communication and collaboration among students and instructors in a technology-enhanced environment may be fostered and researched in relation to CAS or DGS, the next subsection offers insights to possibilities for development of these competencies (K6) from a bit different aspect.

### 3.4.1 Cyber Learning, Communication and Collaboration

One of the oldest functions of technology is the collection of data required for teaching and learning processes. New Web 2.0 and 3.0 technologies allow for the exchange of collected data, as well as time and place independent communication and collaboration. Nevertheless, a first reaction to the teaching and learning of a specific mathematical content, including linear algebra, through *social networks* for example, is a dose of skepticism. Social media may involve inaccurate information, biased comments and unreceptive responses, yet an effortless search shows that hundreds of groups called "Linear algebra" or similar, already exist and have thousands of members on Facebook and Twitter. It is predictable that the number of such groups will continue to grow. Research Gate, the largest academic social network (Matthews, 2016), is another type of virtual resource that may or may not serve teaching and learning beyond research purposes. There are also numerous online *forums* and *blogs*, specifically related to linear algebra, YouTube *tutorials* and *courses* with over millions of views (e.g. thirty four videos of the MIT OpenCourseWare, Linear Algebra, Strang, 2005). Virtual teaching and learning environments, for example, *online classrooms, flipped classrooms, wikis* (e.g. GeoGebra Wiki and Wikispaces) could also be part of the repertoire for organizing linear algebra courses. An *inverted* or *flipped classroom* used as a "one-time class design to teach a single topic, as a way to design a recurring series of workshops, and as a way of designing an entire linear algebra course" (Talbert, 2014, p. 361). Love et al. (2014) found that students in a flipped linear algebra classroom had "a more significant increase between the sequential exams compared to the students in a traditional lecture section, while performing similarly in a final exam" (Love et al., 2014, p. 317) and expressed conceptual understanding and joy (Love et al., 2014, p. 323). A teaching and learning platform for linear algebra for engineers, created according to the *blended-learning*-concept, has significantly improved students' performance (Roegner & Seiler, 2012). The participating students in the project had direct access to an online-script with visualizations, individualized homework problems with an interactive training environment and automatic corrections. The sustainability, further expansion, and dissemination of these projects, as well as the development of other such platforms remain ongoing.

These plentiful and diverse educational innovations have the potential to become a part of quality teaching and (in)formal learning of linear algebra after careful research and documentation has been undertaken. There is still a lack of satisfactory evidence that these innovative forms of instruction and learning guarantee development of subject matter competencies. Even development of other non-content specific competencies such as networked debating, blogging, and chatting, as well as socio-cultural and anthropological aspects of formal and informal education in a pure context of teaching and learning linear algebra need to be further examined and documented. In the era of global digitalization, computer technologies have high social legitimacy, but their educational legitimacy, remains an open research question.

## 4 Discussion and Challenges Ahead

Alongside the evolution of technologies and the enlargement of classroom accessibility, critical research suggests cautious implementation. Even in the late 90s, Guin and Trouche (1998) pointed out complexities in teaching and learning

**Table 1** Technological tools for facilitating development of competencies in linear algebra

| Competency | Technology based environment | Connecting theories for research |
|---|---|---|
| Defining, proving, understanding (K1) | CAS, DGS | (1) General theories about mathematics education<br>(2) Theories about technologies in mathematics education<br>(3) Specific theories for the teaching and learning of linear algebra |
| Computation, symbol manipulation, programming (K5) | CAS | |
| Visualization, representation, exploration, generalization (K4) | DGS | |
| Communication, collaboration (K6) | CAS, DGS, virtual environments | Online classroom, inverted (flipped) classroom, blended learning |

situations which are brought by implementation of calculators. The above analysis shows that the learning and understanding of the abstract nature of linear algebra through axioms, definitions, theorems and structures does not become straightforward by a simple use of a CAS or a DGS. Though a rigorous systematization regarding the research question would be difficult to establish, an attempt to show which digital tool may facilitate the development of which competency and how it may be researched is proposed in Table 1.

The initial ideas illustrated in Table 1 must be considered with some flexibility. For example, a CAS environment may also be helpful for visualization and representation, (K4), though the DGS with embedded algebraic features is seen as having more potential for this purpose due to mutual dependence and invariant properties which can be simultaneously investigated (are not always brought by the software a priory but are additionally designed). This shows that a whole linear algebra course does not necessarily have to be designed in a single, either CAS or DGS environment. It could be the case that combining different technological tools for facilitating the learning of particular concepts in a single course may also be useful.

What has been considered as "smarter technologies, like computer software or symbolic calculators [and] emerging technologies (Internet, etc.)" (Lagrange et al., 2001, p. 3) fifteen years ago may seem history now. New emerging technological devices such as touch and multi-touch (iPads, iPhones, etc.) open new questions for further investigations.

## 5 Conclusion

This chapter has surveyed current literature on both technology-facilitated teaching and the learning of linear algebra, taking the discussions in two ICME13 working groups as starting viewpoints. The analysis focused on whether and how technology-enhanced environments could facilitate the development of students' competency in linear algebra. From this analysis and regarding the research question, it seems that CAS are more suitable for the development of competencies such as symbol manipulation and programming (K5) (in Sect. 3.2) and DGS are better for competencies such as visualization, representation, exploration, and generalization (K4) (in Sect. 3.3). Both types of environments may be appropriate for defining, proving and understanding (K1) (Table 1). In order to show how a digitally based environment may be considered for fostering K1, K4 and K5, I have suggested a *nested model* (Fig. 2). This model presents all three *modes of description and thinking* that I consider important in instruction and in the design of tasks or teaching environments. It is a visual presentation showing that the modes are not dispersed one from another, rather connected. In particular, specifically designed technologically-based environments may enable easier and more efficient shifts between the modes, facilitating the development of competencies for defining, representing and understanding concepts in linear algebra.

## References

Bagley, S., Rasmussen, C., & Zandieh, M. (2015). Inverse, composition, and identity: The case of function and linear transformation. *The Journal of Mathematical Behavior, 37*, 36–47.

Britton, S., & Henderson, J. (2009). Linear algebra revisited: An attempt to understand students' conceptual difficulties. *International Journal of Mathematical Education in Science and Technology, 40*(7), 963–974.

Bussi, M. B., & Mariotti, M. A. (2008). Semiotic mediation in the mathematics classroom: Artifacts and signs after a Vygotskian perspective. *Handbook of international research in mathematics education, New York*, 746–783.

Buteau, C., Muller, E., Marshall, N., Sacristán, A. I., & Mgombelo, J. (2016). Undergraduate Mathematics Students Appropriating Programming as a Tool for Modeling, Simulation, and Visualization: A Case Study. *Digital Experiences in Mathematics Education, 2*(2), 142–166.

Caridade, C. M. R., Encinas, A. H., Martín-Vaquero, J., & Queiruga-Dios, A. (2015). CAS and real life problems to learn basic concepts in Linear Algebra course. *Computer Applications in Engineering Education, 23*(4), 567–577.

Carlson, D., Johnson, C. R., Lay, D. C., & Porter, A. D. (1993). The Linear Algebra Curriculum Study Group recommendations for the first course in linear algebra. *The College Mathematics Journal, 24*(1), 41–46.

Day, J. M., & Kalman, D. (1999). Teaching linear algebra: What are the questions. *Department of Mathematics at American University in Washington DC*, 1–16.

De Villiers, M. (1998). To teach definitions in geometry or teach to define? In *Proceedings of the 22nd Conference of the International Group for the Psychology of Mathematics Education* (Vol. 2, pp. 2–248).

Díaz, A., García López, A., & Villa Cuenca, A. D. L. (2011). An example of learning based on competences: Use of Maxima in Linear Algebra for Engineers. *International Journal For Technology in Mathematics Education, 18*(4), 177–181.

Dios, A. Q., Martínez, V. G., Encinas, A. H., & Encinas, L. H. (2012). The computer as a tool to acquire and evaluate skills in math courses. In *4th International Conference on computer research and development, IPCSIT* (Vol. 39).

Donevska-Todorova, A. (2015). Conceptual Understanding of Dot Product of Vectors in a Dynamic Geometry Environment. *Electronic Journal of Mathematics & Technology, 9*(3).

Donevska-Todorova, A. (2016). Procedural and Conceptual Understanding in Undergraduate Linear Algebra. In Krainer, K. & Vondrova, N. (Eds)., Proceedings INDRUM2016.

Donevska-Todorova, A. & Trgalova, J. (2017). Learning Mathematics with Technology. A Review of Recent CERME Research. In Dooley, T. & Gueudet, G. (Eds.). Proceedings of the Tenth Congress of the European Society for Research in Mathematics Education (CERME10, February 1 – 5, 2017). Dublin, Ireland: DCU Institute of Education and ERME.

Donevska-Todorova, A. (2017a). Utilizing Technology to facilitate the transition between the Upper Secondary- to Tertiary Level of Linear Algebra. (unpublished PhD Thesis).

Donevska-Todorova, A. (2017b). Recursive Exploration Space for Concepts in Linear Algebra. In Tabach, M. & Siller, S. (Eds). *Uses of Technology in K-12 mathematics Education: Tools, topics and Trends*. Springer (in press).

Dorier, J. L. (2000). Epistemological Analysis of the Genesis of the Theory of Vector Spaces. In: Dorier, J. L. (Ed.). *On the Teaching of Linear Algebra*. Mathematics Education Library, vol 23. Springer, Dordrecht.

Dray, T., & Manogue, C. A. (2008). The geometry of the dot and cross products. AMC, 10, 12.

Dreyfus, T., Hillel, J., & Sierpinska, A. (1998). Cabri-based linear algebra: Transformations. *European Research in Mathematics Education I*, 209–221.

Drijvers, P., Ball, L., Barzel, B., Heid, M. K., Cao, Y. & Maschietto, M. (2016). *Uses of Technology in Lower Secondary Mathematics Education*. ICME-13 Topical Survey, pp. 1–34. Springer International Publishing. https://doi.org//10.1007/978-3-319-33666-4.

Dubinsky, E., & McDonald, M. A. (2001). APOS: A constructivist theory of learning in undergraduate mathematics education research. In *The teaching and learning of mathematics at university level* (pp. 275–282). Springer Netherlands.

García, A., García, F., Del Rey, Á. M., Rodríguez, G., & De La Villa, A. (2014). Changing assessment methods: New rules, new roles. *Journal of Symbolic Computation*, *61*, 70–84.

García López, A., García Mazario, F., & Villa Cuenca, A. D. L. (2011). Could it be possible to replace DERIVE with MAXIMA? *The International Journal for Technology in Mathematics Education*, *18*(3), 137–142.

Guin, D., & Trouche, L. (1998). The complex process of converting tools into mathematical instruments: The case of calculators. *International Journal of Computers for Mathematical Learning*, *3*(3), 195–227.

Falcade, R., Laborde, C., & Mariotti, M. A. (2007). Approaching functions: Cabri tools as instruments of semiotic mediation. *Educational Studies in Mathematics*, *66*(3), 317–333.

Francis, K., Khan, S., & Davis, B. (2016). Enactivism, spatial reasoning and coding. *Digital Experiences in Mathematics Education, 2*(1), 1–20.

Hannah, J., Stewart, S., & Thomas, M. O. J. (2016). Developing conceptual understanding and definitional clarity in linear algebra through the three worlds of mathematical thinking, *Teaching Mathematics and its Applications: An International Journal of the IMA*, *35*(4), 216–235.

Hegedus, S., Dalton, S., & Moreno-Armella, L. (2007). Technology that mediates and participates in mathematical cognition. *Proceedings of CERME5, WG 9 Tools and technologies in mathematical didactics 1331*, pp. 1419–1428.

Heid, M. K., & Edwards, M. T. (2001). Computer algebra systems: revolution or retrofit for today's mathematics classrooms? *Theory into Practice, 40*(2), 128–136.

Hillel, J. (2000). Modes of description and the problem of representation in linear algebra. In *On the teaching of linear algebra* (pp. 191–207). Springer Netherlands.

Janetzko, H.-D. (2016). The GUI CATO – how natural usage of CAS with CATO modified the mathematical lectures and the interface itself. In *the Proceedings of the 22nd Conference on Applications of Computer Algebra*, ACA, August 2016, Kassel, Germany.

Jin, L., & Bi, C. (2011, July). Application of matlab software for linear algebra. In *Circuits, Communications and System (PACCS), 2011 Third Pacific-Asia Conference on* (pp. 1–3). IEEE.

Konyalioglu, A. C., Isik, A., Kaplan, A., Hizarci, S., & Durkaya, M. (2011). Visualization approach in teaching process of linear algebra. *Procedia-Social and Behavioral Sciences, 15*, 4040–4044.

Konferenz der Kultusminister der Länder in der Bundesrepublik Deutschland [KMK] (2012). Bildungsstandards im Fach Mathematik für die Allgemeine Hochschulreife (Beschluss der Kultusministerkonferenzvom 18.10.2012). Available at: http://www.kmk.org/fileadmin/Dateien/veroeffentlichungen_beschluesse/2012/2012_10_18-Bildungsstandards-Mathe-Abi.pdf.

Lagrange, J. B., Artigue, M., Laborde, C., & Trouche, L. (2001). A meta study on IC technologies in education. Towards a multidimensional framework to tackle their integration. In *Proceedings of the 25th Conference of the International Group for the Psychology of Mathematics Education* (Vol. 1, pp. 1–111).

Love, B., Hodge, A., Grandgenett, N., & Swift, A. W. (2014). Student learning and perceptions in a flipped linear algebra course. *International Journal of Mathematical Education in Science and Technology, 45*(3), 317–324.

Matthews, D. (2016). "Do academic social networks share academics' interests?". *Times Higher Education. Retrieved 2016-04-22.*

NCTM - National Council of Teachers of Mathematics (2000). *Principles and Standards for School Mathematics.* Reston, Virginia, USA.

Oldenburg, R. (2016). A Transparent Rule Based CAS to support Formalizationof Knowledge. *In the Proceedings of the 22nd Conference on Applications of Computer Algebra*, ACA, August 2016, Kassel, Germany.

Roegner, K., & Seiler, R. (2012). Das multimediale Lehr-und Lernsystem MUMIE/TUMULT in der universitären Mathematikausbildung. *Hochschuldidaktik–Mathematik und Informatik. Symposiumsband „Verbesserung der Hochschullehre in Mathematik und Informatik"*, 115–122.

Sinclair, N., Bartolini Bussi, M. G., de Villiers, M., Jones, K., Kortenkamp, U., Leung, A. & Owens, K. (2016). Recent research on geometry education: an ICME-13 survey team report *ZDM Mathematics Education 48*(5), pp 691–719.

Sierpinska, A. (2000). On Some Aspects of Students' Thinking in Linear Algebra. In Dorier, J.-L. (Ed.). *On the teaching of linear algebra*. Mathematics Education Library, vol 23. Springer, Dordrecht.

Stewart, S., & Thomas, M. O. J. (2004). The learning of linear algebra concepts: Instrumentation of CAS calculators. Proceedings of the 9th Asian Technology Conference in Mathematics (ATCM), Singapore, 377–386.

Strang, G. (2005). Linear Algebra. Video Lectures. MIT OpenCourseWare. https://ocw.mit.edu/courses/mathematics/18-06-linear-algebra-spring-2010/video-lectures/. Last access 25.10.2017.

Talbert, R. (2014). Inverting the linear algebra classroom. *Primus, 24*(5), 361–374.

Tall, D. (2004). Thinking through three worlds of mathematics. In *Proceedings of the 28th Conference of the International Group for the Psychology of Mathematics Education* (Vol. 4, pp. 281–288).

Tall, D., & Vinner, S. (1981). Concept image and concept definition in mathematics with particular reference to limits and continuity. *Educational studies in mathematics, 12*(2), 151–169.

Trouche, L. (2005). Instrumental genesis, individual and social aspects. In *The didactical challenge of symbolic calculators* (pp. 197–230). Springer US.

Turgut, M. & Drijvers, P. (2017). Students' Thinking Modes and the Emergence of Signs in Learning Linear Algebra. In the Proceedings of ICME 13 Topical Survey (to appear).

Varbanova, E. & Durcheva, M. (2016). Developing Competences in Higher Mathematics in a CAS Supported Learning Environment. *In the Proceedings of the 22nd Conference on Applications of Computer Algebra,* ACA, August 2016, Kassel, Germany.

# Part IV
# Challenging Tasks with Pedagogy in Mind

# Linear Algebra—A Companion of Advancement in Mathematical Comprehension

**Damjan Kobal**

**Abstract** Linear algebra is considered a core subject with its specific cognitive and teaching challenges at the very start of university mathematics teaching. By views and experience of many students and teachers linear algebra 'defines the change of culture between secondary and university teaching'. Lots of educational research has explored productive transitions to 'higher levels of conceptualization' symbolized by linear algebra. We propose a rather different and simple perspective: linear algebra might be motivated and its basics successfully taught if presented as a tool for mastering diverse mathematical problems. Basic linear algebra concepts can be used for a smooth transitions from intuitive to abstract cognition and to deepen student's understanding. 'Scholar-teacher' can use rich linear algebra contents for guided learning through exploration and discovery. We will present a few samples of challenging mathematical problems where 'linear algebra reasoning intuitively comes to rescue' and gradually develops into a powerful and beautiful subject of its own value.

**Keywords** Intuitive · Abstract · Visualization · Linear · Geometric · Challenge

## 1 Introduction

Linear algebra is an important branch of mathematics, but, as for other subjects, it is impossible to strictly define what is and what is not linear algebra. Maybe most of the problems related to "change of 'culture' between secondary and university teaching" (Dorier, Robert, & Sierpinska, 2000, p. 276) are a consequence of the definition of linear algebra as a branch of mathematics that 'explores vector spaces and linear mappings between them'. As a study of vector spaces and linear mappings linear algebra is surely completely new and an abstract subject to students who are only used to high school (intuitive) mathematical thinking. The above definition of linear algebra is appropriate for *Mathematics Subject Classification—MSC* and

D. Kobal (✉)
Faculty of Mathematics and Physics, Department of Mathematics,
University of Ljubljana, Ljubljana, Slovenia
e-mail: damjan.kobal@fmf.uni-lj.si



S. Stewart et al. (eds.), *Challenges and Strategies in Teaching Linear Algebra*,
ICME-13 Monographs, https://doi.org/10.1007/978-3-319-66811-6_13

surely, the question of how to smartly teach these advanced subjects to students is legitimate. But a global challenges of mathematical learning should point to another direction. Namely, advanced linear algebra concepts can only be developed gradually. A teacher should not be afraid to see and present linear algebra as 'the study of line-like relationships', which only gradually evolve into a sophisticated concepts of 'vector spaces' and 'linear mappings between them'.

A lot has been written about the myths of good (and bad) teaching ('paradigms')—for ex. (Claudi, 2002, p. 3). Comprehensive views on 'good teaching' remain understandably general and talk in terms of 'scholar-teacher' (Fung & Siu, 2005, p. 3). Good teaching practice always comprises of good mathematics understanding as well as of profound sensitivity, skill and devotion for 'complex aspects of teaching'. These provide means for guided learning and motivate explorations leading to discoveries and new knowledge.

In the following samples we describe some interesting mathematical and mathematics teaching ideas which combine visual and analytical thinking, offer opportunities for motivated explorations and better comprehension, while providing cognitive means for soft transition from concreteness and intuition to abstraction. As such, the samples are not 'carefully designed teaching units' but mere backbone ideas, which a 'scholar-teacher' will implement by filling up many fine details of the specific teaching environment. Presented mathematical ideas and mathematics teaching accents are derived from empirically successful implementations through decades of teaching experience.

If the given ideas and its almost trivial connection between 'geometric meaning' and simple 'vector concepts' seem clear, it is much harder to fit these ideas within formal concepts of linear algebra teaching. And maybe this is one of the important causes for difficulties in transition to abstract (linear algebra) teaching. Namely, teachers are (or have to be) more preoccupied by formal concepts and strict following of the syllabus, then by the beauty and understanding of interconnected mathematical concepts. It is a shame, that there are mathematics students, who have passed tests of advanced linear algebra courses, but find themselves surprised and amassed by simple and meaningful connections between linear algebra concepts and simple intuitive geometric meanings (like those presented). Therefore, as we will try to suggest the use of presented ideas within formal teaching concepts, we could start with the suggestion, that it is never too late to emphasize simple and nice ideas even to students, who are already familiar with advanced linear algebra concepts. All ideas presented in this chapter of the book are similar in the sense that they show elegant, powerful and easily understood meanings of (elementary linear algebra) concepts, which, when understood, bring elegant solutions just by taking the right (or alternative) perspective. As such they could probably be best used as motivation for introduction of new (more abstract) concepts. Namely, rich concepts like for example 'length', 'direction', 'perpendicular' should first be motivated and understood through intuitive and geometrical comprehension in order to later successfully build up the abstract notion of vector space.

Presented samples could be especially useful with prospective teachers to enhance their sensitivity for comprehension and mathematical beauty.

Linear algebra with its wide content spanning from very elementary geometric insights to its deep abstract formulations is a challenge and a teaching opportunity *per se*. Contemplating about teaching linear algebra, we should very seriously consider famous Felix Klein sentence: *I believed that the whole sector of Mathematics teaching, from its very beginnings at elementary school right through to the most advanced level research, should be organized as an organic whole* (Klein, 1923).

## 2 Practitioner's Agenda

The goal of this paper is to provide some concrete ideas and attitudes to practitioners, that is to teachers of linear algebra, which have the potential to make the teaching and learning of linear algebra a challenge that motivates rather than frustrates. At the same time educational researchers could explore and compare the potential such an approach has in relation to the pedagogical questions of teaching of linear algebra and its applications.

We shall start with some very elementary problems, where only very basic (high school) linear algebra concepts are used—or maybe we should say: introduced.

### 2.1 Four Points

The task is nicely presented and illustrated by the use of dynamic geometry software (Kobal, 2016, Chap. 1.1.). Classical abstract formulation of the problem would be:

- Given any point $A$, how can we describe (in terms of point $A$) the (coordinates of) points $B$, $C$ and $D$, if $ACBD$ is a square as in Fig. 1—left?

Quite standard solution of a university student, who is familiar with vector algebra and knows some geometry, would be given by intersections of a circle and appropriate perpendicular lines, all centered at the (invisible) origin.

The 'linear algebra solution', which, by a given point $A = (x, y)$, simply defines the points $B = (-x, -y), C = (-y, x)$ and $D = (y, -x)$ is far more straightforward and conveys important content of coordinate system and vectors as well as it confirms (or motivates) the basic scalar product property (Fig. 1—right) $\vec{a} \perp \vec{b} \iff \vec{a} \cdot \vec{b} = 0$.

The idea could be used at several different levels of student's mathematical background:

- If we start with dynamic interaction using 'computer applet' (Kobal, 2016, Chap. 1.1.) students can intuitively grasp important mathematical concepts even before acquiring abstract notions of coordinate system with its origin, length, perpendicularity, vectors, ... By the help of a sensitive teacher, already a primary school pupil will be able to observe all the essential properties of 'dependent moves...' and thus

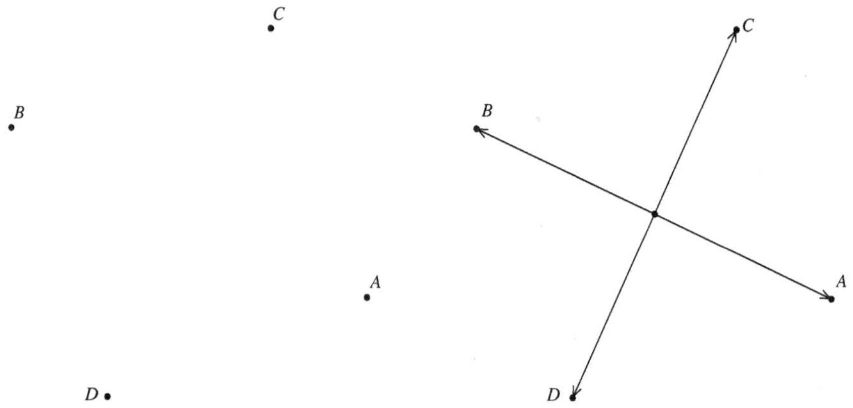


**Fig. 1** Points $A$, $C$, $B$ and $D$ are vertices of a square

build up motivation for rigorous definitions. By such dynamic explorations students gain valuable intuitive insights and comprehension of 'independent' versus 'dependent' point, which leads to the understanding of abstract notion of 'independent' versus 'dependent' variable. By visually attractive dynamic interaction an abstract problem becomes an intuitive challenge.

In such a form the problem could be used as an example to illustrate the use of $xy$-coordinate system and to better understand geometrical interpretation of the position of a point $(x, y)$ in the $(\mathbb{R}^2)$ plane. Also, the same approach could be used just before introducing the most elementary notion of 'plane vectors' in $xy$-coordinate system. In both cases, understanding the geometrical content of $x$ and $y$ coordinates shifting and/or adding 'minus', while considering a position of a point, is essential for good initial comprehension of coordinate system concept.

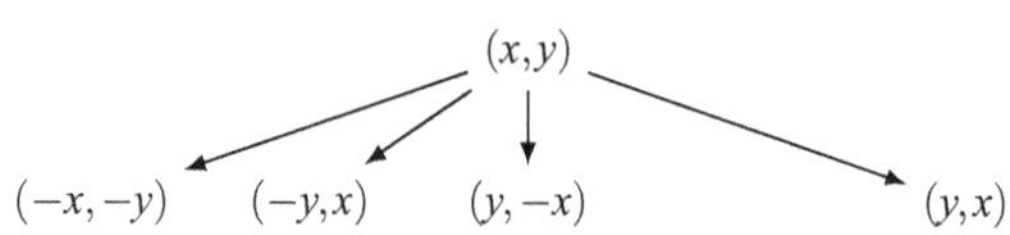


Students will gain essential understanding that coordinates have in defining 'direction', which is a prerequisite for vector comprehension.

- In many places around the globe already high school students learn the basics of 'plane and space vectors' together with its coordinate (i.e. $\mathbb{R}^2$ and $\mathbb{R}^3$ vectors) and simple 'scalar product' notion. Others do that in basic college mathematics classes. The sample could be used at this stage as a nice example and illustration of these basic vector and scalar product notions. Namely, it shows the power, that already primitive notion of high school 'plane vectors' and scalar product formula $(a, b) \cdot (c, d) = a \cdot c + b \cdot d$ holds.
  The idea also nicely illustrates the 'intuitive protocol' of constructing (two) perpendicular vectors (in $\mathbb{R}^2$) by 'shifting the two coordinates and adding one minus':

$$(x,y) \longrightarrow (y,x) \begin{array}{l} \nearrow (y,-x) \\ \searrow (-y,x) \end{array}$$

- As mentioned, the idea could also be used as almost trivial but meaningful ($\mathbb{R}^2$) illustration of vectors and its relation to ($\mathbb{R}^n$, Euclidean) coordinates and scalar product in a more advanced abstract vector space setting.
- Because of its intuitive accessibility and abstract meaning the problem is especially suitable for prospective teachers. Firstly, to help them understand these concepts themselves and secondly, to suggest and give them ideas to be used in their later teaching practice.

## 2.2 *Triangle on the Top of a Square*

The task is nicely presented by the use of dynamic geometry software (Kobal, 2016, Chap. 1.2.).

- Having a square and an equilateral triangle sitting on a top of a square (Fig. 2—left), what is the radius and where is the center of the circle going through the bottom two and the top corner of the obtained 'house' shape?

As in problem 2.1, the same idea can be formulated differently. Above formulation is as intuitive as possible. Teachers should be aware, that formal notations, as given below, too often distract weaker students. Thus, in order to develop also the necessary rigorous vocabulary, formal notations should only be developed gradually. This

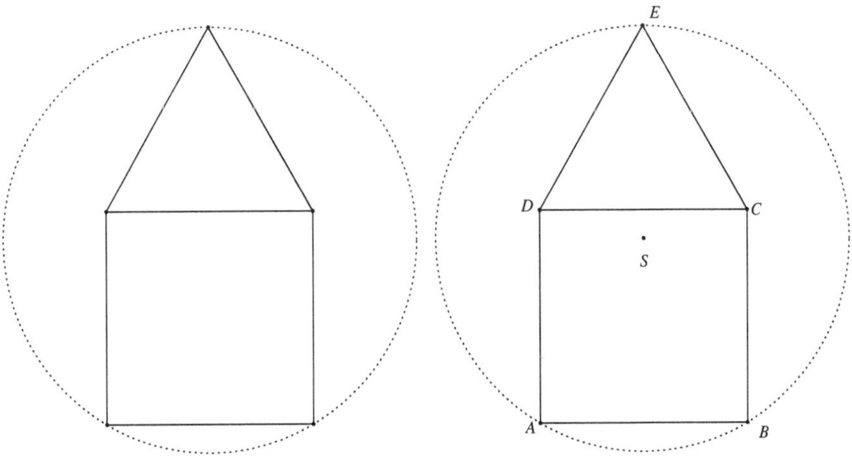


**Fig. 2** An equilateral triangle on a top of a square

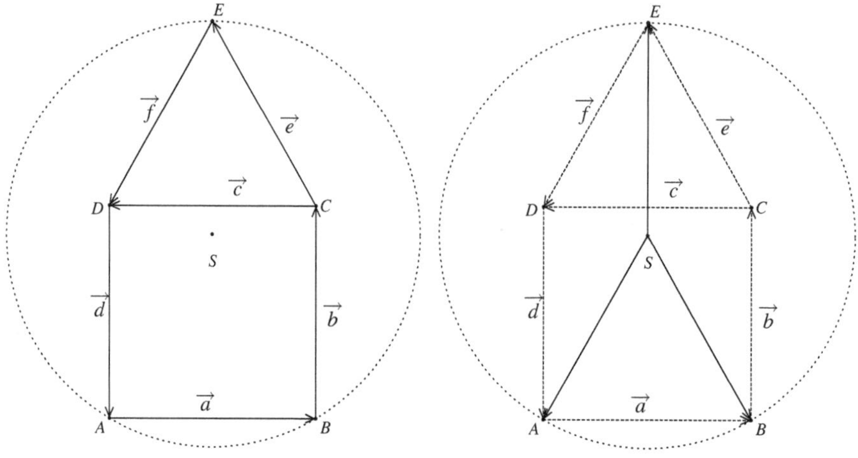


**Fig. 3** Vectors of an equilateral triangle on a top of a square

gradual transition from 'intuitive to formal' is essential for a successful transition from secondary to university teaching.

Given a square *ABCD* with a side $a$ and an equilateral triangle *DCE* sitting on the top of the given square *ABCD* (Fig. 2—right), where is the center *S* and what is the radius of the circle circumscribed to the triangle *ABE*?

Even advanced students often find complicated solutions (using trigonometry or advanced elementary geometry calculations), but very few use the most simple and intuitive insight, which give an immediate and aesthetic solution:

Focused on Figs. 2 and 3, we might ask several questions leading to the solution:

- Observing the paths between corners of the shape on Fig. 2, how many different paths (vectors) do you get?
- What can you say about the vectors? How different are they?
- What can you tell about the vectors $\overrightarrow{a}$, $\overrightarrow{b}$, $\overrightarrow{c}$, $\overrightarrow{d}$, $\overrightarrow{e}$ and $\overrightarrow{f}$ on Fig. 3—left?
- Can you draw a triangle, which you get, if triangle *DCE* is moved (pushed down) for vector $\overrightarrow{d}$?
- If *S* is a point, where *E* is moved by vector $\overrightarrow{d}$, what can you say about the vectors $\overrightarrow{SE}$, $\overrightarrow{SA}$ and $\overrightarrow{SB}$ (Fig. 3—right)?

Therefore, the circle circumscribed to *ABE* has its center in *S* and radius $a$.

The idea could be best used:

- As a challenge and task just after the (elementary) geometric notion of a vector (defined by direction and length) is introduced.
- Or, if and when we consider geometric transformations (i.e. translation) of a plane.
- Or, in a yet more intuitive setting, when students are only familiar with very basic geometric notions like 'equilateral triangle'. We can use just informal 'translation' in the form of 'push down' as visualized in dynamic applet (Kobal, 2016, Chap. 1.2.). Obtaining such an elegant and easy solution can be a good motivation to

introduce the notion of 'a push in a certain direction for a certain distance', which is a prelude to a more formal definition of a vector.

## 2.3 *Midpoints of a Quadrilateral*

The task is nicely presented by the use of dynamic geometry software (Kobal, 2016, Chap. 1.3.).

- The midpoints of the sides of a quadrilateral form a parallelogram.

True! The statement is often used and proved at different levels, but too often the problem is not adequately motivated and not the right emphasis is given. In mathematics teaching too often the proofs are given without students reaching the point of cognitive puzzlement. In this problem, as well as in many others, students should be challenged, maybe alongside tasks and questions as follows:

- Draw a quadrilateral. Draw the quadrilateral with vertices at the midpoints of the sides of your quadrilateral.
- Oh..., you drew a special one. You seem to obtain a parallelogram. Can you draw another one, so that you would get a more general quadrilateral?
- Is this a good/bad luck or ...?

After students reach 'a parallelogram hypothesis' themselves, we should lead them step by step to essential realizations of the proof. For example, we could proceede as follows: Drawing any quadrilateral *ABCD*, means choosing points (sequentially) $A$, $B$, $C$ and $D$. Choosing a point $A$ does not tell anything yet about the quadrilateral shape. In perspective, it only means we chose an initial point (the origin). After having a point $A$, choosing a point $B$ is equivalent of choosing a vector $\vec{a} = \overrightarrow{AB}$. Choosing a point $C$ is the same as choosing (for example) a vector $\vec{b} = \overrightarrow{BC}$. And finally, choosing a point $D$ is the same as choosing (for example) a vector $\vec{c} = \overrightarrow{AD}$. Therefore, a choice of any quadrilateral 'is equivalent' to a choice of three vectors (Fig. 4).

It is now a straightforward calculation to obtain $\overrightarrow{EF} = \overrightarrow{HG} = \frac{1}{2}(\vec{a} + \vec{b})$. It is a review task to also get $\overrightarrow{EH} = \overrightarrow{FG} = \frac{1}{2}(\vec{c} - \vec{a})$. But at this point students should be challenged to understand, that $\overrightarrow{EF} = \overrightarrow{HG}$, suffices for the conclusion of the quadrilateral $EFGH$ to be a parallelogram. At this initial stage of vector understanding, this is exactly what should be emphasized, namely, a 'vector' carries much more information than a 'segment'.

The idea could be best used:

- As a challenge and task just after the (elementary) geometric notion of a vector (defined by direction and length) is introduced.

## 2.4 *Regular Octagon?*

The task is nicely presented by the use of dynamic geometry software (Kobal, 2016, Chap. 1.4.).

- Draw a square and midpoints of its sides. Connect the four midpoints to its opposite corners. Is the octagon, which we obtain (see Fig. 5—left), regular?

Prove/disprove of the 'regular octagon hypothesis' is a nice example of advancement from 'intuitive/geometric linearity' to 'algebraic linearity' and can be obtained by two conceptually different but in essence similar ways of elementary 'linear algebra thinking'.

### 2.4.1 Linear Functions in a Standard Plane Coordinate System

It is a nice high school (if not primary school) task to position the given square within a standard coordinate system (for example as seen in Fig. 5—middle). The two sketched lines passing through the two pairs of points $(-2, 2)$, $(2, 0)$ and $(0, 2)$, $(2, -2)$ are obtained as a routine task. The two lines are given by equations $y = 1 - \frac{1}{2}x$ and $y = 2 - 2x$ (linear functions $f(x) = 1 - \frac{1}{2}x$ and $f(x) = 2 - 2x$) respectively. The intersection $B$ of the two lines is calculated as $(\frac{2}{3}, \frac{2}{3})$. By similar arguments we obtain $A = (1, 0)$. Since the two vertices $A$ and $B$ are not equidistant from the octagon's center of symmetry, which is obviously at the origin, the octagon is NOT regular.

The idea and such a solution could and should be used:

- As a challenge and task soon after the (elementary) concept of linear function is introduced. It could easily be done at the secondary mathematics level.

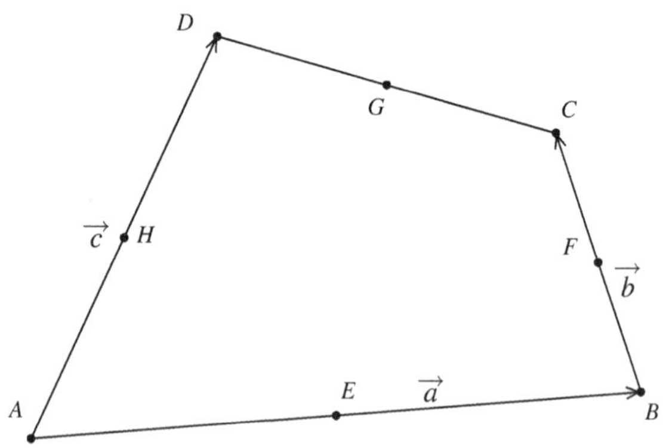


**Fig. 4** Midpoints of a quadrilateral

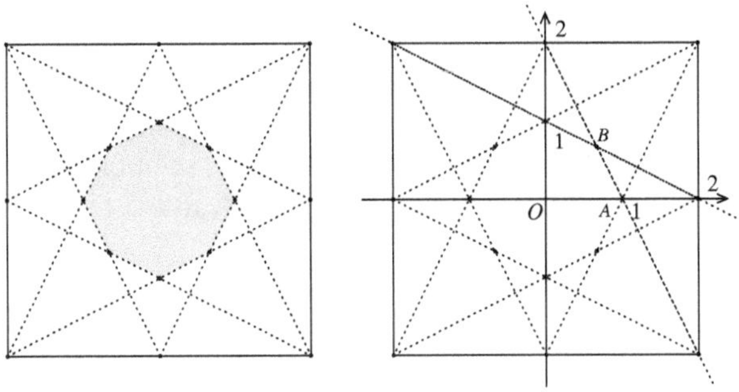


**Fig. 5** Regular octagon?

### 2.4.2 Vectors

Of course, we could proceed by the use of position vectors in a standard plane coordinate system, or, as we sketch in the next lines, by choosing two (independent) vectors $\vec{a}$ and $\vec{b}$, starting at the center $O$, as seen on Fig. 5—right.

This approach requires some very basic vector routine to define the correct linear combination of vectors associated to appropriate paths. We can reach point $A$ by going different paths from $O$. For example:

$$\lambda\vec{a} = -\vec{b} + \mu(\vec{a} + 2\vec{b})$$

By the use of linear independence of vectors $\vec{a}$ and $\vec{b}$ we find $\lambda = \mu = \frac{1}{2}$ and we know that point $A$ can be reached by $\frac{1}{2}\vec{a}$ from point $O$. Similarly we travel two different paths to reach point $B$:

$$\frac{1}{2}\vec{a} + \lambda(2\vec{b} - \vec{a}) = \vec{a} + \mu(\vec{b} - 2\vec{a})$$

Since $\vec{a}$ and $\vec{b}$ are linearly independent, we get a linear system of two equations with two unknowns:

$$\begin{aligned} -\lambda + 2\mu &= \frac{1}{2} \\ 2\lambda - \mu &= 0 \end{aligned}$$

Finding solutions $\lambda = \frac{1}{6}$ and $\mu = \frac{1}{3}$ we get $B$ as $\frac{1}{3}(\vec{a} + \vec{b})$ from $O$. Since $|\vec{a}| = |\vec{b}|$, it is a simple high school vector exercise to conclude that $|\overrightarrow{OB}| = \frac{\sqrt{2}}{3}|\overrightarrow{OA}|$. Therefore, the two octagon's vertices $A$ and $B$ are not equidistant from the center and the octagon is NOT regular.

The idea and such a solution could and should be used:

- As a challenge and task soon after we introduce 'linearly independent vectors'. This is usually done in the first year of university studies or already in more advanced secondary schools. It is useful if students are reminded of the other solution, which is based on linear function understanding. Students should be able to see, that both approaches basically describe the same 'geometric facts'.

## 3 Basics and Advanced Topics

We proceed by using quite elementary linear algebra concepts within intuitive settings of some more advanced mathematical topics.

### *3.1 Fixed Point Theorem*

It seems quite a naive and ambitious idea to try to (non-trivially) visualize concepts as advanced as *Brouwer fixed point theorem* with the most elementary mathematical knowledge. With very basic linear algebra thinking, which formally only requires the understanding of linear function $f(x) = k \cdot x + n$, one can solve a highly non-trivial challenge. The below presented problem offers motivation and visual understanding, which easily engages students of all levels and offers many interesting issues to discuss. The problem is well illustrated by dynamic geometry software (Kobal, 2016, Chap. 2.1.) and by 'hands on task' as follows:

- Take any picture or a map and its identical but sized down copy, putting the small over the big one (see Fig. 6—left).
- Is there a point where the same place on both maps coincide?
- Can you find a positioning of the small map on the big one, so that one could find such a point? Could there be more than one such point?

'Proportional vertical lines' define points on the two maps with identical 'longitudes' (Fig. 6—right). At the points, where the two 'vertical lines' intersect, the longitude ($x$-coordinate) of the two maps coincide (Fig. 7—left). Similarly, 'proportional horizontal lines' define points on the two maps with identical 'latitudes' ($y$-coordinates), and where the two 'horizontal lines' intersect, the latitudes ($y$-coordinate) of the two maps coincide. The intersection of the 'line defining equal longitudes' and 'line defining equal latitudes' gives the unique point where the same place on both maps overlap (Fig. 7—right).

If above visualizations and especially the use of dynamic geometry software (Kobal, 2016, Chap. 2.1.) is intuitively persuasive, it can be a nice and useful exercise to explicitly calculate the fixed point and to demonstrate the power of (very basic) linear algebra. In essence it is just a routine task of finding lines (linear functions)

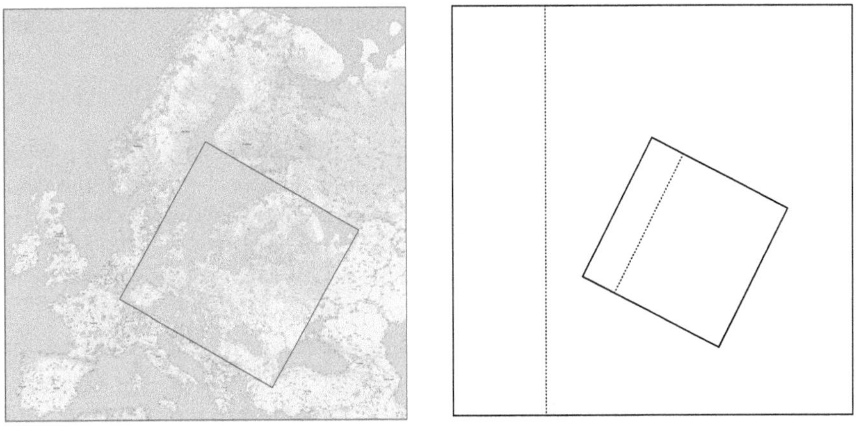

**Fig. 6** Small over a big map

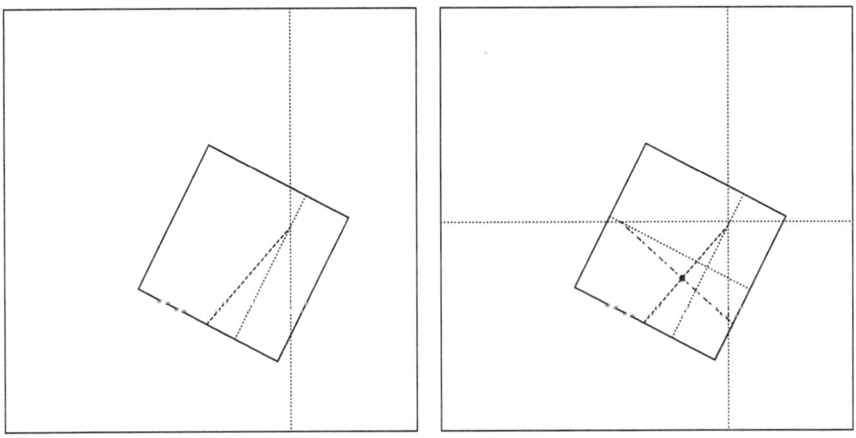

**Fig. 7** Points with equal longitudes and latitudes

defined by points and slopes and calculating their intersections. We suggest for example the following concrete calculations (within standard coordinate system).

Let the 'big square map' be defined by the three vertices $(0, 0)$, $(10, 0)$, $(10, 10)$. Find the coinciding point, if the small (scaled down copy) square map is positioned to have three points at

1. $(1, 1)$, $(5, 1)$, $(1, 5)$
2. $(2, 1)$, $(5, 2)$, $(1, 4)$
3. $(u, v)$, $(u, v) + (k, h)$, $(u, v) - (h, -k)$

As said, we just need some basic 'linear algebra' thinking that involves proportionality, coordinate system and linear functions understanding.

1. In the first case we have the line $x = a$ on the big map corresponding to the line $x = \frac{2a}{5} + 1$ on the small map. Therefore the two lines overlap for $x = \frac{5}{3}$, and since the positioning of the small map on the big one is $x - y$ symmetric, the coinciding point is $(\frac{5}{3}, \frac{5}{3})$.
2. In the second case we see that a line $x = a$ passing through point $(a, 0)$ on the big map, corresponds to the line with slope $-3$ passing through the point $(2, 1) + (\frac{3a}{10}, \frac{a}{10})$, which is $y = -3x + 7 + a$. The two lines intersect at $(a, 7 - 2a)$. Therefore, all the points on the line $y = 7 - 2x$ have the same longitude on the big and on the small map. Similarly, a line $y = b$ passing through point $(0, b)$ corresponds to the line with slope $\frac{1}{3}$ passing through the point $(2, 1) + (-\frac{b}{10}, \frac{3b}{10})$, which is $y = \frac{1}{3}x + \frac{1}{3} + \frac{b}{3}$. The two lines intersect at $(2b - 1, b)$. Therefore, all the points on the line $y = \frac{x}{2} + \frac{1}{2}$ have the same latitude on the big and on the small map. The intersection of the two lines $(\frac{26}{10}, \frac{18}{10})$ is therefore the overlapping point on both maps.
3. General case requires careful calculations with analogous argumentation.

The idea could be used:

- As a challenge and task for motivated students with very little formal mathematical knowledge. Namely, above formal expressions like 'longitude' and/or $x$-coordinate could be exchanged with very intuitive descriptions like 'how far to the East'. And above arguments, especially if illustrated by dynamic applet (Kobal, 2016, 2.1.) are very persuasive and basically use no formal mathematical knowledge. As such they might be a good motivational tool to introduce 'coordinates' and 'lines', which provide means for calculating exact position of the 'fixed point'.
- As a challenge and task for motivated students already familiar with 'linear function'. For them the above described solution would be a good use of the new linear function concept, giving them interesting application of 'formal linear (function) equations'. Maybe we should emphasize, that linear functions (see also Sect. 2.4) are usually not classified as 'linear algebra chapter'. Nevertheless, linear function and especially the presented use of it, surely belongs to 'the study of line-like relationships', which are the essence of linear algebra.
- As an illustration of more advanced linear algebra concepts. Positioning the origin into the center of the (big) square, we could define the appropriate map by rotation composed by scalar multiplication and translation. Finding the fixed point, would then mean finding the fixed point of the obtained composition (affine) map.

In a way, the third approach would be a simplification of the second one. But sticking to the main emphasis of our chapter, the first two given approaches are more intuitive and build up on geometric understanding. As such they present a desired transition from elementary geometric and intuitive to more abstract linear algebra concepts.

### 3.2 An Introduction to Projective Spaces

The idea of a projective plane and its intuitive visualization of extended $z = 1$ plane within three dimensional space $\mathbb{R}^3$ is a wonderful (linear algebra) subject to introduce students to more advanced mathematical content. At the same time students acquire deeper understanding of basic linear algebra concepts.

Let us sketch the main cognitive steps in building up a well motivated (and well known) intuitive visualization of a projective plane.

- Simple discussion about the visualization of a plane $z = 1$ in regular $\mathbb{R}^3$ space.
- Every point in the plane $z = 1$ can be associated with the line including the point and the $\mathbb{R}^3$ origin.
- The concept of (one dimensional) vector (sub)space is over-viewed and associated with a line corresponding to a point.
- Avoiding 'horizontal vectors' $(x, y, 0)$, every one dimensional subspace $\mathcal{L}\{(x, y, z)\}$ gives a unique point in our plane $z = 1$.
- Every point in a plane $z = 1$ can be represented by one dimensional subspace. Which one? (Uniformity!)
- Every point in a plane $z = 1$ can be represented by a vector in $\mathbb{R}^3$. Which one? (Not uniform choice!)
- Every one dimensional subspace in $\mathbb{R}^3$ represents a unique point in 'extended plane $z = 1$'. 'Directions'. i.e. $\mathcal{L}\{(x, y, 0)\}$ with no intersection with plane $z = 1$ are added as 'intuitively self-explanatory points in given directions' at infinity.
- Points in 'extended plane $z = 1$' (projective plane) are given by homogeneous coordinates $(x, y, z)$.

Such an intuitive visualization of a concept can surely be a nice introduction to a robust and abstract definition of a 'projective space' as ' ... the space of one-dimensional vector subspaces of a given vector space'. As in other given examples, such a discussion could lead to enhancement of 'spacial orientation' and improve the understanding of elementary $\mathbb{R}^3$ linear algebra concepts (points, lines, planes), while providing means to characterize previously only intuitive points 'at the edge of a plane'—as direction. We believe such intuitive and inter-connecting ideas are inspiring for students and are essential for good (teaching of) mathematics. It is to make students say: *J'ai compris, c'est ça la mathématique: c'est d'imaginer!* (Now I understand, that is the real mathematics: to imagine!). Such was a sigh of a 14-year old student understanding the beauty and the power of well presented concept of barycentric coordinates and its application, within Emma Castelnuovo presentation of her teaching at ICME-1 in Lyon in 1969 (Castelnuovo, 1969, p. 192).

Therefore, the idea could be used:

- As a project-exercise after students are familiar with basic $\mathbb{R}^3$ (vector) linear algebra. It could be done even in a very elementary level without using the terminology of 'one dimensional vector spaces' (using just vectors/lines—passing through the origin).

- 'Projective spaces' as otherwise classical chapter of linear algebra is usually introduced later in the university studies within more advanced mathematics classes or for mathematics majors. Presented idea is a very useful exercise and motivation to bring students from elementary and intuitive $\mathbb{R}^3$ linear algebra to the more abstract subject of projective spaces.
- The idea could be used in combination with 'barycentric' and/or 'trilinear' coordinates (discussed in the following).

### 3.3 *Barycentric Coordinates*

It is a wonderful 'abstract visualization' challenge to introduce the idea of a triangle (equilateral for start) as a form of coordinate system in a sense of carefully and intuitively introduced barycentric coordinates. The concept we need, is just (an extension of) 'linearity'. The readers are warmly invited to carefully read the above mentioned presentation of Emma Castelnuovo (Castelnuovo, 1969). The presentation is an inspirational proof of how a charismatic teacher with a profound understanding of mathematical subject ('scholar-teacher') can bring average students to unimaginably deep understanding and what is even more important, can make students enjoy mathematics. The barycentric coordinates, even if proved by Emma Castelnuovo to be accessible already to 14-years old students, might be an extensive chapter of study with many interesting 'synapses' to other linear algebra and mathematical 'neurons'.

Let us just state the very intuitive definition of barycentric coordinates and some very basic but beautiful ideas, which, when presented to students, should be introduced gradually and with great sensitivity. Starting with a given triangle as the basis of orientation, an (ordered) triple of numbers represents the point in a plane (of a triangle) defined as a 'center of mass', where the three 'masses/numbers are placed to the vertices'... It is very intuitive to understand, that the new coordinates are 'homogeneous'. Intuitive meaning of so described barycentric coordinates has also very easy to understand geometric description. Starting with an equilateral triangle and any point within a triangle (this notion can of course be extended to any triangle and to points 'outside' of a triangle), we draw parallel lines to the triangle sides, passing through the given point. We obtain three 'smaller equilateral triangles', which we denote by $p$, $q$ and $r$ (which shall mean lengths of sides of respective triangles) and three parallelograms as seen on Fig. 8—left.

Putting some notation on our figure and drawing a segment from a vertex $A$ over our point $P$ to the opposite side as seen on Fig. 8—middle, we get some very important conclusions. Sides of a triangle are nicely 'decomposed' as $AB = q + r + p$, $BC = r + p + q$ and $CA = p + q + r$ and by using similarity, we easily conclude that $\frac{BD}{DC} = \frac{r}{q}$ and $\frac{AT}{TD} = \frac{q+r}{p}$. Therefore, point $D$ is a center of gravity of masses $q$ and $r$ positioned in vertices $B$ and $C$ respectively, and our point $T$ is the center of gravity of masses $p$, $q$ and $r$ positioned in vertices $A$, $B$ and $C$ respectively.

Therefore, considering the obvious nature of 'homogeneity' of barycentric coordinates, and placing any (positive) values $p$, $q$ and $r$, for which $p + q + r = 1$, at

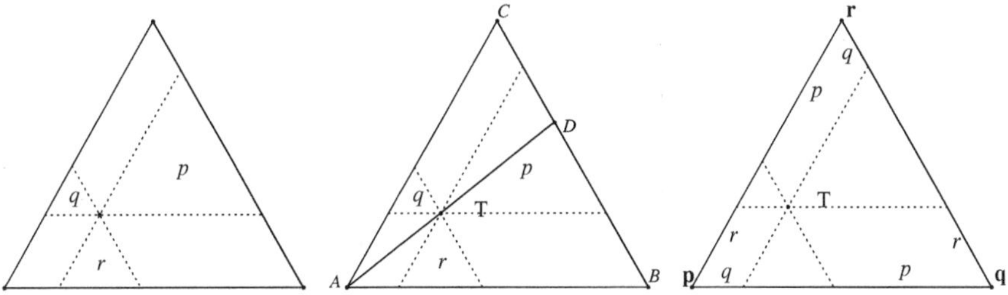


**Fig. 8** Geometric visualization of barycentric coordinates

the vertices of an equilateral triangle with a side of length 1, we see, that the center of gravity is given by point $P$ with barycentric coordinates $(p, q, r)$ and with a very intuitive position in a triangle, as seen on Fig. 8—right.

This nice notion of barycentric coordinates which is based solely on a very simple (primary school) 'linearity' and/or 'proportionality', has a wonderful 'extension' into the 'high school level linearity'.

The idea could be used:

- As an interesting problem-exercise for smart kids. Have only to be familiar with 'basic triangle geometry'. Pupils use 'linear algebra' thinking before they know that 'linear algebra' exists.
- As an interesting problem-exercise for high school students. Gives an alternative understanding of a concept of 'coordinate system', where the position of a point is determined relatively to a given triangle. Students use 'linear algebra' thinking without knowing it.
- As a simple and intuitive example of otherwise abstract concept of projective coordinates/spaces, for advanced university students.

### 3.4 *Trilinear Coordinates*

Focusing our attention on Fig. 8—right, it is quite straightforward to notice, that a position of a point $P$ within our triangle could also be described as a position in a triangle, of which the distances to (appropriate) sides of a triangle are in the ratio of $(p, q, r)$. Therefore, the position within a triangle could be given by the distances from the triangle's sides. It is obvious, that only two such distances suffice, and that the same information can be given by three homogeneous coordinates.

But don't these coordinates of points in plane look like 'projective coordinates' of a projective plane discussed in Sect. 3.2? Might this construction of coordinates by the use of a triangle in fact give some kind of 'projective plane coordinates'? If so, how could these $(x, y, z)$ coordinates (giving the ratio of distances to triangle's sides) be connected to one dimensional vector spaces $\mathscr{L}\{(x, y, z)\}$ spanned over $\mathbb{R}^3$ vectors?

Answers to these questions are quite easy and give rise to the use of (slightly more advanced) vector part of linear algebra. As mentioned for barycentric coordinates, it is not too difficult to do it in general for any triangle. In the case of trilinear coordinates it is even more natural and easy to understand, that a position in a plane is well defined, if we give its two distances to the appropriate triangle's sides (or lines containing them). It is also more natural to talk about negative distances, where positive distance from a line is defined by the 'triangle's side'. And as we will see, the proof that 'trilinear coordinates', given by homogeneous distances from lines containing triangle's sides (i.e. the triple $(x, y, z)$ give the ratio of the distances), are in fact projective coordinates in a 'projective plane', is an inspiring and easy exercise in basic vector linear algebra.

We start with any triangle $\Delta ABC$ with standard notation: $A, B, C$ are the vertices and $a, b, c$ are triangle's (opposite) sides. With $v_a$, $v_b$ and $v_c$ we denote the respective heights. Let $T$ be any point within our triangle.

Drawing lines parallel to the sides of a triangle, passing through point $T$, we obtain three triangles, which are similar to triangle $\Delta ABC$ (Fig. 9—left):

$$\Delta ABC \sim \Delta C_1C_2T \sim \Delta B_2TB_1 \sim \Delta TA_1A_2$$

If we denote the distances of a point $T$ from respective triangle's sides by $d_a$, $d_b$ and $d_c$, it is easy to use similarity to obtain the following equalities:

$$\overline{AC_1} = \frac{c \cdot d_b}{v_b}; \quad \overline{C_1C_2} = \frac{c \cdot d_c}{v_c}; \quad \overline{C_2B} = \frac{c \cdot d_a}{v_a} \quad \text{and} \quad \overline{C_1T} = \frac{b \cdot d_c}{v_c} \tag{1}$$

which are equivalent to

$$\frac{\overline{AC_1}}{c} = \frac{d_b}{v_b}; \quad \frac{\overline{C_1C_2}}{c} = \frac{d_c}{v_c}; \quad \frac{\overline{C_2B}}{c} = \frac{d_a}{v_a} \quad \text{and} \quad \frac{\overline{C_1T}}{b} = \frac{d_c}{v_c} \tag{2}$$

Of course, by $\overline{XY}$ we denote the length of a respective segment $XY$. Since $c = \overline{AC_1} + \overline{C_1C_2} + \overline{C_2B}$ and by our equations (1) we have

$$1 = \frac{d_a}{v_a} + \frac{d_b}{v_b} + \frac{d_c}{v_c} \tag{3}$$

Now let us place our triangle $\Delta ABC$ on a plane within our regular $\mathbb{R}^3$ space. Imagine the heights $v_a$, $v_b$ and $v_c$ are 'rigidly dropped down' below the triangle $\Delta ABC$ plane as straight links, but flexibly attached at the respective vertices of the triangle. Conclude, that there is a unique point/node in $\mathbb{R}^3$ space below our plane where the three 'hanging links' can be joined. We name this point $O$ (for origin). Switching to the basic vector concept we can now say that vectors $\overrightarrow{OA}$, $\overrightarrow{OB}$ and $\overrightarrow{OC}$ are uniquely defined and are of respective lengths $v_a$, $v_b$ and $v_c$.

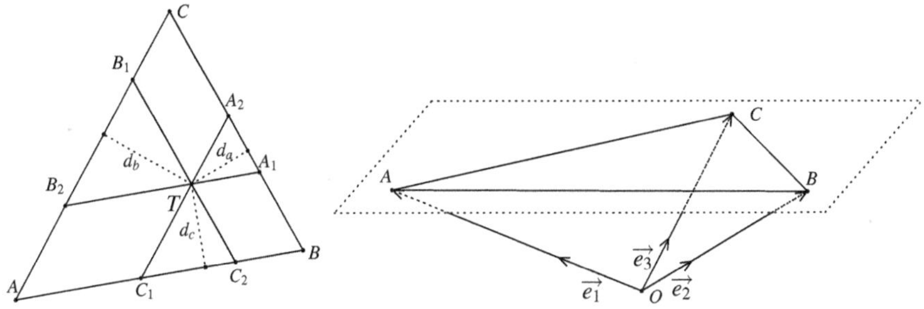


**Fig. 9** Trilinear coordinates as projective plane coordinates

If we define (Fig. 9—right) $\overrightarrow{e_1}$, $\overrightarrow{e_2}$ and $\overrightarrow{e_3}$ to be the unit vectors in respective directions by $\overrightarrow{OA} = v_a \cdot \overrightarrow{e_1}$, $\overrightarrow{OB} = v_b \cdot \overrightarrow{e_2}$ and $\overrightarrow{OC} = v_c \cdot \overrightarrow{e_3}$, we have a basis $\{\overrightarrow{e_1}, \overrightarrow{e_2}, \overrightarrow{e_3}\}$ for $\mathbb{R}^3$ vector space (with $O$ as origin).

Now let us express the point $T$, i.e. the vector $\overrightarrow{OT}$ with basis $\{\overrightarrow{e_1}, \overrightarrow{e_2}, \overrightarrow{e_3}\}$.

$$\begin{aligned}
\overrightarrow{OT} &= \overrightarrow{OA} + \overline{AC_1} \cdot \frac{\left(\overrightarrow{OB} - \overrightarrow{OA}\right)}{c} + \overline{C_1T} \cdot \frac{\left(\overrightarrow{OC} - \overrightarrow{OA}\right)}{b} \\
&= \left(1 - \frac{\overline{AC_1}}{c} - \frac{\overline{C_1T}}{b}\right) \cdot \overrightarrow{OA} + \frac{\overline{AC_1}}{c} \cdot \overrightarrow{OB} + \frac{\overline{C_1T}}{b} \cdot \overrightarrow{OC} \\
&= \left(1 - \frac{d_b}{v_b} - \frac{d_c}{v_c}\right) \cdot \overrightarrow{OA} + \frac{d_b}{v_b} \cdot \overrightarrow{OB} + \frac{d_c}{v_c} \cdot \overrightarrow{OC} \\
&= \frac{d_a}{v_a} \cdot \overrightarrow{OA} + \frac{d_b}{v_b} \cdot \overrightarrow{OB} + \frac{d_c}{v_c} \cdot \overrightarrow{OC} \\
&= \frac{d_a}{v_a} \cdot v_a \cdot \overrightarrow{e_1} + \frac{d_b}{v_b} \cdot v_b \cdot \overrightarrow{e_2} + \frac{d_c}{v_c} \cdot v_c \cdot \overrightarrow{e_3} \\
&= d_a \cdot \overrightarrow{e_1} + d_b \cdot \overrightarrow{e_2} + d_c \cdot \overrightarrow{e_3}
\end{aligned}$$

In rows 3. and 4. we used equations (2) and (3). Obviously, the point $T$, given by trilinear coordinates $(d_a, d_b, d_c)$, is the the same as the point given by projective plane coordinates $(d_a, d_b, d_c)$ rising from $\mathscr{L}\{(d_a, d_b, d_c)\}$ (over the basis $\{\overrightarrow{e_1}, \overrightarrow{e_2}, \overrightarrow{e_3}\}$).

**Note**: Point $T$ could well be outside the given triangle. In such case some distances of $T$ to the lines containing triangle sides would be negative.

The idea could be used:

- As an interesting problem-exercise for smart kids. We could avoid 'sophistication' of projective spaces and vector algebra and just intuitively introduce 'orientation in a plane relative to a given triangle', where points are determined by distances to the

given triangle's sides. 'Homogeneous coordinates' could be introduced intuitively as a triple determining the ratio of appropriate distances from triangle's sides.
- The idea could be best used with students already familiar with projective space, or at least with students understanding projective plane and the concept of homogeneous coordinates. Otherwise, the presented proof that trilinear coordinates are just a special case of (projective) homogeneous coordinates is only a nice practical exercise, which will enhance students understanding of elementary $\mathbb{R}^3$ vector algebra.

### 3.5 *Barycentric Versus Trilinear Coordinates and Applications*

We conclude the discussion about 'projective/barycentric/trilinear' coordinates by rather simple observation, that in the case of equilateral triangle barycentric and trilinear coordinates are the same and by giving two challenging (application) problems with inspiringly beautiful solutions. These might be an encouragement for teachers and students. Namely, sophisticated problems are solved by smart use of simple but contentwise sophisticated linear algebra thinking. This was wonderfully illustrated by already mentioned Emma Castelnuovo presentation (Castelnuovo, 1969).

**Challenge 1**: A fragile glass stick, when dropped, randomly breaks into (exactly) three pieces. If many such sticks are dropped, and of each stick's three pieces we try to form a triangle, of how many a triangle could be formed? One quarter? One third? One half? Of more than half? (In high school or college vocabulary: What is the probability that a stick breaks into three pieces that could form a triangle?)

It is an enlightening exercise to see, that 'different breaks' of a stick are well presented by trilinear (or barycentric) coordinates of a triangle, and that exactly points with all three coordinates smaller than $\frac{1}{2}$ (within an equilateral triangle of height 1) represent 'breaks' that allow triangle forming. Therefore only points within a triangle with vertices at the midpoints of the original triangle's sides are acceptable and we conclude that we will be able to form a triangle only 'every fourth time'.

**Challenge 2**: We have three (unmeasured) containers. The first, which is full of water, of 12 $l$, the second of 9 $l$ and the third of 5 $l$ are empty. Could (without any additional containers) the 12 $l$ of water be split on half?

The problem is elegantly described and solved by trilinear coordinates. We start with an equilateral triangle with a side 12 (total quantity of water) and with sketched 'coordinate lines' (Fig. 10—left). The distance from 'C-12' side denotes the quantity of water in the 12 $l$ container. The distance from 'C-9' and 'C-5' sides respectively denote the quantities of water in 9 $l$ and 5 $l$ containers. For example, the top vertex of our triangle has coordinates $(12, 0, 0)$ and tells us, that all the water is in the 'C-12' container. Since the other two containers only hold 9 $l$ and 5 $l$, we get the 'permissible region' sketched with pentagon on Fig. 10—left. Since the containers are not measured, we can control the moves within the pentagon only by 'hitting

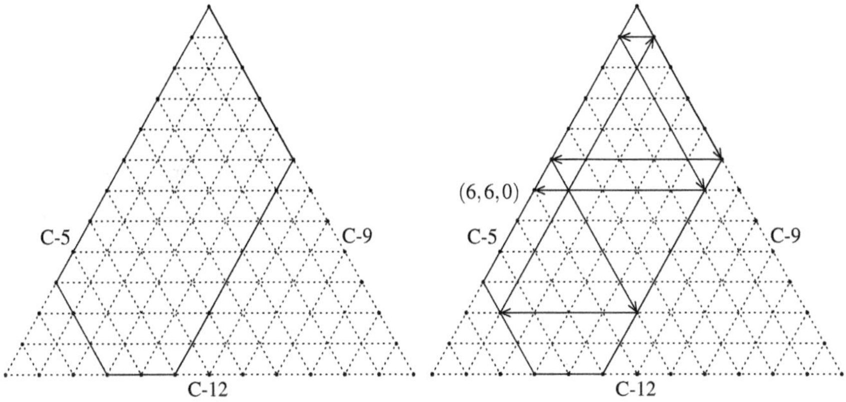


**Fig. 10** Quantities in three containers represented by trilinear coordinates

the edge of the pentagon', which means 'filling up' one of the containers. From the starting point $(12, 0, 0)$ we can only move to points $(3, 9, 0)$ or to $(7, 0, 5)$, which means we fill up the second or the third container respectively.

Moving in direction 'top down-right' means pouring from 'C-12' to 'C-5' container, moving in direction 'top down-left' means pouring from 'C-12' to 'C-9' container and moving in direction 'left-right' means pouring from 'C-9' to 'C-5' container. The solution to the problem is therefore given by sequence of eight pouring indicated by vectors on Fig. 10—right: $(12, 0, 0)$ to $(7, 0, 5)$ to $(7, 5, 0)$ to $(2, 5, 5)$ to $(2, 9, 1)$ to $(11, 0, 1)$ to $(11, 1, 0)$ to $(6, 1, 5)$ to $(6, 6, 0)$.

The ideas could be used:

- As interesting problems-exercises for smart kids, who only have to be familiar with 'basic triangle geometry'. If barycentric and trilinear coordinates were introduced intuitively, as mentioned above, it is a revealing recognition, that in the case of equilateral triangle the two coordinates yield the same 'orientation in a triangle's plane'. And the two challenges provide an appealing use of the concept.
- As examples for students already familiar with projective spaces, showing a nice elementary and intuitive use of the concept.

## 4 Conclusion

Teaching mathematics has always been about good presentation of good ideas. Linear algebra offers many wonderful and smart mathematical ideas, which combine visual and analytical thinking, which offer a smooth transition from concreteness to abstraction and which appear on crossroads of all mathematical fields. This is a great challenge and a valuable opportunity, which a devoted teacher should not miss.

## References

Castelnuovo, E. (1969). Différentes Représentations Utilisant la Notion de Barycentre. In T. E. B. of Educational Studies in Mathematics (Ed.), Proceedings of the First International Congress on Mathematical Education, Lyon, 24–30 August 1969 (pp. 175–200).
Claudi, A. (2002). Why the Professor Must be a Stimulating Teacher: Towards a New Paradigm of Teaching Mathematics at University Level. In H. Derek (Ed.), The Teaching and Learning of Mathematics at University Level (pp. 3–12). Springer, Netherlands.
Dorier, J., Robert, A., & Sierpinska, A. (2000). Conclusion. In J. Dorier (Ed.), On the Teaching of Linear Algebra (pp. 273–276). Springer, Netherlands.
Fung, C., & Siu, M. (2005). Mathematics in Teaching and Teaching of Mathematics [Private correspondence]. Fourth Asian Mathematical Conference, Singapore.
Klein, F. (1923). Retrieved 24 October 2017, from http://www.icmihistory.unito.it/
Kobal, D. (2016). Basic Linear Algebra Samples—Visualisations; GeoGebra book. Retrieved 24 October 2017, from https://www.geogebra.org/m/CTXPCC5x

# A Computational Approach to Systems of Linear Equations

**Franz Pauer**

**Abstract** We present a purely algebraic/computational approach to systems of linear equations. It requires only a little previous knowledge. We define a system of linear equations as a task, discuss how to specify the input and output data and elucidate the basic ideas to solve the system. Geometric interpretations of systems of linear equations (for the case of two or three unknowns) are postponed to the last section. Firstly, because precise geometric reasoning is not simple but quite demanding, secondly, most applications need systems of linear equations with more than three variables.

**Keywords** System of linear equations · Elementary transformation Echelon form Gaussian elimination · Implicit form of a line

## 1 Introduction

The main topic of a basic linear algebra course is systems of linear equations. Despite several advantages of geometric reasoning in the plane or in the space, there are two objections to the geometric approach:

- Interesting applications in science, engineering, and economics need systems of linear equations with (much) more than 3 unknowns (e.g. electric circuits, interpolation by polynomials of degree $\geq 3$, linear optimization, etc.).
- Geometry might appear simple, but precise geometric reasoning is quite demanding.

We propose an algebraic/computational approach to systems of linear equations which requires only a little previous knowledge. We define a system of linear equations as a task, discuss how to specify the input and output data and elucidate

F. Pauer (✉)
Institut für Mathematik, University of Innsbruck, Technikerstr. 13,
A-6020 Innsbruck, Austria
e-mail: franz.pauer@uibk.ac.at

F. Pauer
Institut für Fachdidaktik, University of Innsbruck, Innrain 52f,
A-6020 Innsbruck, Austria

 
S. Stewart et al. (eds.), *Challenges and Strategies in Teaching Linear Algebra*,
ICME-13 Monographs, https://doi.org/10.1007/978-3-319-66811-6_14

the basic ideas to solve the system. The question of how we can describe an infinite solution set by a finite set of data leads naturally to the consideration of other central topics of linear algebra: vector spaces, linear combination, bases, dimension,. . .. These results can subsequently be used for a geometrical interpretation of systems of linear equations.

The theory of (systems of) linear equations is particularly suitable to introduce algorithmic thinking. First the problem must be defined, this includes the specification of the input and output data. Then the basic strategy to solve the problem has to be exhibited and the structure of the algorithm must be explained. Finally the details of the algorithm should be presented.

In Sect. 2 systems of linear equations with at most 2 unknowns are defined and solved. For this section no particular previous knowledge is necessary. In Sect. 3 we define general systems of linear equations and introduce their matrix form (which was already used in Chinese mathematics more than 2100 years ago, see for example Gabriel (1996), Chapter A.2). Then we show by two examples that linear equations (with more than 3 unknowns) are very useful in different fields of application of mathematics. In Sect. 5 we indicate how to describe an infinite solution set by finitely many data. In Sects. 6–9 the ideas of how to solve systems of linear equations are described. The basic idea is the strategy of equivalence transformations. Then we look for tools to transform a system without changing its solution set. After having identified a sufficiently wide class of systems of linear equations which can be solved without a single calculation we explain how an arbitrary system can be transformed to an equivalent one in this class. In the last section we give a geometric interpretation of systems of linear equations and their solution sets.

I have used this approach for many years in a course of linear algebra for first-year students of computer science, physics, and mathematics (including prospective teachers), see Pauer (2007). By this contribution I want to share with the readers my experience of teaching.

## 2 Prelude: Linear Equations in High School

A *linear equation with one unknown* is the following task: Given are (real) numbers $a$ and $c$. Find all numbers $x$ such that $a \cdot x = c$. Such a number is a *solution* of the linear equation.

If $a = 0$ and $c = 0$ then all numbers are solutions.

If $a = 0$ and $c \neq 0$ then there are no solutions.

Now we assume that $a \neq 0$. If two numbers are equal, then so are their products with the same number. Hence $a \cdot x = c$ iff $x = a^{-1} \cdot a \cdot x = a^{-1} \cdot c$. Thus $a^{-1} \cdot c$ is the unique solution.

A *linear equation with two unknowns* is the following task: Given are numbers $a$, $b$ and $c$. Find all pairs of numbers $(x, y)$ such that $a \cdot x + b \cdot y = c$. Such a pair of numbers is a *solution* of the linear equation. The equation is homogeneous iff $c = 0$.

If $a = 0,\ b = 0$ and $c = 0$ then all pairs of numbers are solutions.
If $a = 0,\ b = 0$ and $c \neq 0$ then there are no solutions.
Now we assume that $a \neq 0$ or $b \neq 0$.
First we consider the homogeneous equation given by $a$ and $b$. Since $a \cdot b + b \cdot (-a) = 0$ the pair $(b, -a)$ is a solution. For all numbers $t$ we have

$$a \cdot (t \cdot b) + b \cdot (t \cdot (-a)) = t \cdot (a \cdot b + b \cdot (-a)) = t \cdot 0 = 0$$

hence all pairs $(t \cdot b, t \cdot (-a)) = t \cdot (b, -a)$ are solutions too.

On the other hand, if $(x, y)$ is a solution and $a \neq 0$ resp. $b \neq 0$ then $(x, y) = -(y/a) \cdot (b, -a)$ resp. $(x, y) = (x/b) \cdot (b, -a)$. Hence

$$\{\, t \cdot (b, -a) \mid t \in \mathbb{R} \,\}$$

is the set of all solutions of the homogeneous equation.

Note that the set of all solutions is infinite but we can describe it by a single pair of numbers. Note also that we can solve the equation without a single calculation.

If the equation is not homogeneous we observe that

- $a \cdot x + b \cdot y = c$ and $a \cdot x' + b \cdot y' = c$ imply $a \cdot (x' - x) + b \cdot (y' - y) = 0$. The difference of two solutions is a solution of the corresponding homogeneous equation.
- $a \cdot x + b \cdot y = c$ and $a \cdot u + b \cdot v = 0$ imply $a \cdot (x + u) + b \cdot (y + v) = c$. The sum of a solution and a solution of the corresponding homogeneous equation is again a solution.

Therefore if $(x, y)$ is a solution then any other solution $(x', y')$ can be written as a (componentwise) sum of $(x, y)$ and a solution of the corresponding homogeneous equation: $(x', y') = (x, y) + (x' - x, y' - y)$.

If $a \neq 0$ then obviously $(c/a, 0)$ is a solution. If $b \neq 0$ then obviously $(0, c/b)$ is a solution. Therefore the set of all solutions is

$$\{\, (r, s) + t \cdot (b, -a) \mid t \in \mathbb{R} \,\},$$

where $(r, s)$ is any solution of the equation, for example $(c/a, 0)$ or $(0, c/b)$. Note that the set of all solutions is infinite but we can describe it by only two pairs of numbers: one arbitrarily chosen solution and one non-zero solution of the corresponding homogeneous equation. Note also that we needed only one calculation to solve the equation, namely the division $c/a$ or $c/b$.

*Example* Find all pairs of numbers $(x, y)$ such that $1.28 \cdot x + 4.17 \cdot y = 1.97$! Since $1.97/1.28 = 1.5390625$ the set of all solutions is

$$\{\, (1.5390625, 0) + t \cdot (4.17, -1.28) \mid t \in \mathbb{R} \,\}.$$

A *system of two linear equations in two unknowns* is the following task: Given are numbers $a_1, a_2, b_1, b_2$ and $c_1, c_2$. Find all pairs of numbers $(x, y)$ such that $a_1 \cdot x + b_1 \cdot y = c_1$ *and* $a_2 \cdot x + b_2 \cdot y = c_2$. Such a pair of numbers is a *solution* of the system of linear equations.

The system is homogeneous iff $c_1 = 0$ and $c_2 = 0$.

We shortly say: Solve the system

$$\begin{aligned} a_1 \cdot x + b_1 \cdot y &= c_1 \\ a_2 \cdot x + b_2 \cdot y &= c_2. \end{aligned}$$

We say that two systems of linear equations are *equivalent* iff they have the same set of solutions. It is easy to verify that the following systems are equivalent to the system above:

- $$\begin{aligned} a_2 \cdot x + b_2 \cdot y &= c_2 \\ a_1 \cdot x + b_1 \cdot y &= c_1 \end{aligned}$$

  "We can swap the two equations."
- Let $s, t$ be non-zero numbers.

  $$\begin{aligned} s \cdot a_1 \cdot x + s \cdot b_1 \cdot y &= s \cdot c_1 \\ t \cdot a_2 \cdot x + t \cdot b_2 \cdot y &= t \cdot c_2 \end{aligned}$$

  "We can multiply the equations by non-zero numbers."
- $$\begin{aligned} a_1 \cdot x + b_1 \cdot y &= c_1 \\ (a_2 \pm a_1) \cdot x + (b_2 \pm b_1) \cdot y &= c_2 \pm c_1 \end{aligned}$$

  "We can add (subtract) one equation to (from) the other one."

It is easy to verify that by performing a sequence of these actions (swap two equations, multiply one equation by a non-zero number, add (subtract) one equation to (from) the other one) we obtain an equivalent system of one of the following types:

- $$\begin{aligned} a'_1 \cdot x + b'_1 \cdot y &= c'_1 \\ 0 \cdot x + 0 \cdot y &= 1 \end{aligned}$$

  In this case there are no solutions.
- $$\begin{aligned} 1 \cdot x + 0 \cdot y &= c'_1 \\ 0 \cdot x + 1 \cdot y &= c'_2 \end{aligned}$$

In this case there is exactly one solution, namely $(c_1', c_2')$.

•

$$\begin{aligned} a_1' \cdot x + b_1' \cdot y &= c_1' \\ 0 \cdot x + 0 \cdot y &= 0 \end{aligned}$$

In this case there are infinitely many solutions, namely all solutions of the linear equation with two unknowns given by $a_1'$, $b_1'$ and $c_1'$.

Note that the way to reach one of this types is not unique. The order in which the above-mentioned actions are performed depends on the given system and on the taste of the person solving the system of equations.

*Example* We transform the system

$$\begin{aligned} 2 \cdot x + 3 \cdot y &= 4 \\ 5 \cdot x + 6 \cdot y &= 7 \end{aligned}$$

successively to

$$\begin{aligned} 4 \cdot x + 6 \cdot y &= 8 \\ 5 \cdot x + 6 \cdot y &= 7 \end{aligned}, \quad \begin{aligned} 4 \cdot x + 6 \cdot y &= 8 \\ 1 \cdot x + 0 \cdot y &= -1 \end{aligned}, \quad \begin{aligned} 1 \cdot x + \tfrac{3}{2} \cdot y &= 2 \\ 1 \cdot x + 0 \cdot y &= -1 \end{aligned},$$

$$\begin{aligned} 0 \cdot x + \tfrac{3}{2} \cdot y &= 3 \\ 1 \cdot x + 0 \cdot y &= -1 \end{aligned}, \quad \begin{aligned} 0 \cdot x + 1 \cdot y &= 2 \\ 1 \cdot x + 0 \cdot y &= -1 \end{aligned}, \quad \text{and} \quad \begin{aligned} 1 \cdot x + 0 \cdot y &= -1 \\ 0 \cdot x + 1 \cdot y &= 2 \end{aligned}.$$

Hence all these systems have only one solution, namely $(-1, 2)$.

## 3 What is a System of Linear Equations?

A system of linear equations is the follwing task: Given are numbers $A_{ij}$ and $b_i$ with indices $1 \le i \le m$ and $1 \le j \le n$. Search for a "good description" of the set of all $n$-tuples $(x_1, \ldots, x_n)$ of numbers fulfilling the constraints

$$\begin{aligned} A_{11}x_1 + A_{12}x_2 + \cdots + A_{1n}x_n &= b_1 \\ A_{21}x_1 + A_{22}x_2 + \cdots + A_{2n}x_n &= b_2 \\ \vdots \qquad\qquad &\ \vdots \ \ \vdots \\ A_{m1}x_1 + A_{m2}x_2 + \cdots + A_{mn}x_n &= b_m \end{aligned}.$$

This set is called the *solution set* of the system of linear equations, its elements are *solutions*.

The system of linear equations is *homogeneous* iff $b_1 = \cdots = b_m = 0$. The $n$-tuple $(0, \ldots, 0)$ is always a solution of a homogeneous system of linear equations.

The solution set of a inhomogeneous system of linear equations may be empty or may contain only one $n$-tuple or may be an infinite set. Clearly, in the latter case we cannot write down all solutions. We show in the next section how to overcome this problem and define what we expect as a "good description" of the solution set.

Many lectures on linear algebra start with matrix algebra, a topic which is on the one hand easily comprehensible for first-year-students and which on the other hand facilitates speaking about systems of linear equations. As soon as students are familiar with basic notions and notations of matrix algebra (i.e. addition and multiplication of matrices and the corresponding calculation rules like distributive laws, associative laws, ...), we can define systems of linear equations in the following very short form:

Given are a matrix $A$ with $m$ rows and $n$ columns and a column $b$ with $m$ entries. Search for a "good description" of the set of all columns $x$ with $n$ entries such that

$$A \cdot x = b\,.$$

This set is called the *solution set* of the system of linear equations, its elements are *solutions*. We write $S(A, b) := \{x \mid A \cdot x = b\}$ for the solution set of the system of linear equations given by $A$ and $b$.

This "matrix form" of systems of linear equations is useful to implement systems of linear equations in a programming language, since the input data are just a matrix and a column. A "good description" of the solution set means in this context to specify reasonable output data.

It is little-known that the matrix form of a system of linear equations is historically the earlier one. More than 2100 years ago in China systems of linear equations in matrix form (motivated by problems arising from commerce) were solved by calculations with matrices. Hence Peter Gabriel calls "Fang-Cheng-Algorithm" what nowadays is known as "Gaussian Elimination", see Gabriel (1996, Chapter A.2).

## 4 Two Examples of Systems of Linear Equations

Systems of linear equations occur in nearly every field of application of mathematics, see e.g. Meyer (2000). Here are two examples:

*Example 1* Melting alloys
An alloy containing $b_i$ grams of the metal $M_i$ for $1 \leq i \leq m$ shall be produced. The alloy is produced by combining appropriate quantities of the alloys $L_1, \ldots, L_n$ of the metals $M_1, \ldots, M_m$, where 1 gram of the alloy $L_j$ contains $A_{ij}$ grams of the metal $M_i$. How many grams of $L_1, \ldots, L_n$ have to be melted to obtain the desired alloy? (We assume that the mass of the metals does not change during the melting process).

Melting $x_1, \ldots, x_n$ grams of the alloys $L_1, \ldots, L_n$ yields an alloy with

$$A_{i1}x_1 + A_{i2}x_2 + \cdots + A_{in}x_n$$

grams of $M_i$, $1 \leq i \leq m$. Hence we have to find $n$-tuples $(x_1, \ldots, x_n)$ such that

$$A_{i1}x_1 + A_{i2}x_2 + \ldots + A_{in}x_n = b_i, \;\; 1 \leq i \leq m\,.$$

*Example 2* Electric circuits

In the following electric circuit the voltages $U_1$ and $U_2$, and the electric resistances $R_1, R_2, R_3, R_4, R_5$ are known. Compute the electric currents $I_1, I_2, I_3, I_4, I_5$ through the resistances $R_1, R_2, R_3, R_4, R_5$.
Voltages are given in Volts, currents in Ampères and resistances in Ohms. Hence we consider $R_i, I_i, U_i$ as numbers.

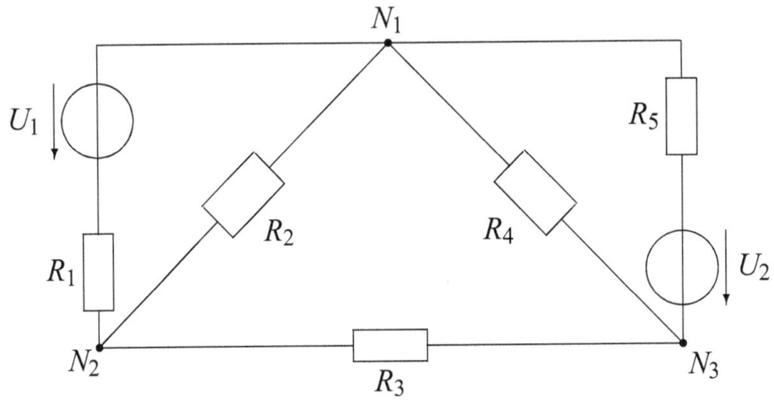


We have to choose (arbitrarily) a direction of the electric current in each wire between two vertices.

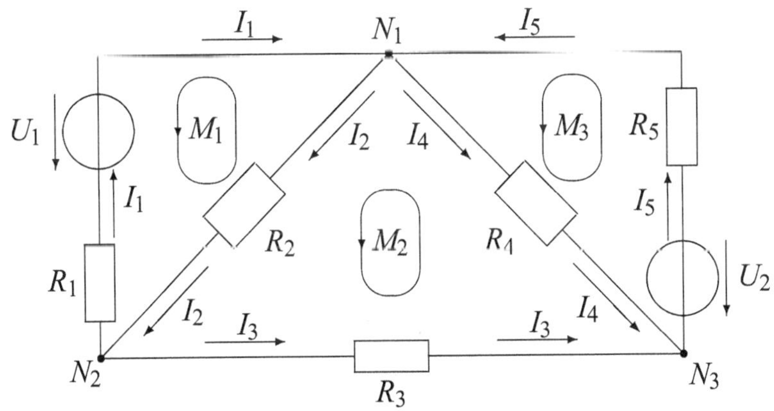


Then, using Ohm's law, Kirchhoff's law for nodes, and Kirchhoff's law for meshes, we transform this problem into a system of linear equations:

Kirchhoff's law for nodes yields
$I_1 + I_5 = I_2 + I_4$
$I_2 \quad = I_1 + I_3$
$I_3 + I_4 = I_5$

Kirchhoff's law for meshes yields
$I_1 \cdot R_1 + I_2 \cdot R_2 = U_1$
$I_2 \cdot R_2 + I_3 \cdot R_3 = I_4 \cdot R_4$
$I_4 \cdot R_4 + I_5 \cdot R_5 = U_2$
Thus we search for all 5-tuples $(I_1, \ldots, I_5)$ such that

$$\begin{pmatrix} 1 & -1 & 0 & -1 & 1 \\ -1 & 1 & -1 & 0 & 0 \\ 0 & 0 & 1 & 1 & -1 \\ R_1 & R_2 & 0 & 0 & 0 \\ 0 & R_2 & R_3 & -R_4 & 0 \\ 0 & 0 & 0 & R_4 & R_5 \end{pmatrix} \cdot \begin{pmatrix} I_1 \\ I_2 \\ I_3 \\ I_4 \\ I_5 \end{pmatrix} = \begin{pmatrix} 0 \\ 0 \\ 0 \\ U_1 \\ 0 \\ U_2 \end{pmatrix}.$$

## 5 How to Describe an Infinite Solution Set by Finitely Many Solutions

It is not possible to describe any infinite set by finitely many of its elements. This is only possible, if there is some additional information or structure on this set. Solution sets of systems of linear equations have two important properties:

1. The problem of describing the solution set of any system of linear equations can be reduced to that of describing the solution set of homogeneous ones. More precisely: if $v$ is any solution of the system given by a matrix $A$ and a column $b$, then its solution set $S(A, b)$ consists of all sums of $v$ with a solution of the corresponding homogeneous system of linear equations (given by $A$ and 0), i.e.

$$S(A, b) = \{v + w \mid w \in S(A, 0)\}.$$

   This easily follows from
   $A \cdot (v + w) = A \cdot v + A \cdot w = b + 0 = b.$
2. If $x$ and $y$ are solutions of the homogeneous system of linear equations given by $A$, then for all numbers $c$ and $d$ the column $cx + dy$ is a solution too.
   This follows easily from calculation rules for matrices:
   $A \cdot (cx + dy) = c(A \cdot x) + d(A \cdot y) = 0 + 0 = 0.$
   More generally, if the columns $w_1, \ldots, w_k$ are solutions and $c_1, \ldots, c_k$ are arbitrary numbers, then the linear combination $c_1 w_1 + \cdots + c_k w_k$ is a solution too.

Now we are able to state more precisely what a "good description" of the solution set means: compute one solution $v \in S(A, b)$ and a basis of $S(A, 0)$, i.e.

finitely many solutions $w_1, \ldots, w_k$ of the corresponding homogeneous equation such that any element of $S(A, 0)$ can uniquely be written as a linear combination $c_1 w_1 + \cdots + c_k w_k$ of $w_1, \ldots, w_k$. Then

$$S(A, b) = \{v + c_1 w_1 + \cdots c_k w_k \mid c_1, \ldots, c_k \text{ numbers}\} .$$

One can show that the number $k$ is uniquely determined, it is called the *dimension* of the solution set.

This could be the starting point to speak about vector spaces, vectors, linear combinations, bases, dimension, ... in a course on linear algebra. In the context of linear equations vectors are columns and vector spaces are non-empty sets of columns which are closed under addition and multiplication by numbers, for example solution sets of systems of homogeneous linear equations.

In general, vector spaces are sets together with an "addition" and a "multiplication with numbers (or scalar multiplication)", such that certain calculation rules hold. Vectors are (by definition) elements of vector spaces. Note that for the following it is not necessary to know that any vector space has a basis, since there is an algorithm which effectively computes a basis of the solution set of homogeneous systems of linear equations and thus shows (in the best way) that a basis exists.

The notion *linear combination* enables another interpretation of systems of linear equations. If $A$ is a matrix with $n$ columns and $y$ is a column with $n$ entries, then

$$A \cdot y = y_1 A_{-1} + \cdots + y_n A_{-n} ,$$

i.e. the matrix product $A \cdot y$ is a linear combination of the columns of $A$. Here $A_{-j}$ denotes the jth row of $A$. The task "Find a column $y$ such that $A \cdot y = b$" is therefore equivalent to the task "Write $b$ as a linear combination of the columns of $A$".

## 6 The Strategy to Solve Systems of Linear Equations

*If you cannot solve a problem directly then move on to an easier problem which has the same solution set as the former one. Repeat this until you arrive at a problem which you can solve.*

This is a basic strategy in mathematics. More than 2200 years ago it was successfully applied by Greek mathematicians to compute the greatest common divisor $gcd(a, b)$ of two positive integers $a$ and $b$, i.e. the greatest integer which divides $a$ and $b$. The key observation for the *euclidean algorithm* (for $a > b$) is:

$$gcd(a, b) = gcd(a - b, b) .$$

This follows from the fact that any common divisor of $a$ and $b$ is also a divisor of $a + b$ and of $a - b$. As long as the two numbers are different we replace the greater one by the difference of the greater and the smaller one. As soon as the two numbers are equal we know $gcd(a, b)$. This case occurs after at most $a$ steps. For example:

$$gcd(355, 213) = gcd(213, 142) = gcd(142, 71) = gcd(71, 71) = 71 .$$

The same strategy is used to solve systems of linear equations. We replace the given system by another system of linear equations, *which has the same solution set*. Such a system of linear equations is called *equivalent* to the first one, and the replacement is called an *equivalence transformation*.

## 7 Elementary Transformations

We now try to transform the given system of linear equations by several equivalence transformations to a system which is easy to solve, i.e. where we need no further computations to obtain one solution and a basis of the solution set of the corresponding homogeneous system. But to this end we have to answer two questions:

1. How can we transform a system of linear equations into an equivalent one?
2. What is the goal of the transformation process, i.e. which systems of linear equations are easy to solve?

The answer to question 2 is given in the next section. To answer question 1 we recall that a matrix $P$ with $m$ rows and $m$ columns is *invertible* if there is a matrix $Q$ such that $P \cdot Q = I_m$, where $I_m$ is the identity matrix. Then we write $P^{-1}$ for $Q$ and call it the *inverse matrix* of $P$. Recall that $A$ is a matrix with $m$ rows and $n$ columns and $b$ ist a column with $m$ entries. The key observation is the following: if $P$ is invertible, then

$$S(A, b) = S(P \cdot A, P \cdot b).$$

This is true since $A \cdot x = b$ implies $(P \cdot A) \cdot x = P \cdot b$ and $(P \cdot A) \cdot x = P \cdot b$ implies $A \cdot x = P^{-1} \cdot (P \cdot A) \cdot x = P^{-1} \cdot (P \cdot b) = b$. Hence, if $P$ is an invertible matrix we get an equivalent system of linear equations by replacing $A$ and $b$ by $P \cdot A$ and $P \cdot b$.

The matrices obtained in the following way are obviously invertible:

1. Replace in $I_m$ one row by the sum of this row and another one.
2. Swap two rows of $I_m$.
3. Multiply one row of $I_m$ with a non-zero number.

We call these matrices *elementary matrices*.

*Example* The matrices

$$\begin{pmatrix} 1 & 0 & 1 \\ 0 & 1 & 0 \\ 0 & 0 & 1 \end{pmatrix}, \quad \begin{pmatrix} 0 & 1 & 0 \\ 1 & 0 & 0 \\ 0 & 0 & 1 \end{pmatrix}, \quad \begin{pmatrix} 1 & 0 & 0 \\ 0 & 3 & 0 \\ 0 & 0 & 1 \end{pmatrix}.$$

are elementary matrices, the inverse matrices of them are

$$\begin{pmatrix} 1 & 0 & -1 \\ 0 & 1 & 0 \\ 0 & 0 & 1 \end{pmatrix}, \quad \begin{pmatrix} 0 & 1 & 0 \\ 1 & 0 & 0 \\ 0 & 0 & 1 \end{pmatrix}, \quad \begin{pmatrix} 1 & 0 & 0 \\ 0 & \frac{1}{3} & 0 \\ 0 & 0 & 1 \end{pmatrix}.$$

If $P$ is an elementary matrix, then we obtain the product $P \cdot A$ in one of the following ways:

1. Replace one row of $A$ by the sum of this row and another one.
2. Swap two rows of $A$.
3. Multiply one row of $A$ with a non-zero number.

We say that we have obtained $P \cdot A$ by an *elementary transformation* from $A$. The systems of linear equations given by $A$ and $b$ and by $P \cdot A$ and $P \cdot b$ are equivalent, i.e. elementary transformations are equivalence transformations.

Assume that $m = n$ and suppose that by several elementary transformations we transform $A$ to the identity matrix $I_m$ and $b$ to $b'$. Then the only column $x$ with $I_m \cdot x = b'$ is $b'$. Thus $b'$ is also the only element of $S(A, b)$.

But by far not all matrices can be transformed by elementary transformations to the identity matrix. Thus even in the case $m = n$ it cannot be the unique goal of our transformation process.

## 8 The Easy Case: Systems in Echelon Form

First we reflect which systems of linear equations can be solved without any computation. A matrix $A$ with $m$ rows and $n$ columns has *reduced row echelon form* iff the following conditions hold:

(1) If all entries in a row of $A$ are zero, then the same is true for all rows below.
(2) The first non-zero entry in each row is called *pivot* and is 1.
(3) The pivot in the (i+1)-th row is in a column to the right of the pivot in the i-th row.
(4) A pivot is the only non-zero entry in its column.

A matrix in reduced row echelon form has the shape

$$\begin{pmatrix} 0 \ldots 0 \; 1 * \ldots * \; 0 * \ldots * \; 0 * \ldots \\ 0 \quad \ldots\ldots \quad 0 \; 1 * \ldots * \; 0 * \ldots \\ 0 \quad \ldots\ldots\ldots\ldots \quad 0 \; 1 * \ldots \\ \vdots \end{pmatrix},$$

where the stars can be arbitrary numbers.

Let $e_i$ be the i-th standard column, i.e. the column with 1 in the i-th row and 0 in the other rows.
Suppose that $A$ has reduced row echelon form. Then its columns with pivots are by definition standard-columns. If $A$ has $r$ pivots, then among the columns of $A$ are the standard columns $e_1, e_2, \ldots, e_r$. If $A$ has more than $r$ rows, then all entries in the rows with indices $r+1, \ldots, m$ are 0. Recall from Sect. 5 that finding a solution of the system of linear equations given by $A$ and $b$ means to write $b$ as a linear combination of the columns of $A$. This is not possible if one of the entries $b_i$ of $b$ with $i > r$ is not zero. Hence in this case there is no solution. Otherwise, since $b = b_1 e_1 + \cdots + b_r e_r$ it is easy to write $b$ as a linear combination of columns of $A$: if the column-indices of the pivots are $p_1 < p_2 < \ldots < p_r$ then $e_i$ is the $p_i$-th column of $A$. Hence the column $x$ with $x_{p_i} = b_i$, $1 \leq i \leq r$, and $x_\ell = 0$ for the other indices $\ell$, is a solution.

By similar considerations we get a basis of the solution set of the homogeneous linear equation given by $A$: Let $q_1, q_2, \ldots, q_{n-r}$ be the indices of columns without pivot. The columns

$$w_j := \begin{pmatrix} 0 \\ \vdots \\ 0 \\ -A_{1q_j} \\ 0 \\ \vdots \\ 0 \\ 1 \\ 0 \\ \vdots \\ -A_{rq_j} \\ 0 \\ \vdots \\ 0 \end{pmatrix},$$

$1 \leq j \leq n-r$, where the numbers $-A_{1q_j}, \ldots, -A_{rq_j}$ are in the rows with indices $p_1, \ldots, p_r$ and 1 is in the row with index $q_j$, are a basis of $S(A, 0)$.

*Example*

Find all 5-tuples $(x_1, \ldots, x_n)$ such that

$$\begin{aligned} x_2 + 3x_3 + 2x_5 &= -1 \\ x_4 - x_5 &= 2 \end{aligned}$$

In matrix-form: Find all columns $x$ with 5 entries such that

$$(*) \qquad A \cdot x := \begin{pmatrix} 0 & 1 & 3 & 0 & 2 \\ 0 & 0 & 0 & 1 & -1 \end{pmatrix} x = \begin{pmatrix} -1 \\ 2 \end{pmatrix}$$

Since

$$\begin{aligned} &0 \cdot A_{-1} + (-1) \cdot A_{-2} + 0 \cdot A_{-3} + 2 \cdot A_{-4} + 0 \cdot A_{-5} \\ &= (-1) \cdot \begin{pmatrix} 1 \\ 0 \end{pmatrix} + 2 \cdot \begin{pmatrix} 0 \\ 1 \end{pmatrix} = \begin{pmatrix} -1 \\ 2 \end{pmatrix}, \end{aligned}$$

$\begin{pmatrix} 0 \\ -1 \\ 0 \\ 2 \\ 0 \end{pmatrix}$ is a solution.

Recall that $A_{-j}$ denotes the j-th column of $A$. Then

$$\begin{aligned} 1 \cdot A_{-1} &= 0, \\ -3 \cdot A_{-2} + 1 \cdot A_{-3} &= 0 \quad \text{and} \\ -2 \cdot A_{-2} + 1 \cdot A_{-4} + 1 \cdot A_{-5} &= 0 \end{aligned}$$

hence $\begin{pmatrix} 1 \\ 0 \\ 0 \\ 0 \\ 0 \end{pmatrix}$, $\begin{pmatrix} 0 \\ -3 \\ 1 \\ 0 \\ 0 \end{pmatrix}$ and $\begin{pmatrix} 0 \\ -2 \\ 0 \\ 1 \\ 1 \end{pmatrix}$ are solutions of the system of homogeneous linear equations. Moreover they are a basis of its solution set which has dimension 3.

Therefore the solution set of the system (*) is

$$\begin{aligned} &\{\begin{pmatrix} 0 \\ -1 \\ 0 \\ 2 \\ 0 \end{pmatrix} + c_1 \begin{pmatrix} 1 \\ 0 \\ 0 \\ 0 \\ 0 \end{pmatrix} + c_2 \begin{pmatrix} 0 \\ -3 \\ 1 \\ 0 \\ 0 \end{pmatrix} + c_3 \begin{pmatrix} 0 \\ -2 \\ 0 \\ 1 \\ 1 \end{pmatrix} \mid c_1, c_2, c_3 \in \mathbb{R}\} \\ &= \{\begin{pmatrix} c_1 \\ -1 - 3c_2 - 2c_3 \\ c_2 \\ 2 + c_3 \\ c_3 \end{pmatrix} \mid c_1, c_2, c_3 \in \mathbb{R}\}. \end{aligned}$$

It is important to realize that we can solve systems of linear equations in reduced row echelon form without any calculation! We only have to assign the entries of $A$ and $b$ to the right places in the columns describing the solution set.

## 9 Gaussian Elimination

The section before suggests to try to transform any system of linear equation by elementary transformations to a system which has reduced row echelon form. It is easy to see that this is always possible. Starting with the first column of $A$, the matrix is transformed step-by-step into reduced row echelon form.

For example: If the entry $A_{11}$ is not zero, then multiply the first row by $1/A_{11}$. Then add to the i-th row the $-A_{i1}$-fold (new) first row, $2 \leq i \leq m$. This yields a matrix whose first column is the first standard-column. If $A_{11} = 0$ but there is a non-zero entry in the i-th row of the first column, then swap the first and i-th row of $A$ and continue as above. If all entries of the first column are 0 then move on to the second column.

In several Computeralgebra-Systems there are commands to transform a matrix $A$ to a matrix which has reduced row echelon form. For example, in Maple 18 this can be done by the command *with(LinearAlgebra): ReducedRowEchelonForm(A).* From the reduced row echelon form we can directly read off the solutions.

Instead of a detailed description of the process of transformation we present an example. Since the elementary transformations for $A$ and $b$ must be the same ones it is reasonable to write $A$ and $b$ as one (extended) matrix $(A|b)$.

$$A := \begin{pmatrix} 0 & 2 & 3 & -1 \\ 2 & 4 & 6 & 0 \\ 2 & 6 & 9 & -1 \end{pmatrix}, \quad b := \begin{pmatrix} 2 \\ 4 \\ 6 \end{pmatrix}$$

$$(A|b) = \left(\begin{array}{cccc|c} 0 & 2 & 3 & -1 & 2 \\ 2 & 4 & 6 & 0 & 4 \\ 2 & 6 & 9 & -1 & 6 \end{array}\right) \to \left(\begin{array}{cccc|c} 2 & 4 & 6 & 0 & 4 \\ 0 & 2 & 3 & -1 & 2 \\ 2 & 6 & 9 & -1 & 6 \end{array}\right) \to$$

$$\to \left(\begin{array}{cccc|c} 1 & 2 & 3 & 0 & 2 \\ 0 & 2 & 3 & -1 & 2 \\ 2 & 6 & 9 & -1 & 6 \end{array}\right) \to \left(\begin{array}{cccc|c} 1 & 2 & 3 & 0 & 2 \\ 0 & 2 & 3 & -1 & 2 \\ 0 & 2 & 3 & -1 & 2 \end{array}\right) \to$$

$$\to \left(\begin{array}{cccc|c} 1 & 0 & 0 & 1 & 0 \\ 0 & 2 & 3 & -1 & 2 \\ 0 & 2 & 3 & -1 & 2 \end{array}\right) \to \left(\begin{array}{cccc|c} 1 & 0 & 0 & 1 & 0 \\ 0 & 2 & 3 & -1 & 2 \\ 0 & 0 & 0 & 0 & 0 \end{array}\right) \to$$

$$\to \left(\begin{array}{cccc|c} 1 & 0 & 0 & 1 & 0 \\ 0 & 1 & \frac{3}{2} & -\frac{1}{2} & 1 \\ 0 & 0 & 0 & 0 & 0 \end{array}\right)$$

One solution is $\begin{pmatrix} 0 \\ 1 \\ 0 \\ 0 \end{pmatrix}$. A basis of $S(A, 0)$ is $\left( \begin{pmatrix} 0 \\ -\frac{3}{2} \\ 1 \\ 0 \end{pmatrix}, \begin{pmatrix} -1 \\ \frac{1}{2} \\ 0 \\ 1 \end{pmatrix} \right)$. Hence

$$S(A,b) = \left\{ \begin{pmatrix} 0 \\ 1 \\ 0 \\ 0 \end{pmatrix} + c_1 \begin{pmatrix} 0 \\ -\frac{3}{2} \\ 1 \\ 0 \end{pmatrix} + c_2 \begin{pmatrix} -1 \\ \frac{1}{2} \\ 0 \\ 1 \end{pmatrix} \mid c_1, c_2 \in \mathbb{R} \right\}.$$

## 10 Geometrical Interpretation

A solution of an equation with 2 (resp. 3) unknowns is a pair (resp. a triple) of numbers and not a point in the plane (resp. space). In order to describe points by pairs (resp. triples) of numbers we have to choose a coordinate system in the plane (resp. space). Then the points can be identified with their pairs (resp. triples) of coordinates, the plane can be identified with $\mathbb{R}^2$ and the space with $\mathbb{R}^3$.

Of course we have an intuitive conception of lines in $\mathbb{R}^n$. But we have to give an exact definition (which corresponds to our intuition) too.

A *line through* 0 and through a point $P \neq 0$ in $\mathbb{R}^n$ is the set $\mathbb{R} \cdot P := \{t \cdot P \mid t \in \mathbb{R}\}$ of all scalar multiples of $P$. Formulated in the language of vector spaces: A line through 0 in $\mathbb{R}^n$ is a one-dimensional vector subspace of $\mathbb{R}^n$.

The *line* through two different points $P$ and $Q$ in $\mathbb{R}^n$ is the set

$$Q + \mathbb{R} \cdot (Q - P) := \{ Q + t \cdot (Q - P) \mid t \in \mathbb{R} \}.$$

Hence a line is given by one of its points and by a non-zero point of its *parallel line through* 0. Such a representation of a line is called a *parameter form*. It is not unique, we may choose instead of $Q$ any other point of the line and instead of $Q - P$ any non-zero scalar multiple of $Q - P$.

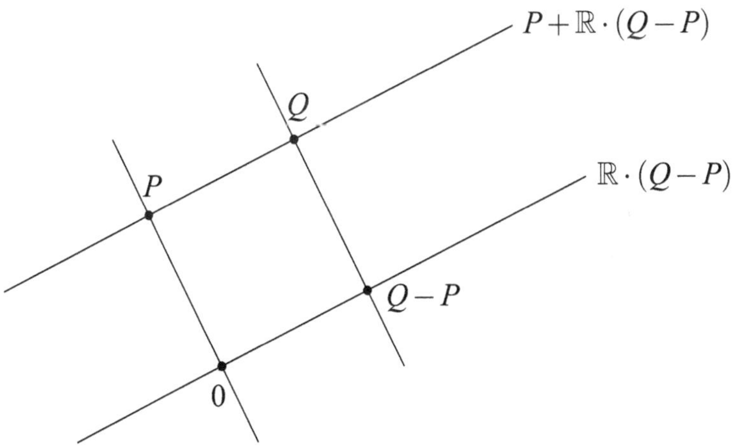


This definition admits a geometric interpretation of one-dimensional solution sets of systems of linear equations in $n$ unknowns: if this system is homogeneous its set of solutions is a line through 0. If the system is not homogeneous, its solution set is a line parallel to the solution set of the corresponding homogeneous system.

In particular, if $(a, b) \neq (0, 0)$ the set of solutions of the linear equation $a \cdot x + b \cdot y = c$ is the line through $(\frac{c \cdot a}{a^2+b^2}, \frac{c \cdot b}{a^2+b^2})$ (or through $(\frac{c}{a}, 0)$ if $a \neq 0$) parallel to the line $\mathbb{R} \cdot (-b, a)$.

A system of linear equations whose solution set is a line is called an *implicit form* of this line. Implicit forms are not unique: all systems of linear equations equivalent to the given one are implicit forms of this line too.

If the implicit form of a line in $\mathbb{R}^n$ is given, we can write it in parameter form by solving this system of linear equations.

If $(q_1, \ldots, q_n) + \mathbb{R} \cdot (r_1, \ldots, r_n)$ is a parameter form of a line, an implicit form can be written down directly (we assume for simplicity that $r_n \neq 0$): Obviously

$$r_n x_1 - r_1 x_n = 0, \ldots, r_n x_{n-1} - r_{n-1} x_n = 0$$

is an implicit form of the line $\mathbb{R} \cdot (r_1, \ldots, r_n)$. Since $(q_1, \ldots, q_n)$ must be a solution we get the system

$$r_n x_1 - r_1 x_n = r_n q_1 - r_1 q_n, \ldots, r_n x_{n-1} - r_{n-1} x_n = r_n q_{n-1} - r_1 q_n$$

as an implicit form of $(q_1, \ldots, q_n) + \mathbb{R} \cdot (r_1, \ldots, r_n)$.

*Example* A parameter form of the line through $(0, 1, 2)$ and $(-2, 3, 1)$ is $\{(0, 1, 2) + t \cdot (2, -2, 1) \mid t \in \mathbb{R}\}$. An implicit form is the system

$$2x_3 - x_1 = 2 \cdot 2 - 0 = 4, -2x_3 - x_2 = -2 \cdot 2 - 1 = -5 \cdot$$

Consider points $P_1$, $P_2$ which are not both elements of a line through 0. A *plane in* $\mathbb{R}^n$ *through* 0 and through points $P_1$, $P_2$ is the set $\{ t_1 \cdot P_1 + t_2 \cdot P_2 \mid t_1, t_2 \in \mathbb{R} \}$ of all linear combinations of $P_1$ and $P_2$. Formulated in the language of vector spaces: A plane through 0 in $\mathbb{R}^n$ is a two-dimensional vector subspace of $\mathbb{R}^n$.

The *plane in* $\mathbb{R}^n$ through three points $P$, $Q$ and $R$, which are not all elements of one line, is the set

$$\{ R + t_1 \cdot (R - P) + t_2 \cdot (R - Q) \mid t_1, t_2 \in \mathbb{R} \}.$$

Hence a plane is given by one of its points and by its *parallel plane through* 0. This representation of a plane is called a *parameter form*. Hence a two-dimensional set of solutions of systems of linear equations in $n$ unknowns is a plane in $\mathbb{R}^n$ which is parallel to the set of solutions of the corresponding homogeneous system.

The solution set of one linear equation with three unknowns

$$a \cdot x + b \cdot y + c \cdot z = d$$

with $(a, b, c) \neq (0, 0, 0)$ (we assume $c \neq 0$) is the plane (we assume $c \neq 0$)

$$\{ (0,0,d/c) + t_1 \cdot (-c,0,a) + t_2 \cdot (0,-c,b) \mid t_1, t_2 \in \mathbb{R} \}.$$

If a plane in $\mathbb{R}^3$ is given in parameter form

$$\{ (r_1,r_2,r_3) + t_1 \cdot (u_1,u_2,u_3) + t_2 \cdot (v_1,v_2,v_3) \mid t_1, t_2 \in \mathbb{R} \},$$

then an implicit form of this plane is

$$a \cdot x + b \cdot y + c \cdot z = a \cdot r_1 + b \cdot r_2 + c \cdot r_3 ,$$

where $(a, b, c)$ is a solution (unique up to a non-zero scalar multiple) of the homogeneous system of linear equations with three unknowns

$$u_1 \cdot x_1 + u_2 \cdot x_2 + u_3 \cdot x_3 = 0, \quad v_1 \cdot x_1 + v_2 \cdot x_2 + v_3 \cdot x_3 = 0.$$

In general, an affine subspace of $\mathbb{R}^n$ is a non-empty subset $Z := \{P + U \mid U \in V\}$, where $P$ is a point and $V$ is a vector subspace of $\mathbb{R}^n$. The dimension of $Z$ is by definition the dimension of $V$, i.e. the number of vectors in a basis of $V$. Affine subspaces of dimension 0, 1, 2 are points, lines and planes respectively. In the previous sections we have shown that solution sets of systems of linear equations are affine subspaces. The geometric formulation of "solve a system of linear equations" therefore is "determine a parameter form of an affine subspace given in implicit form".

The geometric problem "Three planes in $\mathbb{R}^3$ are given. Compute their intersection!" can easily be solved if the three planes are given in implicit form, i.e. if we know linear equations whose solution sets are the given planes. Then we solve a system of three equations in three unknowns and get either no solution or one of the following affine subspaces as solution set: a point, a line, or a plane.

It would not be reasonable to solve a system of linear equations in the following way: Determine first the solution sets of each single equation and then determine their intersection. Then we would get much more information than we need, since we are only interested in *common* solutions.

It is easily seen that this type of approach is not efficient if we consider another problem: consider two polynomials $f := x^9 - 3x^8 - 5x^6 + 4x^5 - 2x^4 - 7x^3 + x^2 - 3$ and $g := x^8 - 3x^7 + 2x^5 - 2x^4 - 2x^3 + 8x - 1$. There is no chance to determine in reasonable time the zero set of $f$ or of $g$ alone. But it is easy to determine the set of common zeros of $f$ and $g$. Using the Euclidean algorithm for polynomials, computeralgebra-systems easily compute $gcd(f, g) = 1$. Thus we see that there are no common zeros.

There is one special case for which systems of linear equations can be solved graphically: a system of two equations with two unknowns. Since we already know that the solution of one equation with two unknowns is a line, it is sufficient to determine two solutions of each equation to draw the solution sets in the plane (after the choice of a coordinate system). If there is exactly one intersection point of these two lines, its coordinates can be read off. But this approach yields not much insight in the general case and has no practicable generalization to three or more variables.

## References

Gabriel, P. (1996). *Matrizen, Geometrie, Lineare Algebra*. Basel: Birkhäuser.
Meyer, C. (2000). *Matrix Analysis and Applied Linear Algebra*. Philadelphia: SIAM.
Pauer, F. (2007). *Lineare Algebra 1*. Lecture notes (in german). Innsbruck: University of Innsbruck. (9th edition 2016). https://www.uibk.ac.at/mathematik/personal/pauer/la-2016/skriptum-la1-2016.pdf

# Nonnegative Factorization of a Data Matrix as a Motivational Example for Basic Linear Algebra

**Barak A. Pearlmutter and Helena Šmigoc**

**Abstract** We present a motivating example for matrix multiplication based on factoring a data matrix. Traditionally, matrix multiplication is motivated by applications in physics: composing rigid transformations, scaling, sheering, etc. We present an engaging modern example which naturally motivates a variety of matrix manipulations, and a variety of different ways of viewing matrix multiplication. We exhibit a low-rank non-negative decomposition (NMF) of a "data matrix" whose entries are word frequencies across a corpus of documents. We then explore the meaning of the entries in the decomposition, find natural interpretations of intermediate quantities that arise in several different ways of writing the matrix product, and show the utility of various matrix operations. This example gives the students a glimpse of the power of an advanced linear algebraic technique used in modern data science.

**Keywords** Nonnegative matrix factorization (NMF) · Topic modeling · Data mining · Matrix multiplication

## 1 Introduction

Examples are an essential part of teaching any mathematical subject. They serve a range of purposes, from checking understanding and deepening knowledge to giving a broader view of the subject and its applications. There are an abundance of examples available in the literature, covering every topic of any basic linear algebra course. However, it is not so easy to find examples that give an insight into the current development of the subject and are at the same time accessible to students. As the applications of linear algebra are rapidly expanding, and several new devel-

B. A. Pearlmutter
Department of Computer Science, Maynooth University, Maynooth, Ireland
e-mail: barak@pearlmutter.net

H. Šmigoc (✉)
School of Mathematics and Statistics, UCD Dublin, Dublin, Ireland
e-mail: helena.smigoc@ucd.ie

 
S. Stewart et al. (eds.), *Challenges and Strategies in Teaching Linear Algebra*,
ICME-13 Monographs, https://doi.org/10.1007/978-3-319-66811-6_15

opments in the subject are motivated by applications, examples showcasing current applications of the subject are of particular interest.

Because of its utility in other domains, linear algebra is a classical subject routinely taught to students not majoring in mathematics. It is a prerequisite not just for advanced mathematics but also for undergraduate degrees in Engineering, Physics, Computer Science, Biology, Chemistry, Business, Statistics, and the like. Those students in particular benefit from learning from examples, and appreciate seeing interesting applications of the material they are learning. The benefits of using models to introduce mathematical concepts has been studied Lesh & English, 2005, and models focusing on different concepts from linear algebra are available (Possani, Trigueros, Preciado, & Lozano, 2010; Salgado & Trigueros, 2015; Trigueros & Possani, 2013).

While very simple examples are essential when introducing a topic, examples of applications presented in classrooms often seem contrived. For example, students' knowledge of economics and agriculture is sufficiently sophisticated that simple linear examples of acreage under cultivation invite criticism. On the other hand, it is impossible to bring to the classroom, for instance, deep applications of linear algebra in genetics (Ponnapalli, Saunders, Van Loan, & Alter, 2011), since most likely neither the instructor nor the students have the necessary background to really understand how they work. To quote Stewart and Thomas (2003),

> While it is true that linear algebra can simplify the solution to many problems, this is only true for those who are very familiar with the subject area. In contrast, the first year university student has a long way to go before being able to see the whole picture.

The press is full of stories about data science: analysis of large corpora of data. Some of these lend themselves naturally to use as motivating examples for various concepts in linear algebra. For example, the Netflix challenge can be viewed as a problem in matrix completion, where a company was highly motivated to recover a low rank decomposition of an almost entirely incomplete matrix of movie ratings.

We present less abstract example, in which matrix multiplication is explicated by examination of a nonnegative decomposition of a term-by-document matrix. This particular example vividly illustrates various views of matrix multiplication (as composition of linear functions; as a sum of outer products of columns with rows; and as a table of inner products of rows with columns), while using only primitive concepts. It also previews and motivates a variety of more advanced concepts (the general algebraic concept of factoring, the notion of rank, approximation and norms, iterative numeric algorithms, constraints like element-wise non-negativity, and column-stochastic matrices), helping sketch the outlines of richer material covered in more advanced courses.

Although intuitive and implemented by a very short algorithm, the technique discussed (NMF) is far from a toy: it has enjoyed a myriad of accessible and engaging applications (Asari, Pearlmutter, & Zador, 2006; Helleday, Eshtad, & Nik-Zainal, 2014; Niegowski & Zivanovic, 2014; O'Grady & Pearlmutter, 2008; Ray & Bandyopadhyay, 2016; Smaragdis & Brown, 2003; Wilson, Raj, Smaragdis & Divakaran, 2008). For this reason, the example we present serves to give a taste of an interesting and accessible application of linear algebra. Although briefly presented in this document for the sake of completeness, we do not suggest attempting to derive the method in the classroom, leaving that too as motivation for the pursuit of more advanced study.

The mathematical notation used below is standard. For example, $e_i$ denotes the vector of appropriate size with $i$-th entry equal to one and other entries equal to zero. The "discussion point" boxes are intended to be illustrative, and can be used for classroom discussion, project-based learning, or as the basis for assignments.

## 2 Term-by-Document Matrix: A Small Example

Numeric data organised in a tabular format is something we are all familiar with in our daily lives. Everyone can understand a spreadsheet whose rows are indexed by products, columns by month, and whose entries contain sales. These are the matrices that students entering a linear algebra course have already seen. In data science, tabular data of this sort is known as a "data matrix".

A data matrix of interest in library science is a tabulation of word frequencies by documents. Rows are indexed by words, columns by documents, and the entries of a matrix are the number of times a given word appears in a given document. This particular kind of data matrix is sometimes called a term-by-document matrix. Although this matrix completely ignores the actual arrangement of words within each document (i.e., it is a bag-of-words model), it still contains sufficient information to allow interesting structure to be discovered.

There are several ways in which matrices and matrix multiplication can be introduced in the classroom. Term-by-document matrices can be one of the examples given to the class, starting with a small example that can be given on a board. In the classroom we can show a pre-prepared example, which can be built on by an assignment in which students have the freedom to chose the documents they want to consider. Since the search function in browsers automatically counts the number of times a word appears on a page, such an assignment is not necessarily time demanding.

Here we present an example where the documents are the Wikipedia entries for the four most venomous animals in the world (*Box Jellyfish*[1], *King Cobra*[2], *Marbled Cone Snail*[3], *Blue-Ringed Octopus*[4]) and we consider only five terms (*venom*, *death*, *danger*, *survive*, *Madagascar*). This gives us Table 1.

[1] https://en.wikipedia.org/wiki/Box_jellyfish.

[2] https://en.wikipedia.org/wiki/King_cobra.

[3] https://en.wikipedia.org/wiki/Conus_marmoreus.

[4] https://en.wikipedia.org/wiki/Blue-ringed_octopus.

**Table 1** Term-by-document matrix of the four most venomous animals

| | | Documents | | | |
|---|---|---|---|---|---|
| | | Jellyfish | Cobra | Snail | Octopus |
| Terms | venom | 32 | 44 | 1 | 18 |
| | death | 9 | 3 | 0 | 2 |
| | danger | 6 | 4 | 0 | 4 |
| | survive | 2 | 0 | 0 | 1 |
| | Madagascar | 0 | 0 | 2 | 0 |

Going from the table to the matrix

$$A = \begin{pmatrix} 32 & 44 & 1 & 18 \\ 9 & 3 & 0 & 2 \\ 6 & 4 & 0 & 4 \\ 2 & 0 & 0 & 1 \\ 0 & 0 & 2 & 0 \end{pmatrix}$$

we can lead the discussion in several directions. A representative set of questions is given below. The questions are of course trivial to answer without referring to matrices. The simplicity of the questions makes it easy for students to understand the corresponding matrix operations and motivates them to think about extensions to more involved tasks.

**Discussion Point**: Determine the frequency of terms appearing in the first document, in the third document, in the first or third document. What is the frequency of terms in all the documents together?

The above questions can all be answered using multiplication of a matrix by a column vector.

$$A \begin{pmatrix} 1 \\ 0 \\ 0 \\ 0 \end{pmatrix} = \begin{pmatrix} 32 \\ 9 \\ 6 \\ 2 \\ 0 \end{pmatrix} \qquad A \begin{pmatrix} 0 \\ 0 \\ 1 \\ 0 \end{pmatrix} = \begin{pmatrix} 1 \\ 0 \\ 0 \\ 0 \\ 2 \end{pmatrix}$$

$$A \begin{pmatrix} 1 \\ 0 \\ 1 \\ 0 \end{pmatrix} = \begin{pmatrix} 33 \\ 9 \\ 6 \\ 2 \\ 2 \end{pmatrix} \qquad A \begin{pmatrix} 1 \\ 1 \\ 1 \\ 1 \end{pmatrix} = \begin{pmatrix} 95 \\ 14 \\ 14 \\ 3 \\ 2 \end{pmatrix}$$

On this small example matrix multiplication, while illustrative, does not help with efficiency of obtaining an answer. However, we can lead the students to think further.

**Discussion Point:** Can you think about other questions about the set of documents that can be answered using matrix multiplication? (E.g., differences in word frequencies.) How would one extract information from very large datasets? (This is for the computer science students in the class: strategies for assembling, representing, storing, and operating upon a very large data matrix.)

At this point the students can appreciate that in order to get the information[5] about the terms in the $i$-th document we need to multiply $A$ by $e_i$, to find the information about the terms in documents $i$, $j$ and $k$ we need to multiply $A$ by $e_i + e_j + e_k$, or equivalently, add $Ae_i$, $Ae_j$, and $Ae_k$. We can view the matrix $A$ as a transformation that takes information about documents $(i, j, k)$ to information about terms ($Ae_i$, $Ae_j$, $Ae_k$).

$$\text{words} \xleftarrow{A} \text{documents}$$

Furthermore, we can remark that this transformation obeys certain rules

$$A(e_i + e_j + e_k) = Ae_i + Ae_j + Ae_k$$

which can be developed into the definition of linearity. We continue the discussion by presenting the transpose matrix.

$$A^T = \begin{pmatrix} 32 & 9 & 6 & 2 & 0 \\ 44 & 3 & 4 & 0 & 0 \\ 1 & 0 & 0 & 0 & 2 \\ 18 & 2 & 4 & 1 & 0 \end{pmatrix}$$

**Discussion Point:** Which documents contain the third term, the fifth term, the third or the fifth term?

$$A^T \begin{pmatrix} 0 \\ 0 \\ 1 \\ 0 \\ 0 \end{pmatrix} = \begin{pmatrix} 6 \\ 4 \\ 0 \\ 4 \end{pmatrix} \qquad A^T \begin{pmatrix} 0 \\ 0 \\ 0 \\ 0 \\ 1 \end{pmatrix} = \begin{pmatrix} 0 \\ 0 \\ 2 \\ 0 \end{pmatrix} \qquad A^T \begin{pmatrix} 0 \\ 0 \\ 1 \\ 0 \\ 1 \end{pmatrix} = \begin{pmatrix} 6 \\ 4 \\ 2 \\ 4 \end{pmatrix}$$

[5]The "information" here can be viewed as histograms over either documents or terms, which is something the students should be comfortable with.

Students can see the similarity with the discussion above. To find out how the $i$-th term is featured in documents we need to multiply $e_i$ by $A^T$.

$$\text{documents} \xleftarrow{A^T} \text{words}$$

Further questions can be discussed in this framework, touching upon elementary ideas not routinely discussed in the first course on linear algebra, such as non-negativity and sparsity.[6]

**Discussion Point:** Note that the term Madagascar only appears in the third document. Can we draw any conclusions from this?

**Discussion Point:** If we were to make a table that includes all the terms that appear in at least one of the four documents, would we expect most of the entries in the matrix to be equal to zero? Why?

**Discussion Point:** Note that all the elements in the matrix are non-negative integers. Can you think of any other tables with only non-negative integers? How about tables containing only non-negative real elements? Can you think about any other conditions on the entries that are imposed naturally in a particular setting?

**Discussion Point:** In class, we usually label the rows and columns of a matrix with successive integers: $1, \ldots, n$. These are generally used as "nominal numbers", meaning only their identities are important—like building numbers, course numbers, or social security numbers. And when we write $\sum_{i=1}^{n}$, what we usually mean is really $\sum_{i \in \text{rows}}$. We can change most of our formulas to use this convention. But in actual applications, as in the example here, often the rows and columns have natural labels: names of chemicals, words, documents, people, months, cities, *etc.* When this holds, we can use these labels instead of numbers as indices. And we can freely rearrange the rows and columns, keeping their labels, while still representing the same underlying mathematical object: the same matrix.

This point is illustrated by a term-by-document matrix, which has rows labeled by terms and columns labeled by documents. Let us look at another example. The two tables below contain movie ratings given by four users to five movies:

[6] Sparsity is of particular importance in computer science, where it impacts the representation and manipulation of both matrices and graphs.

**Table 2** Labeling rows and columns

| | Alice | Becky | Cindy | Dora |
|---|---|---|---|---|
| *Alien* | 4 | 1 | 4 | 5 |
| *Animal House* | 1 | 5 | 4 | 2 |
| *Beetlejuice* | 2 | 2 | 5 | 3 |
| *Jaws* | 5 | 1 | 5 | 5 |
| *Life of Brian* | 1 | 5 | 5 | 1 |

| | Becky | Cindy | Dora | Alice |
|---|---|---|---|---|
| *Animal House* | 5 | 4 | 2 | 1 |
| *Life of Brian* | 5 | 5 | 1 | 1 |
| *Beetlejuice* | 2 | 5 | 3 | 2 |
| *Jaws* | 1 | 5 | 5 | 5 |
| *Alien* | 1 | 4 | 5 | 4 |

**Discussion Point:** Compare the two tables. Do they contain the same information? Can you figure out the principle behind the ordering of rows and columns on the left and on the right?

# 3 Matrix Factorization

Students are familiar with the idea of factoring an integer as a product of prime numbers. Writing $6 = 3 \times 2$ gives us some information about the number 6. Another example is factoring a polynomial. Writing $x^4 - 10x^3 + 35x^2 - 50x + 24$ as $(x - 1)(x - 2)(x - 3)(x - 4)$ uncovers useful information. Both prime factor decomposition and factoring a polynomial are in general hard to do. Given two integers it is straightforward to find their product, but there is no known efficient algorithm for integer factorization.

This concept can, in some sense, be extended to matrices. Given a matrix, we want to write it as a product of two (or more) "simpler" matrices. There are several ways this can be done. A wide range of factorizations of matrices are used in applications, where—depending on the application—different properties of the factors are desired. An example that can be presented in the classroom is given below. The matrix

$$A = \begin{pmatrix} 2 & -1 & 1 & 2 \\ -1 & 1 & -2 & -1 \\ 1 & -2 & 5 & 1 \\ 2 & -1 & 1 & 2 \end{pmatrix}$$

can be factored in several ways:

$$A = \frac{1}{21}\begin{pmatrix} 1 & 1 & -1 & 1 \\ -1 & 0 & 0 & 3 \\ 2 & -1 & 0 & 1 \\ 1 & 1 & 1 & 0 \end{pmatrix}\begin{pmatrix} 7 & 0 & 0 & 0 \\ 0 & 3 & 0 & 0 \\ 0 & 0 & 0 & 0 \\ 0 & 0 & 0 & 0 \end{pmatrix}\begin{pmatrix} 3 & -3 & 6 & 3 \\ 7 & 0 & -7 & 7 \\ -10 & 3 & 1 & 11 \\ 1 & 6 & 2 & 1 \end{pmatrix}$$

$$= \begin{pmatrix} 1 & 1 \\ -1 & 0 \\ 2 & -1 \\ 1 & 1 \end{pmatrix}\begin{pmatrix} 1 & -1 & 2 & 1 \\ 1 & 0 & -1 & 1 \end{pmatrix}$$

$$= \begin{pmatrix} 2 & -1 & 0 & 0 \\ -1 & 1 & 0 & 0 \\ 1 & -2 & 1 & 0 \\ 2 & -1 & 0 & 1 \end{pmatrix}\begin{pmatrix} 1 & 0 & -1 & 1 \\ 0 & 1 & -3 & 0 \\ 0 & 0 & 0 & 0 \\ 0 & 0 & 0 & 0 \end{pmatrix}$$

The students may not have the mathematical tools to develop the factorizations above, but we can ask them to check their correctness, and to explore properties of the factors.

Demands from applications frequently impose conditions on the factors that are too strong to be satisfied exactly. For example, not every matrix can be written as a product of a column by a row. Or more generally, not every matrix can be written as a product of two low rank matrices. If we are unwilling to relax the conditions, we need to resort to approximate factorizations. Let us consider the matrix

$$A = \begin{pmatrix} 1 & 1 & 1 & 1.01 \\ 1 & 1 & 1.01 & 1 \\ 1 & 1.01 & 1 & 1 \\ 1.01 & 1 & 1 & 1 \end{pmatrix}.$$

Using elementary tools one can check that $A$ cannot be written as a product of a column and a row.

**Discussion Point:** Can we find a matrix that is close to the matrix $A$ that can be written as a product of a column by a row?

Students are likely to come up with the following solution:

$$A_1 = \begin{pmatrix} 1 & 1 & 1 & 1 \\ 1 & 1 & 1 & 1 \\ 1 & 1 & 1 & 1 \\ 1 & 1 & 1 & 1 \end{pmatrix} = \begin{pmatrix} 1 \\ 1 \\ 1 \\ 1 \end{pmatrix}\begin{pmatrix} 1 & 1 & 1 & 1 \end{pmatrix},$$

and we may present another one:

$$A_2 = \begin{pmatrix} 1.0025 & 1.0025 & 1.0025 & 1.0025 \\ 1.0025 & 1.0025 & 1.0025 & 1.0025 \\ 1.0025 & 1.0025 & 1.0025 & 1.0025 \\ 1.0025 & 1.0025 & 1.0025 & 1.0025 \end{pmatrix} = \begin{pmatrix} 1 \\ 1 \\ 1 \\ 1 \end{pmatrix} \begin{pmatrix} 1.0025 & 1.0025 & 1.0025 & 1.0025 \end{pmatrix}.$$

**Discussion Point:** Which solutions is better? What does it mean for a matrix $B$ to be close to $A$?

This question could be an introduction to the concept of matrix norms.

Factorizations of matrices have been developed that are used in applications, with the aim of uncovering hidden structure. We present a factorization that requires the factors to be non-negative and of a given low rank $r$. Those conditions are too strong, so the factorization will not be exact. This means that given a matrix $V$, we obtain matrices $W$ and $H$ so that the matrix $\hat{V} = WH$ is in some sense close to $V$.

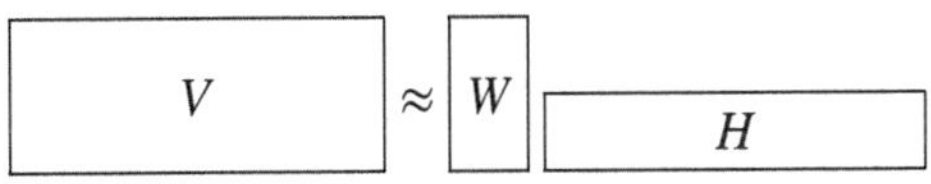


Non-negative matrix factorization, or NMF (Lee & Seung, 1999; Paatero & Tapper, 1994; Wang & Zhang, 2013), is a class of techniques for *approximately* factoring a matrix of non-negative numbers into the product of two such matrices: given an entry wise non-negative $n \times m$ matrix $V$, find two entry wise non negative matrices $W$ and $H$, of sizes $n \times r$ and $r \times m$, such that $V \approx WH$. (Even after a value for $r$ has been chosen, and an appropriate measure of similarity of two matrices has been chosen, there can be many possible solutions. However, popular NMF algorithms empirically usually find good solutions, a phenomenon which has been the subject of considerable analysis (Donoho & Stodden, 2004).)

Let us look at the nonnegative matrix factorization of the matrix that corresponds to the left side of Table 2:

$$A = \begin{pmatrix} 4 & 5 & 4 & 1 \\ 5 & 5 & 5 & 1 \\ 5 & 3 & 2 & 2 \\ 4 & 2 & 1 & 5 \\ 5 & 1 & 1 & 5 \end{pmatrix}$$

First we take $r$ to be equal to one. That means that we want to approximate $A$ by a product of a nonnegative column and a nonnegative row. The algorithm returns the following result:

$$W_1 = \begin{pmatrix} 7.137 \\ 8.214 \\ 6.398 \\ 5.974 \\ 6.155 \end{pmatrix} \qquad H_1 = \begin{pmatrix} 0.6709 & 0.4898 & 0.406 & 0.381 \end{pmatrix}$$

$$A - W_1H_1 = \begin{pmatrix} -0.7885 & 1.504 & 1.102 & -1.72 \\ -0.511 & 0.977 & 1.665 & -2.13 \\ 0.7077 & -0.1334 & -0.5976 & -0.4378 \\ -0.008175 & -0.926 & -1.426 & 2.724 \\ 0.8703 & -2.015 & -1.499 & 2.655 \end{pmatrix}$$

Taking $r = 2$ we get:

$$W_2 = \begin{pmatrix} 6.968 & 1.086 \\ 7.908 & 1.364 \\ 3.763 & 3.558 \\ 0.5117 & 6.448 \\ 0 & 7.197 \end{pmatrix} \qquad H_2 = \begin{pmatrix} 0.5171 & 0.6379 & 0.5707 & 0 \\ 0.6658 & 0.1897 & 0.1095 & 0.7133 \end{pmatrix}$$

$$A - W_2H_2 = \begin{pmatrix} -0.3256 & 0.3496 & -0.09564 & 0.2257 \\ 0.002296 & -0.3033 & 0.337 & 0.02677 \\ 0.6857 & -0.07508 & -0.5374 & -0.5376 \\ -0.5576 & 0.4506 & 0.001623 & 0.4004 \\ 0.2088 & -0.365 & 0.2117 & -0.1333 \end{pmatrix}$$

**Discussion Point:** Compare $A - W_1H_1$ and $A - W_2H_2$. Can you find a non-negative factorization of $A$ for $r = 4$?

Let us have a closer look at $W_2$ and $H_2$. Recall that the rows of $W_2$ correspond to movies, and the columns of $H_2$ correspond to users.

**Discussion Point:** Can we give sensible labels to the columns of $W_2$, or equivalently, the rows of $H_2$.

Note that the highest values in the first column of $W_2$ correspond to movies Alien and Jaws, while the highest values in the second column correspond to movies Animal house and Life of brian. Based on this, we may agree to label the first column "Horror", and the second column "Comedy". The rows of $H_2$ are labeled correspondingly. Values in $H_2$ can now be interpreted in the following way. Cindy likes both horror and comedy movies, Dora and Alice prefer horror movies, and Becky likes comedy, but not horror movies. Matrix factorization uncovered genre for our movies.

In the context of the small example, we can look at $V$ as a transformation from "movies" to "people". Now we have the third notion appearing: "genres". The matrix $H$ can be seen as a transformation that takes movies to genres, and $W$ takes genres to people.

$$\text{people} \overbrace{\xleftarrow{W} \text{genres} \xleftarrow{H}}^{V} \text{movies}$$

We may remark to the students that they are justified in finding this example a bit contrived. The is example is too small (and also made up) to be very convincing. We give a larger example based on term-by-document matrix later in the chapter.

The formula for matrix multiplication, $A = BC$,

$$a_{ij} = \sum_{k=1}^{m} b_{ik} c_{kj}$$

can be intimidating to students.

From the point of view of our example, where the rows of $V$ are indexed by "movies" and the columns by "people", we can write down the same formula in the following way:

$$\hat{v}_{\text{movie,person}} = \sum_{g \in \text{genre}} w_{\text{movie},g} \, h_{g,\text{person}}$$

The entry in the matrix $\hat{V}$ that represents a rating of a chosen "movie" to a chosen "person" is computed by summing up the product of how much the person likes genre $g$ and how much the movie is in genre $g$, over all genres. This process can be depicted graphically:

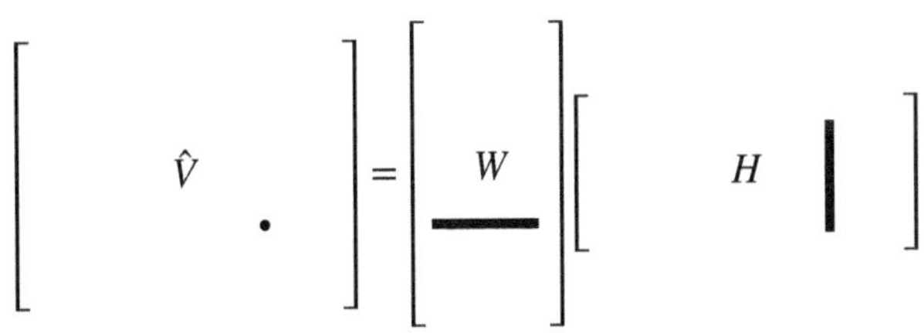


On the other hand we may notice that $\hat{V}$ is the sum of rank one matrices, each of them giving the contribution of a particular genre.

$$\begin{bmatrix} & & \\ & \hat{V} & \\ & & \end{bmatrix} = \sum_g \begin{bmatrix} \\ W_{\bullet,g} \\ \\ \end{bmatrix} \begin{bmatrix} & H_{g,\bullet} & \end{bmatrix}.$$

This is written as:

$$\begin{aligned} V &= \sum_{g=1}^{r} (g\text{-th column of } W) \cdot (g\text{-th row of } H) \\ &= \sum_{g \in \text{genres}} \text{Ratings Matrix for genre } g \\ &= \text{sum of rank-one per-genre matrices} \end{aligned}$$

This interpretation reinforces the power of the nonnegative matrix factorization. From a bundle of documents, it singles out a particular genre in way that agrees with our intuition in a surprisingly strong way.

Let us go back to the example of movie ratings discussed earlier. We have approximated our matrix $A$ by $W_2H_2$. Below this product is written as a sum of two rank one matrices:

$$\begin{aligned} W_2H_2 &= \begin{pmatrix} 6.968 & 1.086 \\ 7.908 & 1.364 \\ 3.763 & 3.558 \\ 0.5117 & 6.448 \\ 0 & 7.197 \end{pmatrix} \begin{pmatrix} 0.5171 & 0.6379 & 0.5707 & 0 \\ 0.6658 & 0.1897 & 0.1095 & 0.7133 \end{pmatrix} \\ &= \begin{pmatrix} 6.968 \\ 7.908 \\ 3.763 \\ 0.5117 \\ 0 \end{pmatrix} \begin{pmatrix} 0.5171 & 0.6379 & 0.5707 & 0 \end{pmatrix} \\ &\quad + \begin{pmatrix} 1.086 \\ 1.364 \\ 3.558 \\ 6.448 \\ 7.197 \end{pmatrix} \begin{pmatrix} 0.6658 & 0.1897 & 0.1095 & 0.7133 \end{pmatrix} \\ &= \begin{pmatrix} 3.603 & 4.445 & 3.977 & 0 \\ 4.089 & 5.045 & 4.514 & 0 \\ 1.946 & 2.4 & 2.148 & 0 \\ 0.2646 & 0.3264 & 0.292 & 0 \\ 0 & 0 & 0 & 0 \end{pmatrix} + \begin{pmatrix} 0.7227 & 0.2059 & 0.1189 & 0.7743 \\ 0.9084 & 0.2588 & 0.1495 & 0.9732 \\ 2.368 & 0.6747 & 0.3897 & 2.538 \\ 4.293 & 1.223 & 0.7064 & 4.6 \\ 4.791 & 1.365 & 0.7883 & 5.133 \end{pmatrix} \end{aligned}$$

**Supplementary Advanced Material**

To minimize the Frobenius norm, $\|V - WH\|_F$, a surprisingly simple method is available. Two updates can be iterated

$$H := H \odot (W^T V \div W^T WH)$$
$$W := W \odot (VH^T \div WHH^T)$$

where $\odot$ denotes elementwise multiplication and $\div$ denotes elementwise division. Typically after each update of $W$ its columns are normalized to unit sum.

An enormous number of variations and embellishments of the basic NMF algorithm have been developed, with applications ranging from astronomy to zoology.

## 4 Example using Module Descriptors

We give an example of factoring a data matrix involving a corpus of documents: module descriptors for 62 mathematics modules that were taught in the School of Mathematics and Statistics, University College Dublin (UCD) in 2015. Module descriptors are relatively short documents that give overviews of the courses. Here are two representative examples of module descriptors.

**Numbers and Functions**

This module is an introduction to the joys and challenges of mathematical reasoning and mathematical problem-solving, organised primarily around the theme of properties of the whole numbers. It begins with an introduction to some basic notions of mathematics and logic such as proof by contradiction and mathematical induction. It introduces the language of sets and functions, including injective surjective and bijective maps and the related notions of left-, right- and 2-sided inverses. Equivalence relations, equivalence classes. It covers basic important principles in combinatorics such as the Principle of Inclusion-Exclusion and the Pigeonhole Principle. The greater part of the module is devoted to number theory: integers, greatest common divisors, prime numbers, Euclid's algorithm, the Fundamental Theorem of Arithmetic, congruences, Fermat's theorem, Euler's theorem, and arithmetic modulo a prime and applications. The module concludes with some topics from elementary coding theory / cryptography such as the RSA encryption system.

**Groups, Rings and Fields**

This course will be an introduction to group theory, ring theory and field theory. We will cover the following topics: definition and examples of groups, subgroups, cosets and Lagrange's Theorem, the order of an element of a group, normal subgroups and quotient groups, group homomorphisms and the homomorphism theorem, more isomorphism theorems, definitions of a commutative ring with unity, integral domains and fields, units, irreducibles and primes in a ring, ideals and quotient rings, prime and maximal ideals, ring homomorphisms and the homomorphism theorem, polynomial rings, the division algorithm, gcd for polynomials, irreducible polynomials and field extensions. Time permitting, we may cover the Sylow theorems, solvable groups and further examples of groups.

This set was chosen because the example was developed for the first linear algebra classes at UCD. All the students in those classes were quite familiar with the chosen set of documents, which they need to navigate each semester when choosing and registering for their modules. The number of documents is large enough that we can make a case for needing a computer to help navigate them, but small enough that we can exert some manual control and can be familiar with the entire corpus.

All the words that appear in any of the 62 documents were collected. So-called stop words (common words like "and", "the", and the like), and all the words that appeared fewer than four times, were removed. Words were also down-cased and stemmed, so for example the terms *Eigenvalue* and *eigenvalues* are deemed equivalent. This resulted in a $290 \times 62$ matrix $V$ of word counts. Below we show some of the data, starting with lists of the most and the least frequent words.

Using the standard Octave (Eaton, Bateman, Hauberg, & Wehbring, 2015) package for NMF, the entry-wise nonnegative matrix $V$ is factored as

$$V \approx WH,$$

where $W$ is an entry-wise non-negative $290 \times r$ matrix and $H$ is an entry-wise non-negative $r \times 62$ matrix. We will see that factoring a matrix in this way reveals a particular structure of the matrix which reveals something about the content of the original documents. Different values for $r$ are used below, and we will see how the information that we obtain changes as we increase $r$.

For easier interpretation, the entries in each column of $W$ have been permuted so that they appear in descending order, and the term corresponding to each row is shown. (Recall the discussion about labelling rows and columns at the end of Sect. 2.) We also present a few columns of the matrix $H$ for $r = 3$.

**Table 3** Most and least frequent words

| **Most Frequent Words** | |
|---|---|
| **word** | **count** |
| function | 117 |
| theorem | 75 |
| linear | 72 |
| matrix | 66 |
| theory | 58 |
| equation | 53 |
| mathematical | 52 |
| mathematics | 52 |
| understand | 49 |
| science | 48 |
| problem | 44 |

| **Least Frequent Words** | | | | |
|---|---|---|---|---|
| addition | advanced | arguments | background | behaviour |
| classify | column | computation | constrained | construct |
| continuous | definite | depth | described | directional |
| double | elimination | engineering | evaluate | expressions |
| flow | foundations | general | importance | independence |
| induction | integrate | interpret | introduces | known |
| manipulate | max | maxima | min | minima |
| nash | nullity | numerical | original | possible |
| prime | quadratic | range | related | riemann |
| row | sample | search | significant | solid |
| special | stock | sum | sylow | together |
| uncountable | variety | | | |

## 5 Discussion

Following on from the discussion around the small example presented above, the students understand how the frequency of the terms across all documents is computed. This gives an easy and automated way to derive the most and least frequent words, given in Table 3. While in small example shown the most and least frequent words could easily be found by hand, this is impractical when the matrix becomes large.

**Discussion Point:** Consider different columns of the matrix $W$ for $r = 2, 3$ given in Tables 4 and 5. What do you observe?

Already in the case when $r = 2$, we can see some regularity in the way the terms are grouped into columns. For example, it makes sense that the terms *function*, *derivative*, *differential*, *integral* appear in the same column. In the second column we see the terms *group*, *ring*, *isomorphism*, *homomorphism*, *sylow*, *subgroups*, *quotient*, *cauchy* appearing together.

The factorization is perhaps the most informative for the choice $r = 3$, so let us take a closer look at this case. The terms in the matrix $W$ are grouped in such a sensible way that we can challenge the students to give them titles. Those students who've read ahead a little may suggest *Abstract Algebra* for the first column, while most should find *Calculus* appropriate for the second, and *Linear Algebra* for the third. Things become a little less clear when we consider the matrix $W$ for $r = 4$ (Table 6).

**Table 4** $W$-matrix, $r = 2$

| | | | |
|---|---|---|---|
| function | 23.0 | group | 17.9 |
| linear | 11.1 | theorem | 15.0 |
| matrix | 11.0 | theory | 6.9 |
| equation | 9.9 | ring | 6.6 |
| derivative | 9.4 | understand | 4.0 |
| calculus | 7.8 | structure | 3.3 |
| differential | 6.5 | example | 3.0 |
| solve | 6.1 | number | 2.9 |
| problem | 5.9 | isomorphism | 2.7 |
| mathematical | 5.3 | concepts | 2.6 |
| science | 5.2 | homomorphisms | 2.5 |
| compute | 5.1 | sylow | 2.3 |
| variable | 4.7 | subgroups | 2.3 |
| applications | 4.7 | quotient | 2.2 |
| integral | 4.6 | cauchy | 2.2 |

**Table 5** $W$-matrix, $r = 3$

| | | | | | |
|---|---|---|---|---|---|
| group | 18.1 | function | 27.1 | matrix | 19.1 |
| theorem | 14.8 | derivative | 11.1 | linear | 14.9 |
| theory | 6.8 | calculus | 9.2 | space | 9.5 |
| ring | 6.7 | equation | 6.3 | vector | 8.7 |
| understand | 3.9 | differential | 5.8 | algebra | 7.1 |
| structure | 3.4 | integral | 5.7 | basis | 6.5 |
| number | 3.0 | problem | 5.5 | equation | 6.4 |
| example | 2.9 | variable | 5.3 | compute | 5.7 |
| isomorphism | 2.6 | graph | 4.8 | system | 4.9 |
| homomorphisms | 2.6 | limit | 4.6 | rank | 3.5 |
| concepts | 2.5 | solve | 4.5 | complex | 3.5 |
| sylow | 2.3 | mathematics | 4.3 | product | 3.4 |
| subgroups | 2.3 | calculate | 3.9 | number | 3.3 |
| applications | 2.2 | applications | 3.8 | mathematical | 3.3 |
| quotient | 2.2 | science | 3.5 | science | 3.3 |
| cauchy | 2.1 | introduction | 3.5 | solve | 3.2 |
| time | 2.1 | mathematical | 3.5 | dimensional | 3.0 |
| finite | 2.1 | method | 3.4 | eigenvalues | 2.9 |
| algebraic | 2.1 | polynomial | 3.4 | set | 2.8 |
| permitting | 2.0 | differentiation | 3.3 | eigenvectors | 2.7 |

**Table 6** $W$-matrix, $r = 4$

| | | | | | | | |
|---|---|---|---|---|---|---|---|
| group | 20.5 | function | 25.2 | matrix | 19.3 | theorem | 9.0 |
| theorem | 11.3 | derivative | 11.7 | linear | 14.8 | understand | 8.7 |
| ring | 6.7 | calculus | 8.4 | space | 9.0 | question | 6.3 |
| theory | 4.8 | equation | 6.7 | vector | 8.7 | complex | 6.1 |
| structure | 3.9 | differential | 6.3 | algebra | 7.1 | number | 5.6 |
| isomorphism | 3.0 | problem | 5.8 | basis | 6.5 | example | 5.6 |
| homomorphisms | 2.9 | variable | 5.5 | equation | 6.4 | concepts | 5.4 |
| applications | 2.7 | graph | 4.9 | compute | 5.7 | mathematical | 5.2 |
| sylow | 2.6 | solve | 4.8 | system | 4.8 | function | 5.1 |
| subgroups | 2.6 | limit | 4.3 | rank | 3.6 | cauchy | 4.9 |
| quotient | 2.6 | applications | 4.2 | product | 3.4 | theory | 4.6 |
| algebra | 2.3 | integral | 4.1 | solve | 3.1 | integral | 3.8 |
| algebraic | 2.2 | calculate | 4.1 | science | 3.1 | demonstrate | 3.8 |
| time | 2.1 | polynomial | 3.7 | complex | 3.0 | correctly | 3.4 |
| permitting | 2.1 | differentiation | 3.4 | eigenvalues | 2.9 | method | 3.4 |
| finite | 1.9 | science | 3.3 | dimensional | 2.9 | series | 3.3 |
| lagrange | 1.8 | mathematics | 3.2 | eigenvectors | 2.8 | write | 3.3 |
| special | 1.7 | introduction | 3.2 | number | 2.7 | set | 3.3 |
| construct | 1.5 | inverse | 3.0 | mathematical | 2.6 | sequence | 3.3 |

**Table 7** $H$-matrix, $r = 3$

| Module Names | | | |
|---|---|---|---|
| Numbers and Functions | Linear Algebra with Applications to Economics | Groups, Rings, and Fields | Differential Equations via Computer Algebra |
| 0.1 | 0.0 | 0.3 | 0.0 |
| 0.1 | 0.0 | 0.0 | 0.1 |
| 0.0 | 0.2 | 0.0 | 0.0 |

**Discussion Point:** What are the advantages and disadvantages of choosing $r$ to be small or big?

While higher values of $r$ will make $\hat{V}$ closer to $V$, they can make it more difficult to interpret the results. An informed choice of $r$, dependent on the needs of the applications, needs to be made. This problem of "model complexity" has been the subject of a great deal of research in Statistics and Machine Learning.

**Discussion Point:** In the $r = 3$ case, we were able to give titles to columns of the matrix $W$. Those titles could be called "topics". The rows of $W$ are indexed by "words" and the columns by "topics". For the multiplication $WH$ to make sense we need to have the rows of $H$ marked by "topics". Let us look at the matrix $H$ given in Table 7 to see if this makes sense.

Representative columns given for the matrix $H$ agree with our prediction that the first row corresponds to Abstract Algebra, the second row to Calculus and the third row to Linear Algebra. For example, the course *Linear Algebra with Applications to Economics* has the only nonzero entry in the third row, while the course *Differential Equations via Computer Algebra* has the only nonzero entry in the second row.

## 6 Conclusion

While this is a black box experiment for the students, they are able to appreciate the result and understand the emergence of the topics in an example. The NMF algorithm yields this topic analysis, helping us appreciate the strengths of the method. If we want to bring the discussion further, it can be pointed out how this class of algorithm is used to decompose speech and music into phonemes and notes (Asari et al., 2006; O'Grady and Pearlmutter, 2008; Smaragdis & Brown, 2003), in speech denoising (Wilson et al., 2008) and recognition (Hurmalainen, 2014), in chemistry (Siy et al., 2008) and biomedical sciences (Helleday et al., 2014; Ortega-Martorell, Lisboa, Vellido, Julià-Sapé, & Arús, 2012; Paine et al., 2016; Ray & Bandyopadhyay, 2016), in the analysis of the cosmic microwave background radiation (Cardoso, Delabrouille, & Patanchon, 2003), etc.

The example presented above can be adapted for classroom needs in various ways. An aspect not discussed here is the potential to turn some of the above ideas into student projects. We are aware that the computational aspects of this may be a big stumbling block, so we are developing a web-based tool to make it easy for students to analyse a set of document in this way. We see a potential for interdisciplinary projects, where students are charged with the task of analyzing a large body of documents on a particular subject, and use linear algebra to reach some conclusions.

In collaboration with Miao Wei,[7] we have created an end-to-end interactive browser-based implementation of the processing pipeline discussed above (taking documents as input and processing them through stemming, the construction of a term-by-document matrix, NMF, and visualization of the resulting factor matrices), which is being made available online.[8]

## References

Asari, H., Pearlmutter, B.A., Zador, A.M.: Sparse representations for the cocktail party problem. Journal of Neuroscience **26**(28), 7477–90 (2006). https://doi.org/10.1523/JNEUROSCI.1563-06.2006

[7]Dept of Computer Science, Maynooth University, Ireland, e-mail: davidweimiao@gmail.com.

[8]http://barak.pearlmutter.net/demo/NMF/

Cardoso, J.F., Delabrouille, J., Patanchon, G.: Independent component analysis of the cosmic microwave background. In: Fourth International Symposium on Independent Component Analysis and Blind Signal Separation, pp. 1111–6. Nara, Japan (2003)
Donoho, D., Stodden, V.: When does non-negative matrix factorization give a correct decomposition into parts? In: Advances in Neural Information Processing Systems 16. MIT Press (2004). http://books.nips.cc/papers/files/nips16/NIPS2003_LT10.pdf
Eaton, J.W., Bateman, D., Hauberg, S., Wehbring, R.: GNU Octave version 4.0.0 manual: a high-level interactive language for numerical computations. Free Software Foundation (2015). http://www.gnu.org/software/octave/doc/interpreter
Helleday, T., Eshtad, S., Nik-Zainal, S.: Mechanisms underlying mutational signatures in human cancers. Nature Reviews Genetics **15**, 585–98 (2014). https://doi.org/10.1038/nrg3729
Hurmalainen, A.: Robust speech recognition with spectrogram factorisation. Ph.D. thesis, Tampere University of Technology, Finland (2014). http://dspace.cc.tut.fi/dpub/bitstream/handle/123456789/22512/hurmalainen.pdf
Lee, D.D., Seung, H.S.: Learning the parts of objects with nonnegative matric factorization. Nature **401**, 788–91 (1999). https://doi.org/10.1038/44565
Lesh, R., English, L.D.: Trends in the evolution of models & modeling perspectives on mathematical learning and problem solving. ZDM Mathematics Education **37**(6), 487–9 (2005). https://doi.org/10.1007/BF02655857
Niegowski, M., Zivanovic, M.: ECG-EMG separation by using enhanced non-negative matrix factorization. In: 2014 36th Annual International Conference of the IEEE Engineering in Medicine and Biology Society, pp. 4212–5 (2014). https://doi.org/10.1109/EMBC.2014.6944553
O'Grady, P.D., Pearlmutter, B.A.: Discovering speech phones using convolutive non-negative matrix factorisation with a sparseness constraint. Neurocomputing **72**(1–3), 88–101 (2008). https://doi.org/10.1016/j.neucom.2008.01.033
Ortega-Martorell, S., Lisboa, P.J., Vellido, A., Julià-Sapé, M., Arús, C.: Non-negative matrix factorisation methods for the spectral decomposition of MRS data from human brain tumours. BMC Bioinformatics **13**(1), 38 (2012). https://doi.org/10.1186/1471-2105-13-38
Paatero, P., Tapper, U.: Positive matrix factorization: A nonnegative factor model with optimal utilization of error estimates of data values. Environmetrics **5**(2), 111–26 (1994). https://doi.org/10.1002/env.3170050203
Paine, M.R.L., Kim, J., Bennett, R.V., Parry, R.M., Gaul, D.A., Wang, M.D., et al.: Whole reproductive system non-negative matrix factorization mass spectrometry imaging of an early-stage ovarian cancer mouse model. PLoS ONE **11**(5), e0154,837 (2016). https://doi.org/10.1371/journal.pone.0154837
Ponnapalli, S.P., Saunders, M.A., Van Loan, C.F., Alter, O.: A higher-order generalized singular value decomposition for comparison of global mRNA expression from multiple organisms. PLOS ONE **6**(12), 1–11 (2011). https://doi.org/10.1371/journal.pone.0028072
Possani, E., Trigueros, M., Preciado, J., Lozano, M.: Use of models in the teaching of linear algebra. Linear Algebra and its Applications **432**(8), 2125–40 (2010). https://doi.org/10.1016/j.laa.2009.05.004. http://www.sciencedirect.com/science/article/pii/S0024379509002523. Special issue devoted to the 15th ILAS Conference at Cancun, Mexico, June 16-20, 2008
Ray, S., Bandyopadhyay, S.: A NMF based approach for integrating multiple data sources to predict HIV-1-human PPIs. BMC Bioinformatics **8**(17) (2016). https://doi.org/10.1186/s12859-016-0952-6
Salgado, H., Trigueros, M.: Teaching eigenvalues and eigenvectors using models and APOS theory. The Journal of Mathematical Behavior **39**, 100–20 (2015). https://doi.org/10.1016/j.jmathb.2015.06.005. http://www.sciencedirect.com/science/article/pii/S0732312315000462
Siy, P.W., Moffitt, R.A., Parry, R.M., Chen, Y., Liu, Y., Sullards, M.C., Merrill Jr., A.H., Wang, M.D.: Matrix factorization techniques for analysis of imaging mass spectrometry data. In: 8th IEEE International Conference on BioInformatics and BioEngineering (BIBE 2008), pp. 1–6 (2008)

Smaragdis, P., Brown, J.C.: Non-negative matrix factorization for polyphonic music transcription. In: IEEE Workshop on Applications of Signal Processing to Audio and Acoustics, pp. 177–180 (2003). https://doi.org/10.1109/ASPAA.2003.1285860

Stewart, S., Thomas, M.O.J.: Difficulties in the acquisition of linear algebra concepts. New Zealand Journal of Mathematics **32**(Supplementary Issue), 207–15 (2003). https://www.math.auckland.ac.nz/~thomas/My%20PDFs%20for%20web%20site/21%20Stewart.pdf

Trigueros, M., Possani, E.: Using an economics model for teaching linear algebra. Linear Algebra and its Applications **438**(4), 1779–92 (2013). https://doi.org/10.1016/j.laa.2011.04.009. http://www.sciencedirect.com/science/article/pii/S0024379511003053. 16th ILAS Conference Proceedings, Pisa 2010

Wang, Y.X., Zhang, Y.J.: Nonnegative matrix factorization: A comprehensive review. IEEE Transactions on Knowledge and Data Engineering **25**(6), 1336–53 (2013). https://doi.org/10.1109/TKDE.2012.51

Wilson, K.W., Raj, B., Smaragdis, P., Divakaran, A.: Speech denoising using nonnegative matrix factorization with priors. In: IEEE International Conference on Acoustics, Speech and Signal Processing (ICASSP), pp. 4029–4032 (2008)

# Motivating Examples, Meaning and Context in Teaching Linear Algebra

**David Strong**

**Abstract** In this chapter I examine how we as authors and instructors can move beyond the *what* and *how* of the ideas that we teach and more effectively address the *why*. I discuss the need for-and give several instances of-relevant and thought-provoking examples that can better motivate the various concepts of linear algebra and simultaneously pique the interest of our students in those concepts.

**Keywords** Motivation · Meaning · Context

## 1 Introduction

Linear algebra textbooks often present and develop a new concept without paying much attention to motivation, real-life meaning or context. It is fairly standard for a section in a textbook to consist of a few definitions, a couple of theorems and their proofs, and a few somewhat contrived examples or applications in which the given ideas are illustrated. Textbooks typically do a good job with the *what* and the *how*, but not as well with the *why*. Often it isn't until subsequent sections that students begin to understand the importance and use of the ideas learned in the previous sections. While this is sometimes the inherent nature of mathematics, it doesn't usually have to be this way. Instead, we (textbook authors and course instructors) have a golden opportunity to simultaneously motivate the need for the ideas and motivate the students to want to learn about those ideas. If students care, they will learn. In this chapter we discuss how we can better address the *why* through relevant and thought-provoking examples to better motivate the need for the ideas taught in the course and to simultaneously pique the interest of the student.

We begin in Sect. 2 by discussing two examples that I typically present at the beginning of an introductory linear algebra course to motivate many of the ideas that my students will learn in the course. In Sect. 3 we discuss examples for motivating a smaller collection ideas, for example as contained within a chapter. We follow this in

D. Strong (✉)
Pepperdine University, 24255 PCH, Malibu, CA 90263, USA
e-mail: David.Strong@pepperdine.edu



S. Stewart et al. (eds.), *Challenges and Strategies in Teaching Linear Algebra*,
ICME-13 Monographs, https://doi.org/10.1007/978-3-319-66811-6_16

Sect. 4 with a discussion on motivating specific concepts by showing those concepts in action before formally introducing them. In Sect. 5 we discuss how including more context for and meaning of linear algebra concepts can increase and enrich learning. Section 6 contains some closing thoughts.

## 2 Examples to Motivate the Entire Course

Mathematics was born and continues to evolve in our attempt to understand and manipulate the world around us. At the heart of mathematics is its usefulness in countless applications, past, present and future.

When I teach a mathematics course that is more naturally application-oriented (linear algebra, calculus, differential equations, numerical analysis, etc.), early in the semester I like to briefly explore with my students five to ten pertinent real world examples to provide some motivation and context for the ideas that they will see in the course. Referring to those same examples throughout the semester also adds some nice continuity to the course and helps students better interconnect the various ideas they learn. Of course, throughout the semester the students will see other applications and examples, as well as more theoretical examples and homework problems in which the ideas are explored at a more abstract level. But I find it very effective to initially motivate the ideas with meaningful and interesting real world problems. The problems are the "hook" that catches a student's attention and gets him or her interested in wanting to learn more. It's like watching a 2 min preview of a film and consequently wanting to see the film in its entirety.

Below I share two examples that my students and I discuss on the first or second day. In Sect. 2.1 is an example of a matrix as an array of coefficients, and in Sect. 2.2 is an example of a matrix as an operator on a vector.

### *2.1 A Matrix as an Array of Coefficients: A Nickel and Dime Problem*

While the following example is a bit contrived to keep it sufficiently simple, it still manages to motivate a plethora of ideas that will arise during a typical introductory linear algebra course. I usually give this example on the first day of class.

*Suppose that you will choose a certain number of nickels and dimes, and that you must satisfy one or two or all three of the conditions below:*

1. *You have six coins.*
2. *You have five times as many dimes as nickels.*
3. *You have 75 cents.*

I ask my students to come up with the number of coins needed to satisfy the seven possible combinations of conditions, as listed below. Of course I don't give my students the solutions listed in the final column.

| Problem | 1 | 2 | 3 | 4 | 5 | 6 | 7 |
|---|---|---|---|---|---|---|---|
| Conditions | 1 | 2 | 3 | 1, 2 | 1, 3 | 2, 3 | 1, 2, 3 |

Students will often initially devote as little time as possible to finding a solution, which in this case means they simply try to guess solutions. This works well enough for the first three problems and typically for the fourth, which has solution $(1, 5)$. Problems 5 and 6 have solutions $(-3, 9)$ and $(\frac{15}{11}, \frac{75}{11})$, respectively. These are solutions only if we allow negative and fractional numbers of coins, thus they are solutions which students probably won't simply guess. Consequently, a systematic way of solving the system of equations is needed. Moreover, once solutions are found, one should determine if they are relevant to the real-life problem.

My students and I begin this exercise at the end of the first day of class, and we complete our discussion at the beginning of the second day. Between class meetings, students have time to try to do more than merely guess solutions. Some do and some don't, but during the first few minutes of Day 2 we collectively realize that we need a more reliable and systematic way of dealing with these problems: language, equations, processes for working with the equations, etc.

Some students come up with the following equations to describe the conditions:

$$\begin{aligned} n + \quad d &= 6 \\ 5n - \quad d &= 0 \\ 5n + 10d &= 75 \end{aligned}, \quad \text{that is,} \quad \begin{bmatrix} 1 & 1 \\ 5 & -1 \\ 5 & 10 \end{bmatrix} \begin{bmatrix} n \\ d \end{bmatrix} = \begin{bmatrix} 6 \\ 0 \\ 75 \end{bmatrix}. \tag{1}$$

(By Day 2 students may not yet have seen the matrix-vector form of this problem, but this is a simple enough transition.) Of course having these equations makes solving Problems 4–7 more doable, but how to do so is still not necessarily obvious. Additionally, there are several other observations and issues that arise in or are motivated by this problem. Too many to discuss on Day 1, they come up throughout the semester as the related ideas are introduced. If I were using this as an introductory example in a linear algebra textbook, I might simply list these issues and observations. When I use this as an in-class example, my students and I discover them together.

- There is more than one way to use the equations to solve this problem. One can solve for one unknown in terms of another, substitute that into another equation, and so on. Or one can add a multiple of one equation to another. Either way, the goal and result are the same: one unknown has been eliminated from an equation.
- If using the substitution approach, it doesn't really matter what form the equations are in, but if adding a multiple of one equation to another, it's more helpful to have the equations in standard form, as above in (1).
- Either of these methods works fine when there is a unique solution, but what if there is no solution—how do we recognize that there is no solution? (Not being able to find a solution of course doesn't necessarily mean there is no solution.)
- Sometimes solutions are valid mathematically, but don't make sense in the real world, such as fractional or negative amounts of coins.

- Visualization is useful, and plotting the three lines makes it clear how many solutions there are: 0, 1 or $\infty$. But what about in higher dimensions, i.e. when there are more than two or three unknowns and we cannot visualize the equations? Is there some process we can use regardless of the number of unknowns?
- The more restrictions (equations) we are trying to satisfy, the more "difficult" it is for there to be a solution. While not completely obvious at this stage, students get an introduction to the typical (albeit with exceptions) relationship between the number of equations $m$ and the number of unknowns $n$:

| $m$ vs. $n$ | Equations | Solutions | Example with $n = 2$ |
|---|---|---|---|
| $m > n$ | "Too many" | 0 | Three lines: do not intersect at any point |
| $m = n$ | "Just right" | 1 | Two lines: intersect at one point |
| $m < n$ | "Not enough" | $\infty$ | One line: every point on line is a solution |

- Another fact that is not yet obvious based on only this problem, but that is first observed while working this problem: if there is more than one solution, then there is an infinite number of solutions.
- A question that arises at this point: if there are "too many" equations, i.e. more equations than unknowns, is it possible to have a solution, or perhaps even an infinite number of solutions?
- When there is no solution, or when there is an infinite number of solutions, or when there is one solution but where the number of equations $\neq$ the number of unknowns, is there a process with which we can determine whether there is a solution, and if so, how many, and if more than one, what those solutions are?

Of course (as much as students sometimes can be annoyed by it) Gaussian Elimination can help us address all of these issues.

Not only does the above coins example motivate and give rise to a number of ideas both simple and deep, other ideas that are introduced later in the course are also motivated or illustrated by this easy first-day example in (1), including:

- Pivots, pivot rows, pivot columns, etc.
- Linear combinations, span, column space, linear independence:
  - In $A\mathbf{x} = \mathbf{b}$ in (1), can we build $\mathbf{b}$ out of the columns of $A$; that is, is $\mathbf{b}$ in the span of the columns of $A$?
  - Do the columns of $A$ span all of $R^3$?
  - What conditions are necessary for the columns of $A$ to span $R^3$?
  - What conditions are sufficient for the columns of $A$ to span $R^3$?
  - If $\mathbf{b}$ can be built out of the columns of $A$, is there more than one way to do so, and if so, is this a "good" thing or "bad" thing?
- A "best" solution, if there is not an exact solution:
  - What does a "best" solution even mean?

  - Building a vector $\hat{\mathbf{b}}$ that is as close to the desired vector $\mathbf{b}$ using the given vectors, e.g. the columns of $A$.
  - Projecting a vector onto the column space of a matrix or onto some other collection of vectors.

- Changing right hand side values possibly changes existence and/or the number (0 or 1 or $\infty$) of solutions and the values of the solution(s):

  - Is it possible to have a solution with three equations and two unknowns? Could we change, for example, the 75 in the third equation in (1) and end up with a modified problem for which there is a solution?
  - In general, is it possible that an overdetermined system ($m > n$) has a solution?
  - Even more generally, is each of the three numbers of solutions 0, 1 and $\infty$ possible for each of the three cases $m < n$, $m = n$ and $m > n$?

Even on Day 1, students already know most of what they need to, and already have a lot of experience and intuition, in order to work with this problem and contemplate many of the above issues. One idea with which they will likely not yet be familiar with is the matrix-vector version of the problem and the corresponding vector version of the problem, that is, that (1) is equivalent to

$$n \begin{bmatrix} 1 \\ 5 \\ 5 \end{bmatrix} + d \begin{bmatrix} 1 \\ -1 \\ 10 \end{bmatrix} = \begin{bmatrix} 6 \\ 0 \\ 75 \end{bmatrix}. \tag{2}$$

As building vectors from other vectors is a core part of linear algebra, I'm a proponent of introducing this view of systems of linear equations as early as possible.

Of course we don't explore all of the above on Day 1 or 2, but the students see that there are many interesting and important questions yet to be answered. My twin goals are to cultivate student curiousity and to create a bit of a challenge for them, both of which ultimately encourage my students to care and to want to learn more. And of course there are several other good motivating first-day examples that could accomplish the same purpose. I also like to return to this example in my end-of-semester review to point out all of the ideas that we've learned that are evident in this one problem.

## 2.2 A Matrix as an Operator: Discrete Dynamical Systems

A second fundamental concept I like to introduce as early in the course as possible is the idea of a matrix as an operator. The discrete predator-prey problem is a great introduction to matrices as operators.

Consider the predator-prey problem

$$W_{k+1} = \frac{\sqrt{3}}{2} W_k + \frac{1}{2} R_k$$
$$R_{k+1} = -\frac{1}{2} W_k + \frac{\sqrt{3}}{2} R_k$$

that is,

$$\begin{bmatrix} W \\ R \end{bmatrix}_{k+1} = \begin{bmatrix} \frac{\sqrt{3}}{2} & \frac{1}{2} \\ -\frac{1}{2} & \frac{\sqrt{3}}{2} \end{bmatrix} \begin{bmatrix} W \\ R \end{bmatrix}_k \approx \begin{bmatrix} 0.866 & 0.500 \\ -0.500 & 0.866 \end{bmatrix} \begin{bmatrix} W \\ R \end{bmatrix}_k \tag{3}$$

in which $W_k$ and $R_k$ are the number of wolves and rabbits in a certain habitat in month $k$. So if initially there are, say in 100s of wolves and rabbits,

$$\begin{bmatrix} W \\ R \end{bmatrix}_0 = \begin{bmatrix} 3 \\ 4 \end{bmatrix},$$

i.e. 300 wolves and 400 rabbits, then after one month there would be

$$\begin{bmatrix} W \\ R \end{bmatrix}_1 \approx \begin{bmatrix} 0.866 & 0.500 \\ -0.500 & 0.866 \end{bmatrix} \begin{bmatrix} 3 \\ 4 \end{bmatrix} \approx \begin{bmatrix} 4.60 \\ 1.96 \end{bmatrix},$$

i.e. 460 and 196 wolves and rabbits, and so on as summarized in the following table.

| $k$ | 0 | 1 | 2 | 3 | 4 | 5 | 6 | ... | 11 | 12 |
|---|---|---|---|---|---|---|---|---|---|---|
| $W_k$ | 3.00 | 4.60 | 4.96 | 4.00 | 1.96 | −0.60 | −3.00 | ... | 0.60 | 3.00 |
| $R_k$ | 4.00 | 1.96 | −0.60 | −3.00 | −4.60 | −4.96 | −4.00 | | 4.96 | 4.00 |

This example doesn't introduce as many ideas as the coins example. But it does motivate the ideas of matrices as operators, linear transformations, rotational matrices and complex eigenvalues/vectors, among others (Figs. 1 and 2).

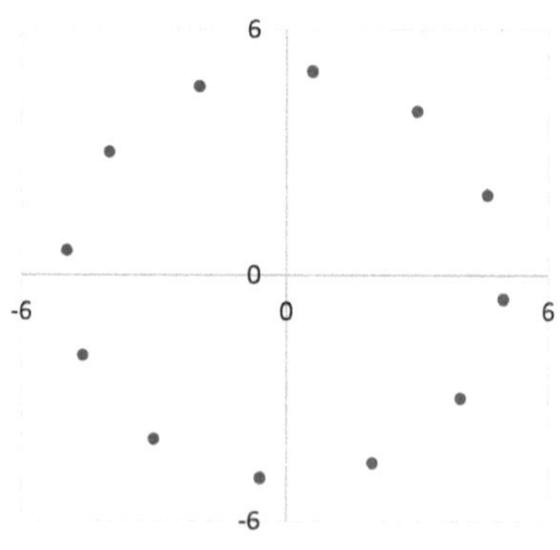


**Fig. 1** The two populations each month, plotted as vectors

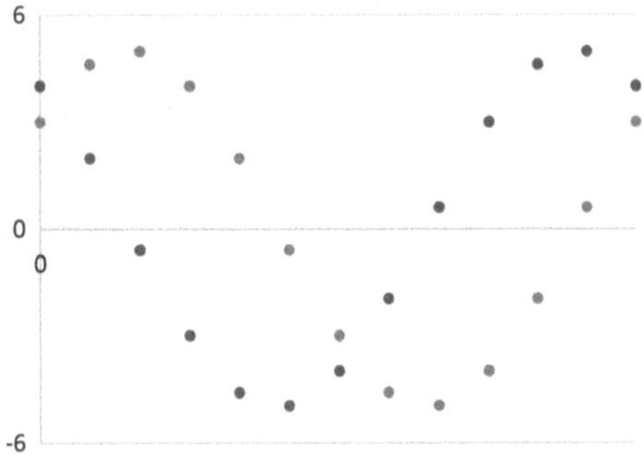


**Fig. 2** The two populations each month, plotted independently as functions of month

Along with the above predator-prey problem, I like to give a second example of a matrix acting on vectors, one in which the parameters don't fit the predator-prey problem, but that more nicely introduce other ideas. Given

$$A = \begin{bmatrix} 1.5 & 1 \\ 0.5 & 1 \end{bmatrix} \quad \text{and} \quad \mathbf{x} = \begin{bmatrix} -3 \\ 7 \end{bmatrix}, \tag{4}$$

if we repeatedly multiply $\mathbf{x}$ by $A$, as done in the above discrete predator-prey problem, what happens to the resulting vectors? Where $\mathbf{x}_k = A^k\mathbf{x}$, we find (rounded to one decimal place).

| $k$ | 0 | 1 | 2 | 3 | 4 | 5 | ... | 50 | 51 | 52 |
|---|---|---|---|---|---|---|---|---|---|---|
| $\mathbf{x}_k$ | −3.0 | 2.5 | 9.3 | 20.6 | 42.3 | 85.2 | ... | 3.0E15 | 6.0E15 | 1.2E16 |
| | 7.0 | 5.5 | 6.8 | 11.4 | 21.7 | 42.8 | | 1.5E15 | 3.0E15 | 6.0E15 |

Some ideas and issues that arise more clearly in this second example include:

- Eigenvectors: What is this vector that seems to be resulting from repeated multiplication by $A$? It seems to be a multiple of the vector (2, 1).
- Eigenvalues: It seems that after a while, each subsequent vector is about twice the previous one. How is that value related to matrix $A$?
- Why is it that in the first "matrix as an operator" example (3) the vectors $\mathbf{x}_0$, $\mathbf{x}_1$, $\mathbf{x}_2$, ... rotate around the origin while the vectors in the second example (4) do not?

Of course most of the beauty and usefulness of eigenvalues and eigenvectors cannot be fully appreciated with an introductory example or two. Still, rather than simply introducing these ideas in a context-less setting, we can give examples in which these ideas naturally arise, thus hopefully making students more curious about understanding what they are observing in those examples.

**Fig. 3** Projecting vector **b** onto the column space of $A$

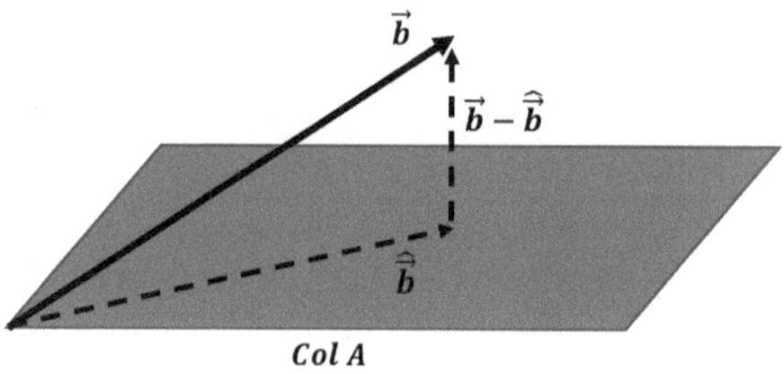


## 3 Examples to Motivate a Chapter

In addition to "bigger" examples that motivate or introduce a wide array of linear algebra ideas, it is also helpful to movitate a smaller collection of ideas, for example, as organized into a chapter, say with a good beginning-of-the-chapter example.

Chapters in textbooks are typically built around a common theme. For example, a chapter on least squares might include the introduction and development of properties of vectors, orthogonality, orthogonal projections, projecting a vector onto the column space of a matrix, the least squares solution, and the extension of these ideas to functions. Rather than simply jumping into the chapter and assuming students care about vectors, orthogonality and projecting a vector onto others, we might begin by giving a simple example or problem to solve, perhaps the coins problem (1) or perhaps a simple data fitting problem, say trying to fit a straight line to three non-collinear points, which would result in a system of the same dimensions as (1).

By this time, students will be familiar with trying to build one vector out of others, such as the right hand side in (1) using the two columns of the coefficient matrix $A$. It is easy to show that for (1) this cannot be done *exactly* (Fig. 3). So how do we do so *as well as possible*?

Because these vectors are from $R^3$, this can all be easily visualized: we are trying to get to the point "on the floor" (in the column space of $A$) that is closest to the point/vector $\mathbf{b}$ in $A\mathbf{x} = \mathbf{b}$. So how can you tell which point/vector "on the floor" is closest to $\mathbf{b}$? It's the point $\hat{\mathbf{b}}$ directly "underneath" point/vector $\mathbf{b}$. What does "underneath" mean? It means we want to find $\hat{\mathbf{b}}$ so that $\mathbf{b} - \hat{\mathbf{b}}$ is perpendicular to Col $A$ (See Fig 3). The rest, as they say, it just details. But without getting into details at this point, we (that is, the students) can see the need to understand certain things in order to work this problem: vectors, orthogonality (including in higher dimensions), projecting one vector onto a collection of others, and so on. They have a reason to care about the ideas about to be explored in the coming chapter.

One caveat for beginning-of-chapter examples: my experience with students is that problems that are too "real-life" and too complex, while interesting, can be a bit overwhelming. Thus, I prefer beginning each chapter with an interesting and important real-world example (or at least a sufficiently simplified version of one) that the student could actually work by the time he/she has completed the chapter, perhaps including the beginning-of-chapter example as a problem in the final section of the chapter or in the chapter review exercises.

# 4 Examples to Motivate Individual Concepts: Using an Idea Before Formally Introducing It

In addition to the "course-motivating" and "chapter-motivating" examples, sometimes the best motivation for why we care about a certain idea is to simply see the idea in action. When appropriate, why not do so before we have formally presented the idea, as a way to motivate the need for that idea, as well as to pique the student's interest in that idea? This doesn't replace any pertinent derivation of those ideas, theorems, proofs, etc. Indeed, hopefully it gives the student a reason to care about that derivation, the theorems and proofs, etc.: "I've seen the usefulness of a particular idea, and now I'm curious to learn more about it." Below I give a few of the ideas that could be shown in action before formally introducing them.

## 4.1 *Matrix Inverse*

Often the idea of an inverse, that $A^{-1}A = I$, is presented as yet another in a long line of ideas to learn and be tested on, rather than something that will help us solve and better understand important and interesting problems. Why not simply demonstrate the usefulness of matrix inverses in order to motivate our desire to learn more about them? For example, given the system

$$\begin{bmatrix} 1 & 2 \\ 3 & 4 \end{bmatrix} \begin{bmatrix} x_1 \\ x_2 \end{bmatrix} = \begin{bmatrix} 5 \\ 6 \end{bmatrix},$$

if we multiply both sides on the left by $\begin{bmatrix} -2 & 1 \\ \frac{3}{2} & -\frac{1}{2} \end{bmatrix}$ then we would have

$$\begin{aligned} \begin{bmatrix} -2 & 1 \\ \frac{3}{2} & -\frac{1}{2} \end{bmatrix} \begin{bmatrix} 1 & 2 \\ 3 & 4 \end{bmatrix} \begin{bmatrix} x_1 \\ x_2 \end{bmatrix} &= \begin{bmatrix} -2 & 1 \\ \frac{3}{2} & -\frac{1}{2} \end{bmatrix} \begin{bmatrix} 5 \\ 6 \end{bmatrix} \\ \Rightarrow \qquad \begin{bmatrix} 1 & 0 \\ 0 & 1 \end{bmatrix} \begin{bmatrix} x_1 \\ x_2 \end{bmatrix} &= \begin{bmatrix} -4 \\ \frac{9}{2} \end{bmatrix} \\ \Rightarrow \qquad \begin{bmatrix} x_1 \\ x_2 \end{bmatrix} &= \begin{bmatrix} -4 \\ \frac{9}{2} \end{bmatrix} \end{aligned}$$

Students' interest in this "mystery" matrix hopefully creates an interest to learn more about it, especially since it appears to be an alternative to Gaussian Elimination for solving $A\mathbf{x} = \mathbf{b}$. Students can see the usefulness of a matrix inverse, as well as of the identity matrix, rather than simply being told to learn and care about them, even though they have not yet been formally introduced to them.

We could give the following analogy to motivate both the idea and notation of $A^{-1}$:

$$\begin{aligned} ax &= b \\ x &= \tfrac{b}{a} \end{aligned} \qquad \begin{aligned} A\mathbf{x} &= \mathbf{b} \\ \mathbf{x} &= \tfrac{A}{\mathbf{b}} \end{aligned}$$

Of course dividing a matrix by a vector is not defined (students may not have previously known this, but will now), so we can correctly rewrite our ideas above as

$$\begin{aligned} ax &= b \\ a^{-1}ax &= a^{-1}b \\ 1x &= a^{-1}b \\ x &= a^{-1}b \end{aligned} \qquad \begin{aligned} A\mathbf{x} &= \mathbf{b} \\ A^{-1}A\mathbf{x} &= A^{-1}\mathbf{b} \\ I\mathbf{x} &= A^{-1}\mathbf{b} \\ \mathbf{x} &= A^{-1}\mathbf{b} \end{aligned}$$

In either the "mystery" matrix or the analogy approach, hopefully the student is curious to know more about inverses, including how to find them. The more textbooks and instructors can show students how mathematics naturally arises (ideas, notation, language, etc.) the better, as students will more naturally learn and care about the ideas.

Even before getting into the details, there are a number of questions that arise from just this simple example. For example: Does every matrix $A$ have an inverse?

- What would that mean if that were true? Every system $A\mathbf{x} = \mathbf{b}$ would have a unique solution. Is this the case? Of course not.
- What if $m < n$? There will be non-pivot columns in $A$ and thus one or more free variables, so if there is a solution it will not be unique; consequently if $m < n$ it would not make sense for a matrix to have an inverse.
- What if $m > n$? There will be non-pivot rows in $A$ and thus there may not be a solution, depending on what $\mathbf{b}$ is; consequently if $m > n$ it would not make sense for a matrix to have an inverse.
- So for a matrix to have an inverse, it must be square. So do all square matrices have inverses? No. So how can you tell if a square matrix has an inverse, and how do you find its inverse? Both are good questions.

## *4.2 Other Examples*

There are other ideas which can be demonstrated before giving their formal definition or getting into their details. I list a few involving matrix factorizations.

- $QR$-factorization. Prior to explaining how to find the $QR$-factorization of a matrix, for example, in a section on the Gram-Schmidt process or in a section on matrix factorizations, it's easy enough to give an example in which we use $A = QR$ to more easily solve the problem $A\mathbf{x} = \mathbf{b}$. Of course there are other uses of the $QR$-factorization, but at this point the student has at least one reason to want to learn more about it and how to find it. This is also a good motivation for orthogonality of vectors, as well as the idea of normalizing a vector.

- $LU$-factorization. Similar to the $QR$-factorization, to illustrate its use and help students appreciate its usefulness, we could give an example using $A = LU$ to solve $A\mathbf{x} = \mathbf{b}$. The $LU$-factorization is also another nice reminder of the usefulness of elementary matrices.
- Diagonalization $A = XDX^{-1}$. There are all sorts of uses, both theoretical and computational, for the diagonalization of a matrix. The easiest initially might simply be to point out, for example, that if

$$\begin{bmatrix} 8 & -1 \\ -2 & 7 \end{bmatrix} = \begin{bmatrix} 1 & 1 \\ 2 & -1 \end{bmatrix} \begin{bmatrix} 6 & 0 \\ 0 & 9 \end{bmatrix} \begin{bmatrix} 1 & 1 \\ 2 & -1 \end{bmatrix}^{-1}$$

then

$$\begin{bmatrix} 8 & -1 \\ -2 & 7 \end{bmatrix}^{10} = \begin{bmatrix} 1 & 1 \\ 2 & -1 \end{bmatrix} \begin{bmatrix} 6 & 0 \\ 0 & 9 \end{bmatrix}^{10} \begin{bmatrix} 1 & 1 \\ 2 & -1 \end{bmatrix}^{-1} = \begin{bmatrix} 1 & 1 \\ 2 & -1 \end{bmatrix} \begin{bmatrix} 6^{10} & 0 \\ 0 & 9^{10} \end{bmatrix} \begin{bmatrix} 1 & 1 \\ 2 & -1 \end{bmatrix}^{-1}.$$

So students have at least one reason to care about this factorization, even if they won't appreciate its more important uses until later. Of course the question then is how exactly we come up with this or any of the other factorizations.

## 5 Context and Meaning

In addition to examples for motivation of ideas and students, context and meaning for those ideas can further increase student interest and add additional layers to their understanding. I discuss two familiar examples below for homogeneous solutions.

### *5.1 Significance of a Homogenous Solution: Traffic Flow*

As illustrated in Fig. 4, suppose you have the following intersections in an area of a certain part of a city, with the given numbers of cars entering and exiting this area at different points. How many cars would be travelling these interior one-way streets in the six specified directions?

From this we get the equations corresponding to traffic flow at each intersection

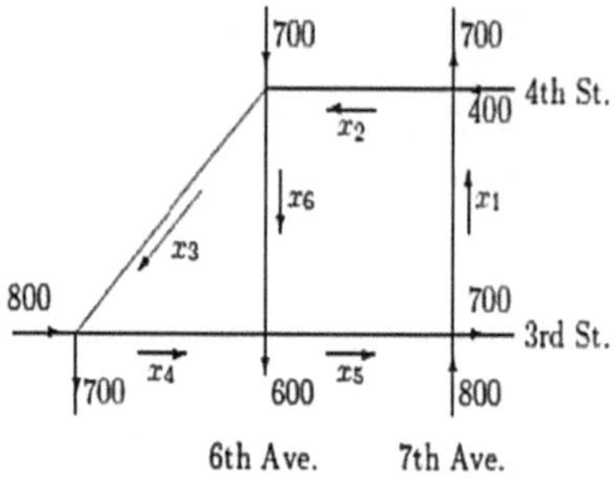


**Fig. 4** The flow of cars within one part of a town, given the number of cars entering and exiting this part of town

$$\begin{aligned} x_1 + 400 &= x_2 + 700 \\ x_2 + 700 &= x_3 + x_6 \\ x_3 + 800 &= x_4 + 700 \\ x_4 + x_6 &= x_5 + 600 \\ x_5 + 800 &= x_1 + 700 \end{aligned}$$

which leads to

$$Ax = \begin{bmatrix} 1 & -1 & 0 & 0 & 0 & 0 \\ 0 & 1 & -1 & 0 & 0 & -1 \\ 0 & 0 & 1 & -1 & 0 & 0 \\ 0 & 0 & 0 & 1 & -1 & 1 \\ 1 & 0 & 0 & 0 & -1 & 0 \end{bmatrix} \begin{bmatrix} x_1 \\ x_2 \\ x_3 \\ x_4 \\ x_5 \\ x_6 \end{bmatrix} = \begin{bmatrix} 300 \\ -700 \\ -100 \\ 600 \\ 100 \end{bmatrix} \tag{5}$$

From this we find one possible way of expressing the general solution

$$\mathbf{x} = \begin{bmatrix} x_1 \\ x_2 \\ x_3 \\ x_4 \\ x_5 \\ x_6 \end{bmatrix} = \begin{bmatrix} 300 \\ 0 \\ 600 \\ 700 \\ 200 \\ 100 \end{bmatrix} + c_1 \begin{bmatrix} 1 \\ 1 \\ 1 \\ 1 \\ 1 \\ 0 \end{bmatrix} + c_2 \begin{bmatrix} 0 \\ 0 \\ -1 \\ -1 \\ 0 \\ 1 \end{bmatrix} \tag{6}$$

There are various observations that we could make, but I want to focus on the homogeneous part of the solution. Let's call the two homogeneous solutions in (6) $\mathbf{v}_1$ and $\mathbf{v}_2$. What does each represent? It's easy (and perhaps lazy, or perhaps our students "can't handle the truth") to simply say that $\mathbf{v}_1$ and $\mathbf{v}_2$ are solutions to the homogeneous problem $A\mathbf{v} = \mathbf{0}$, where $A$ is the coefficient matrix in (5). While true, it doesn't give any meaning to these vectors.

Vector $\mathbf{v}_1$ tells us that if we change $x_5$ in $\mathbf{x}$, the values of $x_1$ through $x_4$ must also change by the same amount in order to satisfy (5), that is, if we still want have the same number of cars entering and exiting this part of town, as given in Fig. 4. So with the same number of cars entering and exiting town, these interior intersections could perhaps handle more cars, in the proportions described by $\mathbf{v}_1$. We could have a similar discussion about $x_6$ and $\mathbf{v}_2$: if we were to *increase* $x_6$, we would need an equal *decrease* in $x_3$ and $x_4$ in order to have the same number of cars entering and exiting this part of town.

We have focused on $x_5$ and $x_6$ because they are the free variables with which $\mathbf{v}_1$ and $\mathbf{v}_2$ naturally arise. However, we can also focus on other unknowns, perhaps equally well called *variables* in this discussion. For example, if $x_1$ were to increase, then an equal increase in $x_2$ through $x_5$ would be needed to still satisfy the given conditions. Thus, one could say that $x_1$ could have been the free variable rather than $x_5$, and indeed this is true. Students usually think that the free variables that naturally arise resulting from a particular process for finding the general solution are *necessarily* the free variables. They think that somehow the unknowns $x_1, x_2, \ldots$ are "designated" to be the dependent variables and that the other unknowns, in this case $x_5$ and $x_6$, are automatically "designated" to be the free variables. Of course this happens because of the way we do Gaussian Elimination with pivoting. In the end, it generally doesn't matter which variables are free. For example, and more interestingly, since the value for $x_3$ is non-zero in both $\mathbf{v}_1$ and $\mathbf{v}_2$, if we were to increase $x_3$ by 1, then we would either need to increase the values of $x_1, x_2, x_4$ and $x_5$ or else increase $x_4$ by 1 and decrease $x_6$ by 1. In other words, there are different ways we could compensate for a change in $x_3$. The same is true for $x_4$.

This observation about $x_3$ (or $x_4$) leads to a more general observation. It is easy to see that $\mathbf{v}_1 + \mathbf{v}_2 = (1, 1, 0, 0, 1, 1)$ is also a homogenous solution. So what does this mean? If we change any one of $x_1, x_2, x_5$ or $x_6$, we need to change the others by the same amounts.

At this point the observent student notices that these three vectors $\mathbf{v}_1, \mathbf{v}_2$ and $\mathbf{v}_1 + \mathbf{v}_2$ are connected to what is shown in the figure: each corresponds to one of the three closed loops seen in the figure. The most natural solutions of the homogenous problem are those that correspond to the closed loops. (Other linear combinations of these are also homogeneous solutions, but they are not as naturally connected to the closed loops in the diagram.)

We're also observing that $\mathbf{v}_1$ and $\mathbf{v}_2$ generate a *vector space*. Depending on what students have learned thus far, this might be their first experience with vector spaces (which tend to seem vague and pointless to students when presented in an abstract, theoretical way) in general and nullspaces in particular. Students are also getting an introduction to the idea that a vector space can have more than one basis: in this example, it is natural to think of $\mathbf{v}_1$ and $\mathbf{v}_2$ forming *the* basis for Nul $A$, but of course any two linearly independent linear combinations of $\mathbf{v}_1$ and $\mathbf{v}_2$ also form a basis for Nul $A$. In the context of certain problems, like this one, this fact might be important.

While it might be simpler and "cleaner" to describe the homogenous solution in a theoretical way, I find it more useful and motivating to have some meaning and context for it. It is not necessary (indeed, it is probably distracting) to have this discussion for every single problem which has a non-trivial homogeneous solution, but it is helpful to do so initially and in other examples where appropriate.

There are other issues and questions that arise in this traffic problem, thus I like to use it as another one of my five to ten recurring beginning-of-the-course examples. A few other issues and questions include:

- Does the number of cars entering and exiting this part of town uniquely determine the number of cars travelling along the six given paths?
- What if fewer or more of the unknown values were specified?
- Would changing certain given values result in there being no solution?
- Is it possible to have more cars within this part of town without having more cars entering and exiting it? That is, it is possible for $x_1 + \ldots + x_6$ to be larger with the same specified right hand side?
- For a different configuration of traffic flow, in particular, with more closed loops, how would the solutions to the homogeneous problem be related to the closed loops?

## 5.2 *Significance of a Homogenous Solution: Investments*

Consider a second example in which the homogenous solution has meaning. Suppose we are to divide \$10,000 into three investments, which have return rates of 5, 10, and 25%, and that we want a total annual return of \$2000. This yields the equations

$$\begin{aligned} x_1 + \quad x_2 + \quad x_3 &= 10{,}000 \\ 0.05x_1 + 0.10x_2 + 0.25x_3 &= 2{,}000 \end{aligned} \tag{7}$$

which has a general solution of

$$\begin{bmatrix} x_1 \\ x_2 \\ x_3 \end{bmatrix} = \begin{bmatrix} 2{,}500 \\ 0 \\ 7{,}500 \end{bmatrix} + c \begin{bmatrix} -3 \\ 4 \\ -1 \end{bmatrix}.$$

In this problem, the homogenous solution $(-3, 4, -1)$ describes the *reallocation of resources* needed in order to meet the given conditions. There are different ways to invest, and $(-3, 4, -1)$ tells us the proportions and directions of change in the three investments needed to still satisfy the two given conditions. For example, a *decrease* in the third investment requires a *decrease* of three times that amount in the first investment and an *increase* in the second investment of four times that amount. Again, the homogenous solution is more than simply a solution to $A\mathbf{v} = \mathbf{0}$.

## 6 Final Thoughts

If a student wants to learn, he or she will. There are many motivations for students in a mathematics course: grades, past success, their own standards of excellence, competition, the fun of the problem solving, the beauty of the ideas, and relevance of what they are learning. People often don't like to do hard things, but they are much more willing (indeed, even excited) to do something difficult if there is a good reason for doing it. We unapologetically expect students to have a certain amount of self-discipline and self-motivation in learning mathematics, and quite rightfully so. Indeed, mathematics courses sometimes necessarily serve as a "weeding out" process to help steer students toward other, less mathematical majors and career paths. But the abstract, sometimes context-less nature of mathematics, while part of its power and beauty, can also be one of the obstacles to student learning, even for mathematically strong students who certainly should be studying mathematics or a related field. This is exasperated by the fact that many students who study linear algebra are "applied mathematicians" from other fields: physics, biology, chemistry, business, computer science, etc. and often don't think like and are not as motivated in the same ways as a pure mathematician. Ultimately, one way in which we can simultaneously increase student interest and enrich their learning is with better motivation for the ideas being taught, including with examples that precede and motivate those ideas, along with more context and richer meaning for those ideas.

# Holistic Teaching and Holistic Learning, Exemplified Through One Example from Linear Algebra

**Frank Uhlig**

**Abstract** Here we discuss ways to teach and learn holistically. The holistic method encourages student curiosity and respects student input fully, however qualified. A teacher's holistic approach leads to open discussions and deep learning in class. Using the students' innate desire to understand drives the course of such a class. The teacher's role is to guide and adjust the course as the subject matter, experience, and—in mathematics courses—as the algebraic, geometric and logic rules of mathematics dictate. Holistic Teaching respects and adheres to the 'necessity principle' of learning and it applies the 'holistic management principle' that is successfully used for many other complex systems to achieve a comprehensive teaching and learning experience for both teachers and students. Teaching holistically is exemplified by one extended in-class study of how to measure angles in $\mathbb{R}^n$ from first principles of both Geometry and Linear Algebra.

**Keywords** Holistic teaching · Necessity principal · Holistic management principle · Mathematics teaching · Linear algebra · Geometry · Proof · Angle Unit vector · Linear transformation · Matrix · Dot product · Trigonometry

## 1 Introduction

This exposition gives an outline of "Holistic Teaching and Learning", philosophically and in practice.

It is based on the author's experience of teaching mathematics to diverse students at all levels over 48 years, at seven universities, on two continents, and specifically at Auburn University for the last 32 years. Early teaching assignments in my 2nd year at the University of Cologne in Germany have given me a precious perspective on the educational process, both of teaching and of learning, combined into one process. As a very young student-teacher, just 20 years old, I experienced that I learned best

F. Uhlig (✉)
Department of Mathematics and Statistics, Auburn University,
Auburn, AL 36849-5310, USA
e-mail: uhligfd@auburn.edu



S. Stewart et al. (eds.), *Challenges and Strategies in Teaching Linear Algebra*,
ICME-13 Monographs, https://doi.org/10.1007/978-3-319-66811-6_17

when I taught and what I taught. I was happiest in mathematics classes that were not taught in the strict 'Definition-Lemma-Proof-Theorem-Proof-Corollary' lecture style, but rather loosely presented and occasionally left questions unanswered. And more specifically, I liked it when teachers went out on limbs of discovery and sharing. Courses that explained the historic and current developments of a subject's notions sat best with me; when and where—in a way—the teacher shared her or his whole knowledge and the desires and gaps therein. Conversely I was slightly uncomfortable in courses that were built on drilling us to remember proofs and theorems, facts, tricks and formulas because I did not feel nurtured or helped with my inner personal needs and my own desire to grasp and understand the subject matter wholly.

The development of this, my so called 'Holistic Teaching and Learning' method, took time. It was influenced by having the chance of teaching in both systems, the hierarchical lectures culture and in more interactive academic settings. Helped by these experiences I have learned to understand and synthesize both into one, namely into the 'Holistic Teaching and Learning' approach to education. First I will explain the general twofold framework for this method which can be used in every educational effort from first grade through graduate school and in driver's education etc. as well. This is followed by one detailed in-class example from sophomore mathematics.

## 2 The Framework of 'Holistic Teaching and Learning'

Holistic Teaching and Learning applies to and is useful in almost any educational environment and for many subject areas. It has two foundations: one lies in satisfying our students' need to know and another is to approach teaching and learning as mirror images of each other and to view the educational process as a whole; thus the name.

Over time I learned of the 'necessity principle' of teaching as it affects our cognitive development and individual learning. For a detailed introduction and history of this principle see e.g. (Harel, 2013). It has become obvious now that my way of teaching had become—unknowingly at first—well aligned with this aspect of modern educational research. As a student I had experienced the big difference between two types of mathematics classes: One where we had to remember definitions, formulas, theorems and their proofs and had to repeat what we were taught; and another that taught concepts, explored proofs and helped us understand the subject matter from within, as well as giving us broad and deep subject knowledge. When using an open minded interactive approach as a teacher, students will engage. The exchange of ideas, insights and questions then becomes bidirectional between me as teacher and my students; information passes back and forth. Even multidirectional if we include the textbook and student to student interactions. By fostering my students' desire to learn and letting the subject matter develop organically in class, students take control of their own intellectual and cognitive growth. Therewith they can take ownership of their field of studies. When students feel free and are encouraged to ask questions, they begin to sort things out in their own personal ways of thinking and seeing, and thereby they truly learn. This questioning, exploring and understanding, including occasional mis-understandings, builds up mature thinkers and knowledgable individuals. Formulas and theorems become appendages to genuine comprehension. We all seemingly gain and retain knowledge best if we are motivated and eager to know and if we are encouraged to ask 'why' and 'how' frequently. When the 'necessity principle' of learning is respected it bears fruit. Every student's natural intellectual need is to try and understand how and why a subject, any subject, and our knowledge about it has come about. True problem solving then gives students an intellectual purpose to engage with mathematics in a new, complete way of learning. Adhering to the necessity principle in teaching thus drives a course and transforms students from learners into knowers.

Over time again, I also realized that incorporating the necessity principle alone into my teaching was not all that had occurred in my classes. A change of attitude inside of me had come about through interactive teaching. I had found little use for intellectual or knowledge supremacy when teaching. As I desired to interact more freely back and forth with students over mathematics, I needed to become nonjudgmental and eagerly follow any student's insight, thought path, or proof as it was shared in class. This called for courage and superb subject knowledge on my part in order to steer the process through any pitfalls of logic, geometry, arithmetic or theory that might occur. Likewise for a student it is highly demanding to discuss mathematics and to present ideas and derivations or a proof of his or her own making and then to share his or her thoughts and explain in public. To succeed with education, I believe that both teacher and students have to bare themselves, to open up their thought processes and minds freely and tussle with the problems that arise, all while considering themselves as (near) equals in spirit. For me the educational process has become one of sharing spirit and soul. Unconsciously I had incorporated this leveling duality, the intertwinement of teaching and learning as one whole into my own teaching: As I shared my knowledge with the students I learned to accept their input on an equal footing, guided by and judging for mathematical and logical correctness only.

This is my personal wholeness principle of education. The holistic approach to teaching and learning extends subject-wise even further, namely to the whole of mathematics and related fields as subject matter and all of their histories as grounds for possible class consideration and questioning.

Teaching holistically is fraught with perils and uncertainties. For example, how does one answer a calculus class student who asks

> *How would anyone ever think of inserting a zero in the form of* $-f(x)g(x+h)+f(x)g(x+h)$ *into the numerator* $f(x+h)g(x+h)-f(x)g(x)$ *of the difference quotient for the product of two functions* $f\cdot g$*?*

other than to obtain a limits-based proof of the product rule of differentiation? As a teacher I am often not prepared, and anyone else may not be fully prepared either for all the questions that may arise. However, this uncertainty is very beneficial to students; yet it only occurs in holistic teaching environments. Complete openness, deep mastery of the subject, refined listening ability all around make for a good learning environment. And when I fail or stumble as a teacher, the class usually comes alive, other minds take over and unconventional deep discussions and learning take place.

Natural systems and nature itself have been recognized for centuries as being complex rather than just complicated systems. Much of human knowledge is complicated, consisting of many part and steps; yet the educational transfer of human knowledge is one intrinsically complex whole. Personal growth and learning cannot be reduced to parts and steps at all, refer to the "Outlook" section at the very end. Complex systems of nature have been recognized as such and described and studied extensively since the Enlightenment period, starting possibly with Goethe. They were first systematized in Alexander von Humboldt's 'Kosmos' (von Humboldt 1845–1862). As human growth in general and our education and teachings in particular are complex problems, a holistic approach as outlined above beckons. Eventually I recognized that I had been influenced in my teachings by Allan Savory's general 'Holistic Management' concept (Savory, 1998) and that I had used holistic methods, unknowingly again, out of necessity all along.

Following both the necessity and the holistic principles of teaching and learning benefits our students and the teacher as well. It requires experience, courage and compassion that are well spent on the young.

## 3 Details of My Methodology, Using an Example from Linear Algebra

One of the simplest ways to channel students into a holistic approach to learning and to life in general is to ask them:

*Who will teach your children?*
*Who of you will know Mathematics well enough to explain it in the future when I am old and retired? Who among you will eventually be able to add to Mathematics through original research and problem solving?*
*How can the entities of math and science be understood and kept alive in our culture and society unless future generations are taught?*
*Who will understand and be able to transmit how and why math's very nature forces certain properties, tools and understandings onto the modern world, such as the math behind cell phones, of google searches, of GPS, of wind turbine design, of molecular chemistry, of surveillance techniques, of large data and so forth?*

Whenever I teach a graduate or undergraduate mathematics class I am guided by the two underlying principles and I approach teaching, learning and math holistically. I teach interactively with constant give and take between all members of the class and during most any class hour. More involved exploratory class sessions occur five to eight times during our 15 week semesters. These sometimes arise spontaneously and at other times are planned in advance.

I personally love to teach pure, applied and numerical math courses at all levels that involve Linear Algebra and Matrices. Matrix Theory is the area of my own research and expertise. To start off a semester of sophomore Linear Algebra, for example, I write out the linearity equation for functions on the very first day:

$$f(au + bv) = af(u) + bf(v)\ .$$

Here $f$ is a function that maps $n$-vectors such as $u, v$ and $au + bv$ to $m$-vectors $f(u), f(v)$ and $f(au + bv)$, respectively, and $a$ and $b$ denote scalars. We discuss real $n$-space and $m$-space naively and I introduce the task of our class on Linear Algebra as studying all linear functions $f$ between vector spaces.

*No more, no less and, deceivingly, sounding simple enough.*

Then I recall the spaces of continuous functions and of differentiable or integrable functions from Calculus. I ask the class how to define these function spaces and to explain that they are spaces and why they are. Soon it becomes clear that differentiation and integration are linear operators or functions on their respective spaces since

$$(af + bg)'(x) = af'(x) + bg'(x) \text{ and}$$

$$\int (af(x) + bg(x))dx = a \int f(x)dx + b \int g(x)dx\ .$$

As I and the students discuss these question, I mention that linear functions are the backbone (jointly with physics, chemistry, biology, computer science and engineering) of our whole modern technological world.

Linear functions and their representation as matrices govern much of our interconnected internet lives: They help in search engines, with GPS computations, in automatic control and on and on as they are well suited and adaptable for numerical computations and inside algorithms. Linear functions act as mathematical 'atoms' for modern industry and technology. Linear functions and their matrix representations are nowadays used everywhere in computer number crunching, i.e., they are indispensable in our modern day computer revolution and for our dependencies thereon.

That piques the interest of the class on day one. I then lead the class—with many open ended questions and the student answers discussed and possibly used—through the discovery and proof finding process that each linear transformation $f$ from $\mathbb{R}^n$ to $\mathbb{R}^m$ can be represented by the action of an $m$ by $n$ matrix $A$ on vectors in $\mathbb{R}^n$.

Thereafter we study the classical subjects of Linear Algebra such as equations, linear dependence, span, bases etc. for many weeks. This first half of my first Linear Algebra course is peppered with quick or extended discovery sessions for the students who, on different days, are asked questions such as

*Are there equations that nobody can solve, not even Einstein, if he were alive?*
*How many solutions can a linear system $Ax = b$ have? How can we tell?*
*What distinguishes a uniquely solvable system of linear equations from one with many solutions or one with none?*
*Which sets of vectors are linearly independent? How can one check?*
*What relations are there between the coordinate vectors of one point in* $\mathbb{R}^n$, but for

different bases?
*How can we translate one basis coordinate vector to another coordinate vector for the same point in space but with respect to a different basis?*
*What is the difference between a foreign language dictionary and a basis change matrix?*
and so forth.

Here students learn to discover and form judgments and reasoned opinions in mathematical terms and they begin to argue openly and freely through mathematical statements. They start to understand the concepts and necessity of 'rigor' and 'proof' in mathematics, as well as of subject specific mathematical concepts.

A chance of teaching important linear algebraic concepts, anew and more deeply opens up in the second half of the course. After basis change effects on matrix representations have been observed, and some, but not all matrices have been found diagonalizable via eigenvector bases, it is time to explore the following question:

***What determines angles and how are they measured in $\mathbb{R}^n$?***

The concept of angle precedes our study of orthogonality and a deeper analysis of bases and of special matrices such as symmetric, hermitian, normal and orthogonal matrices that fills out the rest of the semester. On one hand, the concept of angles and their measurement must be understood for comprehending mathematics and to live and move in this world. And on the other hand, a thorough angle definition itself offers a wonderful means to re-introduce vectors and discuss and learn about vector spaces and linear transformations anew on a deeper, more conceptual and also more concrete level. This helps the class to see and understand special linear transformations and matrices in their own intrinsic light later on and renders them more palatable and useful for analysis and applications. A sure win-win situation for teacher and students alike.
Here my startup questions typically are:

> "*What is an angle in $\mathbb{R}^n$? How does it come about? What geometric objects of $\mathbb{R}^n$ define an angle?*

This usually generates reflected student answers such as:

*Take 3 non-collinear points in space, say O, A and B. Look at the plane spanned by O, A and B. Assume that O is the vertex or corner of the angle.*
*Move, rotate and tilt the plane to make it coincide with the ground plane $\mathbb{R}^2$.*
*Draw the points out on paper, and there it is, the space angle $\angle(AOB)$ contained between the rays OA and OB.*

Students can generally construct the equivalence between an angle in $n$-space and its representation in $R^2$. An angle is a planar object after all, defined by three points $A$, $O$, and $B$ in space (Fig. 1). To visualize that a general two dimensional plane in $\mathbb{R}^n$ is and behaves just like our ordinary drawing paper or the black- or white-board involves mental abstraction. Students are best left alone for a short interval to discuss, develop and perform this labor by themselves through discussions and teamwork with their peers.

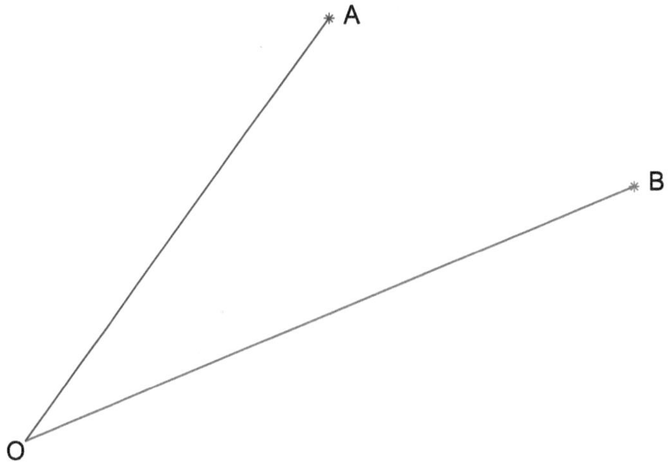


**Fig. 1** A generic angle in $n$-space

To allow for this exchange, I usually step out of class and walk the hallways for a minute or two or three, while they work through the process in accordance with the holistic management approach to teaching. Occasionally I enter back in to learn their progress or to ask further questions until they are done and reasonably clear about the equivalence by and among themselves. My students usually succeed without many prompts due to the mathematical maturity level that they have reached towards the end of the course when they have been holistically taught.

Once the generic $\mathbb{R}^2$ situation of angles in $n$-space is understood, we can use the geometry and trigonometry of $\mathbb{R}^2$ to study the angle between the rays $OA$ and $OB$ that emanate from $O$.

This picture raises further questions:

*How can we measure the three angles $\alpha$, $\beta$ and the difference angle $\alpha - \beta$ in $\mathbb{R}^2$ as labeled in Fig. 2?*
*Does anyone remember trigonometry? What does trigonometry do in this realm? How are the elementary trig functions defined?*

After drawing the coordinate axes onto the angle plot above, my students may recognize the role for sine and cosine here. As the trig functions are based on the unit circle, this leads us to the intersection of the unit circle with the rays from $O$ to $A$ and $O$ to $B$, respectively, as drawn below.

At first, students might be unable to correlate the points $\widetilde{A}$ and $\widetilde{B}$ on the unit circle marked by small + signs in Fig. 3 with the given points $O$, $A$ and $B$. I wait until the class notices that these two points $\widetilde{A}$ and $\widetilde{B}$ are the unit vectors for the rays from $O$ to $A$ and from $O$ to $B$, respectively. Thus as vectors

$$\widetilde{A} = (\cos(\alpha), \sin(\alpha)) = \frac{A}{\|A\|} \qquad \text{and} \qquad \widetilde{B} = (\cos(\beta), \sin(\beta)) = \frac{B}{\|B\|}\,.$$

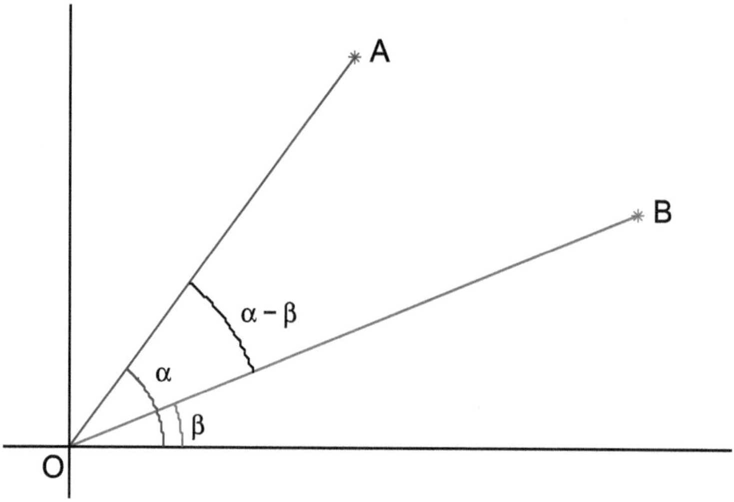


**Fig. 2** An angle in 2-space

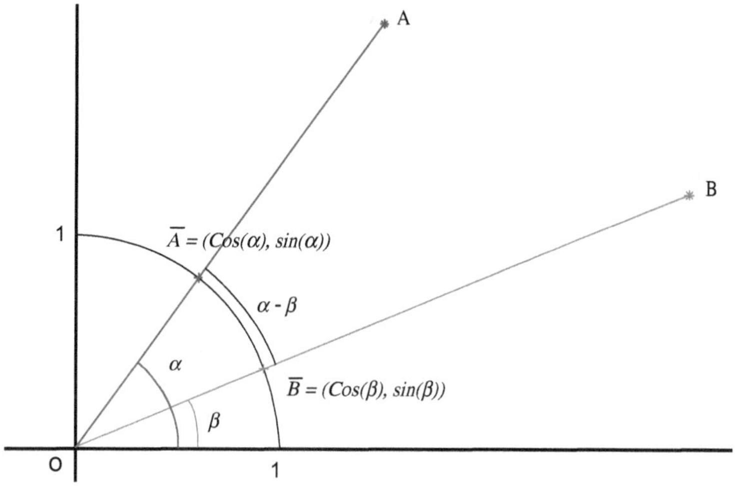


**Fig. 3** The angle with the coordinate axes and the unit circle drawn in the plane

What is really needed, though, is a measure of the angle $\alpha - \beta$ between $OA$ and $OB$, or : *What is* $\cos(\alpha - \beta)$ *given* $\cos(\alpha)$ *and* $\cos(\beta)$*?*

Students now may recall the addition formula for cosine, i.e., $\cos(\alpha + \beta) = \cos(\alpha)\ \cos(\beta) - \sin(\alpha)\ \sin(\beta)$, and also that $\cos(-\gamma) = \cos(\gamma)$ because cosine is an even function, while $\sin(-\gamma) = -\sin(\gamma)$ since the sine function is odd.

Thus simple algebraic manipulations relate the cosine of the desired angle $\alpha - \beta$ between $OA$ and $OB$ to the given sine and cosine coordinates of the two associated unit vectors:

$$\begin{aligned}\cos(\angle(AB)) &= \cos(\alpha - \beta) = \cos(\alpha + (-\beta))\\ &= \cos(\alpha)\ \cos(-\beta) - \sin(\alpha)\ \sin(-\beta)\\ &= \cos(\alpha)\ \cos(\beta) + \sin(\alpha)\ \sin(\beta)\ .\end{aligned}$$

However: *What does this cosine formula have to do with our original angle problem, where the angle is determined by the rays OA and OB and the coordinates of A and B? How does* $\cos(\alpha - \beta)$ *relate to the vector coordinates of A and B and vice versa?* To find an answer we must scrutinize the formula

$$\cos(\alpha - \beta) = \cos(\alpha)\ \cos(\beta) + \sin(\alpha)\ \sin(\beta)$$

while at the same time looking at

$$(\cos(\alpha), \sin(\alpha)) = A/\|A\| \quad \text{and} \quad (\cos(\beta), \sin(\beta)) = B/\|B\|\ .$$

After staring long enough at these three formulas, the cosine of the angle $\alpha - \beta$ between $OA$ and $OB$ appears as a dot product $\cdot$ of two vectors. In fact, $\cos(\alpha - \beta)$ is the dot product of the unit vectors $(\cos(\alpha), \sin(\alpha))$ and $(\cos(\beta), \sin(\beta))$ that point from $O$ to $\widetilde{A}$ and from $O$ to $\widetilde{B}$ in Fig. 3, respectively. Thus

$$\begin{aligned}\cos(\angle(AB)) &= \cos(\alpha - \beta) = \cos(\alpha)\ \cos(\beta)\ +\ \sin(\alpha)\ \sin(\beta)\\ &= (\cos(\alpha), \sin(\alpha)) \cdot (\cos(\beta), \sin(\beta))\\ &= \frac{A}{\|A\|} \cdot \frac{B}{\|B\|} = \frac{A \cdot B}{\|A\|\ \|B\|}\ .\end{aligned} \qquad (*)$$

This is the dot product cosine formula. It measures angles in $\mathbb{R}^2$ in terms of their defining vector coordinates. Formula (*) is the standard coordinates based angle measure and it generalizes verbatim to angles in $\mathbb{R}^n$.

But we are not done yet if we want to understand holistically and appropriately at our knowledge level:

*How does the cosine addition formula* $\cos(\alpha + \beta) = \cos(\alpha)\ \cos(\beta) - \sin(\alpha) \sin(\beta)$ *come about?*

This question can lead us back to our middle or high school days. But instead we can study this question by returning to the very first week of class, namely to linear transformations and their standard matrix representations. Let us follow the following lines of inquiry while viewing Fig. 4:

*Is rotating the plane around the origin O a linear transformation of* $\mathbb{R}^2$ *or not? If so, what is its standard matrix representation?*

Here we use the original linearity condition $f(au + bv) = af(u) + bf(v)$. Consider two nonzero vectors $au$ and $bv \in \mathbb{R}^2$ with $a, b \in \mathbb{R}$ and $u, v \in \mathbb{R}^2$ and the diagonal $au + bv \in \mathbb{R}^2$ of the parallelogram that $au$ and $bv$ form.

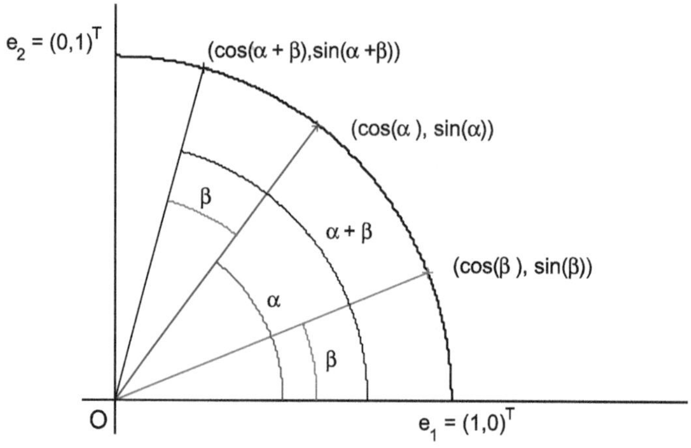


**Fig. 4** Counterclockwise rotation by $\beta$ of the unit vector $(\cos(\alpha), \sin(\alpha))$ around 0 in $\mathbb{R}^2$

*How does a planar rotation $R_\beta$ around the origin $O$ by the angle $\beta$ change this parallelogram?*

The rotated sides and the parallelogram's diagonals form another parallelogram that is congruent to the original one since planar rotation does not change shapes. Therefore $R_\beta(au + bv) = aR_\beta(u) + bR_\beta(v)$, i.e., the linearity condition holds for $R_\beta$. Thus any rotation $R_\beta$ around the origin by an angle $\beta$ is a linear function. And therefore $R_\beta$ can be represented as a 2 by 2 matrix.

*What is the standard matrix representation of the counterclockwise rotation $R_\beta$ of the plane $\mathbb{R}^2$ by $\beta$ around $O$?*

From class week one, the standard matrix representation of any linear transformation $\mathbb{R}^2 \to \mathbb{R}^2$ contains the images of the standard unit vectors $e_1 = \begin{pmatrix} 1 \\ 0 \end{pmatrix}$ and $e_2 = \begin{pmatrix} 0 \\ 1 \end{pmatrix}$ of $\mathbb{R}^2$ in its columns. Thus

$$R_\beta = \begin{pmatrix} \vdots & \vdots \\ R_\beta \begin{pmatrix} 1 \\ 0 \end{pmatrix} & R_\beta \begin{pmatrix} 0 \\ 1 \end{pmatrix} \\ \vdots & \vdots \end{pmatrix}_{2,2} .$$

*What are the images of the two unit vectors $e_i$ of $\mathbb{R}^2$ under counterclockwise rotation by $\beta$?*

By inspection $R_\beta \begin{pmatrix} 1 \\ 0 \end{pmatrix} = \begin{pmatrix} \cos(\beta) \\ \sin(\beta) \end{pmatrix}$ and $R_\beta \begin{pmatrix} 0 \\ 1 \end{pmatrix} = \begin{pmatrix} -\sin(\beta) \\ \cos(\beta) \end{pmatrix}$. And thus

$$R_\beta = \begin{pmatrix} \cos(\beta) & -\sin(\beta) \\ \sin(\beta) & \cos(\beta) \end{pmatrix}.$$

The most recent picture shows the typical situation for angle addition: If we rotate the point $(\cos(\alpha), \sin(\alpha))$ by $\beta$ counterclockwise around the origin $O$, it moves to $(\cos(\alpha+\beta), \sin(\alpha+\beta))$.

*How does the 2 by 2 matrix representation of $R_\beta$ map the vector $(\cos(\alpha), \sin(\alpha))^T \in \mathbb{R}^2$?*

$$\begin{pmatrix} \cos(\alpha+\beta) \\ \sin(\alpha+\beta) \end{pmatrix} = R_\beta \begin{pmatrix} \cos(\alpha) \\ \sin(\alpha) \end{pmatrix} = \begin{pmatrix} \cos(\beta) & -\sin(\beta) \\ \sin(\beta) & \cos(\beta) \end{pmatrix} \begin{pmatrix} \cos(\alpha) \\ \sin(\alpha) \end{pmatrix} =$$
$$= \begin{pmatrix} \cos(\alpha)\cos(\beta) - \sin(\alpha)\sin(\beta) \\ \cos(\alpha)\sin(\beta) + \sin(\alpha)\cos(\beta) \end{pmatrix}.$$

And thus we have derived both trigonometric addition formulas (for sine and cosine) in one formula by using matrix times vector multiplication and elementary Linear Algebra.

---

This angle example takes about 1 hour of holistic, highly interactive class time.

It prepares my students to study orthonormal bases and Gram-Schmidt orthogonalization, as well as orthogonal and unitary matrices such as Householder matrices and Givens rotations.

Then we can apply our freshly gained angle and rotation insights to symmetric and hermitian matrices and—sometimes—we still have time left to deduce the Schur Normal Form of square matrices that is sparse and achieved by orthogonal similarities.

The whole program for teaching elementary Linear Algebra holistically as outlined above, leading us from linear transformations to special matrices assumes a 15 week semester with 3 class hours per week (at 50 min each), with 3 to 4 in-class tests and review sessions before each test. When taught holistically, this class often acts as the transformational class, the springboard that ensures my students' success in their subsequent science and engineering curricula.

## 4 Results and Conclusions

All in all, in my classes I encourage students to think about and discuss all aspects of the subject area. I explain what came before and led to the current class and also what may be important to retain for future learning and in applications. I encourage my students to question and learn responsibly and holistically everything that pertains to the subject, similarly to how I have detailed one example from elementary Linear Algebra here. I teach Calculus, Numerical Analysis, Matrix Theory etc. all guided by the same holistic approach. In all of my classes we use free explorations and interactive discussions. We think jointly and individually and share our minds. I aim for full awareness and mastery of each day's subject and full consciousness of the subject as a whole, both in my students and reflexively in me. On ideal days my students and I learn from each other and teach each other.

In a holistically taught class the actual teaching style adjusts itself from day to day as the subject matter and area warrant and as the students require. In math courses there will be Definition-Theorem-Lemma proving days; as well as practical example solving days; and also days where students, randomly chosen, will solve problems on the board, often four problems with four students on the board at the same time; and at other times interactive and exploratory days; and so forth. This freedom of means to teach and learn is one triple benefit of teaching holistically, namely for students, for the teacher and for spreading the comprehension, general understanding, love and mastery of our subject.

Holistic Teaching and Learning works best with class sizes of around 30 students and in rooms with black- or white-boards. I have also taught large auditorium classes of 80 to over 300 students holistically. These are typically accompanied by smaller group recitation and exercise sessions which again are best built on and taught from both holistic principles. However, in large auditorium classes not all student questions can be answered as easily and comprehensively as they can in smaller classes. Yet for best results, large classes will benefit from adopting the holistic teaching and learning principles.

Generally it takes around 2 weeks until a class becomes comfortable with the two founding principles of holistic teaching and accepts its premises. In holistically taught courses we generally cover and students learn around 10% more subject matter and they do so more deeply than can be achieved in parallel, more hierarchically taught classes on the same subject. The reason is that when studying a subject built on its inherent concepts, student confusion is minimized. A concept once learnt and understood will be recalled and reinforced by applications while an "applications first" approach to teaching often leads to widespread confusion and wasted class time. Not surprisingly, students in my classes tend to be asked and often act as tutors for friends and dorm neighbors that take the same course in parallel, but with different faculty. This indicates that our class is soon ahead of parallel classes and that my students understand the subject matter well enough to be willing to try and test their own understandings and can explain, help and teach their friends how to approach and solve new homework problems. Sometimes they even bring these problems back to our class for us to examine, discuss and solve.

I have never taught where learning assessments or student tracking were established, thus I have no usable statistics data on 'Holistic Teaching and Learning'. Except anecdotally, when I meet former students by chance at the swimming pool, at a restaurant or at graduation time and I repeatedly hear that my class was one of their most demanding, but overall the most valuable lesson of their course of studies.

If interested, I suggest to try and learn more about the 'necessity principle' and the 'holistic principle' of teaching by searching the internet for these terms, as well as look up Jean Piaget (1985, 1977, 1960) and Guershon Harel (2013) for the roots of the necessity principal in teaching, and Allan Savory (1998) and Jan Smuts (2010) or André Voisin (1988) for the holistic principle that originated in nature and agronomy studies. I have been familiar with many of these authors' works over the years, but I have not used nor did I revisit any specific references when writing this paper.

This paper is, however, a continuation and evolution of my earlier thoughts and notes from a decade ago on the need for conceptual teaching and how to structure a holistically taught first Linear Algebra course, see (Uhlig 2002a, b, c, 2003).

## 5 Outlook

This paper is unlike any other paper in the current and recent educational literature. It points to a 'wicked problem' that institutions, corporations and complex human endeavors routinely face when they mature as organizations. For a different but similar assessment of the inherent benefits of approaching complex situations holistically as we have done here for Mathematics, see (Wendell, 2001) for example. I have learned of his writings about science only very recently, but they seem to apply equally well to the 'wicked problem' that education finds itself in today.

Over the last few decades educational research has become more and more reductionally 'scientific' and less a part of the human arts or philosophy. Hypotheses are now proposed and class tested. And assessment statistics and numerical data have taken over the field of educational research and are shaping our ways of teaching. Thoughts and reflections on education, as well as philosophical debates on the principals of teaching have in turn been more and more dismissed, have been forgotten and have become near un-publishable while experiments and reductional fixes fill educational journals.

*'No child left behind'* like test results and analyses have driven our recent educational efforts. In the process, our desired 'scientific' learning outputs have been defined too narrowly and too trivially. Consequently teaching itself has veered off its fundamental task to give students tools for learning and achieving subject mastery and conceptual understanding. In fact, competent teaching and deep learning has become almost secondary in schools and colleges as faculty are trying to obtain positive teaching evaluations for tenure and promotions.

But education is too complex to be solved by popularity contests at the end-of-semester evaluation time.

Narrowly defined tests produce reductional insights into a few educational parameters of learning. Yet the whole of educating our young is a complex natural process that defies being measured or described by 12 or 20 or any finite number of parameters. As long as we continue to declare 'educational success' according to 'good' statistical data as 'good' teaching, unintended social consequences will occur in society, for both the young and the mature. In this sense, how to assess the quality of education is a 'wicked problem'. To name a few of recently lost effects of education, our schools are alienating and losing more and more students, schools and colleges provide less and less life guidance and fewer tools for understanding the world around us and therefore they have become increasingly irrelevant to millennials, for example, if not even noxious. Our GPA statistics and drug and incarceration statistics are following each other nicely and they reflect one devastating image of what is wrong. Could this be mere coincidence or maybe human induced? We are relying on reductionist 'best solutions' with vicious and unintended consequences, where a holistic approach to educating our young could and should be followed instead.

Education as a whole, and Mathematics specifically, is an art and not a technology or a craft that can be learned from formulas or equations. If teaching is treated, measured and applied machine-like, society will logically continue to suffer badly from a slew of unthought-of consequences.

Further and differing thoughts on these issues have also been expressed by Michael Fried (2014).

Leading youths from adolescence to adult life (in Latin the verb 'e-ducere' means 'to lead out of'), i.e., the education of the young as a complex natural process of the society of man and woman. As such, education can only be rightfully assessed and performed holistically. The holistic management context of education is the 'necessity principle'. This is so because a student's desire to learn and understand alone drives her or his educational progress.

No student interest—no education.

The 'holistic management principle' then guides the breadth of our educational efforts in class. And thus it completes the framework of teaching in a wholesome and adequate way for this complex problem that humanity has faced since antiquity and before, namely how to transfer our current knowledge base to the next generations.

## References

Fried, M. N. (2014). Mathematics and Mathematics Education: Beginning a Dialogue in an Atmosphere of Increasing Estrangement, chapter 2 in M. N. Fried, T. Dreyfus (eds.), *Mathematics and Mathematics Education: Searching for Common Ground, Advances in Mathematics Education*. https://doi.org/10.1007/978-94-007-7473-5_2. Springer.

Harel, G. (2013). Intellectual Need. In K.R. Leatham (ed.), *Vital Directions for Mathematics Education Research* (Chapter 6, p. 119–151), Springer Science and Business Media, New York. https://doi.org/10.1007/978-1-4614-6977-3_6.

Piaget, J. (1960). *Child's Conception of Geometry*, (original in French 1948), Basic Books.

Piaget, J. (1985). *The equilibration of cognitive structures: The central problem of intellectual development*, (first published in 1978), University of Chicago Press.

Piaget, J. (1977). *Intellectual evolution from adolescence to adulthood*, (original in French, 1970), Cambridge Univ. Press, 1977.

Savory, A. (1998). *Holistic Management: A New Framework for Decision Making*, 2nd edition, Island Press.

Smuts, J. C. (2010). *Holism and Evolution*, (Macmillan, 1927), Kessinger Publishing.

Uhlig, F. (2002a) *Transform Linear Algebra*, Prentice-Hall, ISBN. 0-13-041535-9, 502 + xx p.

Uhlig, F. (2002b). The role of proof in comprehending and teaching elementary Linear Algebra, *Educational Studies in Mathematics*, vol. 50, 335–346.

Uhlig, F. (2002c) *Author's response to comments on "The role of proof in comprehending and teaching elementary linear algebra" in Educational Studies in Mathematics*, 50 (2002), 335–346, Educational Studies in Mathematics, vol. 53 (2003), 271–274.

Uhlig, F. (2003). A new unified, balanced, and conceptual approach to teaching Linear Algebra, *Lin. Alg. Appl.*, vol. 361, 147–159.

Voisin, A. (1988). *Grass Productivity*, (original in French, 1957), Island Press.

von Humboldt, A. *Kosmos, Entwurf einer physischen Weltbeschreibung*, Cotta, Stuttgart; 5 volumes, 1845–1862.

Wendell, B. (2001). *Life is a Miracle*, Counterpoint, 176 p; ISBN 13: 9781582430584.

# Using Challenging Problems in Teaching Linear Algebra

**Abraham Berman**

**Abstract** We present examples of interesting problems that hopefully make the learning and the teaching of linear algebra enjoyable. The problems are on matrix multiplication, rank, determinants, eigenvalues and eigenvectors, and matrices and graphs. The problem solving strategies used include "look for invariants", "check parity" and "define an energy function".

**Keywords** Teaching through problem solving · Teaching problem solving
Teaching linear algebra

## 1 Introduction

A pleasant way to teach mathematics, in general, and linear algebra, in particular, is **TtPS**—*Teaching through Problem Solving*.

Fi and Degner (2012) define **TtPS** as *"pedagogy that engages students in problem solving as a tool to facilitate students learning of important mathematics subject matter and mathematical practices"*. Teaching through problem solving should not be confused with, the related but different, teaching problem solving, e.g. (Polya, 2004; Schoenfeld, 1985).

The example studied in Fi and Degner (2012) is from high-school mathematics but the idea of **TtPS** is also relevant to university courses. In his book *On linear algebra problems* Zhang (1996) says that "working problems is a crucial part of learning mathematics". Halmos opens his *linear algebra problem* book (Halmos, 1995) by asking "Is it fun to solve problems, and is solving problems a good way to learn?" and replies "the answer seems to be yes provided the problems are neither too hard nor too easy."

In this paper I discuss some of my favorite problems that I have used in teaching linear algebra in more than 40 years. I hope they are neither too hard nor too easy.

A. Berman (✉)
Department of Mathematics, National Technion Institute of Technology,
320000 Haifa, Israel
e-mail: berman@technion.ac.il



S. Stewart et al. (eds.), *Challenges and Strategies in Teaching Linear Algebra*,
ICME-13 Monographs, https://doi.org/10.1007/978-3-319-66811-6_18

Some of them were given, as motivation, before the related material was taught. Some were chosen in order to relate the material to topics that were taught earlier or would be taught later. I must confess that I do not know or do not remember the origin of the problems. They and many other problems appear in my textbook (Berman, 2002) (see also, Carlson, Johnson, Lay, & Porter, 2002; Halmos, 1995; Matousek, 2010; Prasolov, 1994; Zhang, 1996).

## 2 Impossible Tasks

The aim of impossible tasks is to educate the students that the answer to a "find" problem may be "this is impossible".

**Problem 1** Find matrices $A$, $B$ such that

$$AB - BA = I.$$

**Solution** It is impossible to find real (or complex) matrices $A$, $B$ since

$$trace(AB - BA) = 0.$$

It is, however, possible over $\mathbb{Z}_2$, where $-1 = 1$. For example

$$\begin{pmatrix} 0 & 0 \\ 1 & 0 \end{pmatrix} \begin{pmatrix} 0 & 1 \\ 0 & 0 \end{pmatrix} - \begin{pmatrix} 0 & 1 \\ 0 & 0 \end{pmatrix} \begin{pmatrix} 0 & 0 \\ 1 & 0 \end{pmatrix} = \begin{pmatrix} 1 & 0 \\ 0 & 1 \end{pmatrix}.$$

**Remarks**

1. *The natural place for Problem* 1 *is after the concept of a trace has been introduced and it was shown that* $trace(AB) = trace(BA)$. *However, if time permits, it can be given to students that are not familiar with the concept of trace, and used to guide them to develop the concept.*
2. *The fact that* $trace(AB) = trace(BA)$, *even when* $AB \neq BA$, *is important. Later in the course it will be shown that this follows from the fact that if* $AB$ *is* $n \times n$ *and* $BA$ *is* $m \times m, n \geq m, then$

$$\Delta_{AB}(t) = t^{n-m} \Delta_{BA}(t). \tag{1}$$

*(Here* $\Delta_X(t)$ *denotes the characteristic polynomial of a matrix X). Notice that comparing the coefficient of* $t^{n-m}$ *in both sides of* (1) *gives a short and elegant proof of the Cauchy–Binet formula.*

3. *The $\mathbb{Z}_2$ example can be given with appropriate explanation, even if the course does not cover finite fields.*
4. *If we want to consider matrices as finite dimensional operators, it is possible to give an example of infinite dimensional operators A and B such that the commutator $AB - BA$ is the identity:*

$$Au(x) = \frac{d}{dx}u(x), \qquad Bu(x) = xu(x)$$

$$ABu(x) = u(x) + x\frac{d}{dx}u(x), \qquad BAu(x) = x\frac{d}{dx}u(x)$$

$$(AB - BA)u(x) = Iu(x).$$

**Problem 2** Nine coins are arranged in 3 rows and 3 columns. In the middle (2, 2) position the "head" side is up. In all other positions the coins are "tail" side up:

$$\begin{matrix} T & T & T \\ T & H & T \\ T & T & T \end{matrix}.$$

A legal operation is turning over all the coins in one row or in one column. Use only legal operations so that all the coins show "heads" up:

$$\begin{matrix} H & H & H \\ H & H & H \\ H & H & H \end{matrix}.$$

**Solution** Replacing $T$ by $-1$ and $H$ by 1 we want to use only legal operations to replace the rank 2 matrix

$$\begin{pmatrix} -1 & -1 & -1 \\ -1 & 1 & -1 \\ -1 & -1 & -1 \end{pmatrix}$$

by the rank 1 matrix

$$\begin{pmatrix} 1 & 1 & 1 \\ 1 & 1 & 1 \\ 1 & 1 & 1 \end{pmatrix},$$

but the legal operations do not change the rank.

**Remarks**

1. *A solution that does not use the concept of rank and thus can be given to high school students is to consider the four coins at the right upper part:*

$$\begin{matrix} T & T \\ H & T \end{matrix}.$$

   *The number of heads in this part is odd and remains odd after any sequence of legal operations.*
2. *A linear algebraic version of the solution in Remark 1 is to observe that* $det\begin{pmatrix} -1 & -1 \\ 1 & -1 \end{pmatrix} \neq 0$ *but* $det\begin{pmatrix} 1 & 1 \\ 1 & 1 \end{pmatrix} = 0$.
3. *The solution strategy in both solutions is to "look for invariants" i.e., rank in the first solution and parity or nonsingularity, in the second.*

**Problem 3** A legal operation on a real matrix is multiplying a row by $-1$ or multiplying a column by $-1$. Prove that from any real matrix one can get by legal operations, a matrix in which all the row sums and all the column sums are nonnegative.

**Solution** Consider the sum of all the entries at the matrix.

**Remark** *Problem 3 is not an impossible task problem. It is brought here because it looks similar to Problem 2 but is much more challenging. It is also an opportunity to teach "problem solving". A helpful strategy in problems of dynamical systems is to associate a numerical function (called Energy function or Lyapunov function) and see what happens with the function. Here, the sum of all the entries is such a function. If a legal operation is performed on a row or a column where the sum of the entries is negative, the total sum increases. Since the number of matrices that can be obtained using legal operations is finite, this means that one must be able to obtain a matrix in which all row sums and all column sums are nonnegative. A similar strategy is used in Problem 11.*

## 3 Fibonacci Numbers

**Problem 4** The Fibonacci sequence is defined by

$$f_1 = 1, \ f_2 = 1, \ f_n = f_{n-1} + f_{n-2}; \ n \geq 3.$$

$$1, 1, 2, 3, 5, 8, 13, \ldots,$$

Prove that

$$s_n \equiv f_n^2 - f_{n-1} f_{n+1} = (-1)^{n+1}; \ \ n \geq 2.$$

**Solution**

$$s_n = det\begin{pmatrix} f_n & f_{n-1} \\ f_{n+1} & f_n \end{pmatrix} = det\begin{pmatrix} f_n & f_{n-1} \\ f_n + f_{n-1} & f_{n-1} + f_{n-2} \end{pmatrix} = det\begin{pmatrix} f_n & f_{n-1} \\ f_{n-1} & f_{n-2} \end{pmatrix}$$
$$= -det\begin{pmatrix} f_{n-1} & f_{n-2} \\ f_n & f_{n-1} \end{pmatrix} = -s_{n-1}.$$

**Problem 5** Show that $f_n = \frac{1}{\sqrt{5}}\left[\left(\frac{1+\sqrt{5}}{2}\right)^n - \left(\frac{1-\sqrt{5}}{2}\right)^n\right]$.

**Solution**

$$\begin{pmatrix} f_n \\ f_{n-1} \end{pmatrix} = \begin{pmatrix} 1 & 1 \\ 1 & 0 \end{pmatrix}^{n-2} \begin{pmatrix} 1 \\ 1 \end{pmatrix}.$$

To compute $\begin{pmatrix} 1 & 1 \\ 1 & 0 \end{pmatrix}^{n-2}$ we diagonalize the matrix $A = \begin{pmatrix} 1 & 1 \\ 1 & 0 \end{pmatrix}$. Its eigenvalues are $\varphi = \frac{1+\sqrt{5}}{2}$ and $1-\varphi$ and $\begin{pmatrix} \lambda \\ 1 \end{pmatrix}$ is an eigenvector corresponding to $\lambda$, for $\lambda \in \{\varphi, 1-\varphi\}$. Thus,

$$A = \frac{1}{\sqrt{5}}\begin{pmatrix} \varphi & 1-\varphi \\ 1 & 1 \end{pmatrix}\begin{pmatrix} \varphi & 0 \\ 0 & 1-\varphi \end{pmatrix}\begin{pmatrix} 1 & -(1-\varphi) \\ -1 & \varphi \end{pmatrix} \text{ and}$$
$$\begin{pmatrix} f_n \\ f_{n-1} \end{pmatrix} = \frac{1}{\sqrt{5}}\begin{pmatrix} \varphi & 1-\varphi \\ 1 & 1 \end{pmatrix}\begin{pmatrix} \varphi^{n-2} & 0 \\ 0 & (1-\varphi)^{n-2} \end{pmatrix}\begin{pmatrix} \varphi \\ \varphi - 1 \end{pmatrix}$$
$$= \frac{1}{\sqrt{5}}\begin{pmatrix} \varphi & 1-\varphi \\ 1 & 1 \end{pmatrix}\begin{pmatrix} \varphi^{n-1} \\ -(1-\varphi)^{n-1} \end{pmatrix} = \frac{1}{\sqrt{5}}\begin{pmatrix} \varphi^n & -(1-\varphi)^n \\ \varphi^{n-1} & -(1-\varphi)^{n-1} \end{pmatrix}.$$

**Remark** *In this problem the need for diagonalization is clear, since we have to compute a general power of A. This is more convincing than the need to compute, say $A^{10}$, which can be done by calculating $A^2$, $A^4$ and $A^8$.*

**Problem 6** Show that the set of infinite sequences $\{a_i\}$, where $a_1, a_2 \in \mathbb{R}$ and

$$a_n = a_{n-1} + a_{n-2}; \quad n \geq 3,$$

is a two dimensional real vector space.

**Solution** The set is a subspace of the vector space of sequences. Denote $\{a_1, a_2, \ldots; a_k = a_{k-1} + a_{k-2};\ k \geq 3\}$ by $\mathcal{S}(a_1, a_2)$. A possible basis is $\mathcal{S}(1, 1)$—the Fibonacci series; and $\mathcal{S}(1, 3) = \{1, 3, 4, 7, 11, \ldots, \}$—the Lucas sequence.

**Remark** *Using this notation*

$$\mathcal{S}(0, 1) = \{F_{n-1}\}, \qquad \mathcal{S}(1, 2) = \{F_{n+1}\}$$

*it follows that*

$$L_n = F_{n-1} + F_{n+1}.$$

## 4 The Vandermonde Matrix

This section is about the Vandermonde determinant formula and two of its applications.

**Problem 7** Show that the determinant $V(x_1, x_2, \ldots, x_n)$ of the Vandermonde matrix

$$\begin{pmatrix} 1 & x_1 & x_1^2 & \cdots & x_1^{n-1} \\ 1 & x_2 & x_2^2 & \cdots & x_2^{n-1} \\ \vdots & \vdots & \vdots & \cdots & \vdots \\ 1 & x_n & x_n^2 & \cdots & x_n^{n-1} \end{pmatrix}$$

is $\prod_{1 \le i < j \le n}(x_j - x_i)$.

**Solution** The polynomial $V(x_1, x_2, \ldots, x_{n-1}, x)$ is a polynomial of degree $n-1$ in $x$. The coefficient of $x^{n-1}$ is $V(x_1, x_2, \ldots, x_{n-1})$. $V(x_1, x_2, \ldots, x_{n-1}, x_n)$ vanishes at $x = x_1,\ x = x_2,\ \ldots,\ x = x_{n-1}$. So it is equal to $V(x_1, x_2, \ldots, x_{n-1})(x_n - x_1)(x_n - x_2) \ldots (x_n - x_{n-1})$. The result follows by induction on $n$.

**Problem 8** Let $x_1, x_2, \ldots, x_n$ be distinct numbers and let $y_1, y_2, \ldots, y_n$ be $n$ numbers. Prove that there is a unique polynomial $P$ of degree $n-1$ such that $P(x_i) = y_i$; $i = 1, \ldots, n$.

**Solution** Let $P(x) = a_0 + a_1 x + \cdots + a_{n-1}x^{n-1}$. We want to show that the coefficients $a_i$ are defined by a system of $n$ equations in $n$ unknowns $a_i$:

$$\begin{pmatrix} 1 & x_1 & x_1^2 & \cdots & x_1^{n-1} \\ 1 & x_2 & x_2^2 & \cdots & x_2^{n-1} \\ \vdots & \vdots & \vdots & \cdots & \vdots \\ 1 & x_n & x_n^2 & \cdots & x_n^{n-1} \end{pmatrix} \begin{pmatrix} a_0 \\ a_1 \\ \vdots \\ a_{n-1} \end{pmatrix} = \begin{pmatrix} y_1 \\ y_2 \\ \vdots \\ y_n \end{pmatrix}.$$

The system has a unique solution since $V(x_1, x_2, \ldots, x_n) \neq 0$.

**Problem 9** Use the Vandermonde determinantal formula to prove that eigenvectors that correspond to distinct eigenvalues are linearly independent.

**Solution** Let $A\mathbf{v}_i = \lambda_i \mathbf{v}_i$; $\mathbf{v}_i \neq 0$; $i = 1, \ldots, k$, $\lambda_1, \ldots, \lambda_k$ distinct.
We have to show that $\mathbf{v}_1, \ldots, \mathbf{v}_k$ are linearly independent. Let

$$a_1\mathbf{v}_1 + \cdots + a_k\mathbf{v}_k = \mathbf{0}. \tag{2}$$

We want to show that

$$a_1 = \cdots = a_k = 0.$$

Multiply (2) by $A, A^2, \ldots, A^{k-1}$:

$$\lambda_1 a_1 \mathbf{v}_1 + \cdots + \lambda_k a_k \mathbf{v}_k = \mathbf{0}$$
$$\vdots$$
$$\lambda_1^{k-1} a_1 \mathbf{v}_1 + \cdots + \lambda_k^{k-1} a_k \mathbf{v}_k = \mathbf{0};$$

$$\left(a_1\mathbf{v}_1 | a_2\mathbf{v}_2 | \cdots | a_k\mathbf{v}_k\right) \begin{pmatrix} 1 & \lambda_1 & \cdots & \lambda_1^{k-1} \\ 1 & \lambda_2 & \cdots & \lambda_2^{k-1} \\ \vdots & & & \vdots \\ 1 & \lambda_k & \cdots & \lambda_k^{k-1} \end{pmatrix} = \mathbf{0}.$$

Since $V(\lambda_1, \ldots, \lambda_k) \neq 0$, $\left(a_1\mathbf{v}_1 | a_2\mathbf{v}_2 | \cdots | a_k\mathbf{v}_k\right) = 0$. Since $\mathbf{v}_i \neq 0$, $a_i = 0$ for all $i = 1, \ldots, k$.

**Remark** *There are of course other proofs. It is always nice to give several proofs, when possible.*

## 5 Graphs and Matrices

**Problem 10** At each vertex of a graph there is a bulb. The states of the bulbs change every time unit according to the following majority rule.
If at time $t$, a light bulb has more neighbors that are "on", it will be "on" at time $t+1$, if it has more "off" neighbors, it will be "off". In case of a tie, its state does not change.

Prove that for every graph and for any initial states, from some $t$, the states of the lights at time $t+2$ are the same as their states at time $t$ (Fig. 1).

**Solution** Let $N$ be the adjacency matrix of $G$. Let $A = N + \frac{1}{2}I$. Let

$$\mathbf{x}(t)_i = \begin{cases} 1 & \text{if light } i \text{ is on at time } t \\ -1 & \text{if light } i \text{ is off at time } t. \end{cases}$$

The signs vector of $A\mathbf{x}(t)$ is the same as the signs vector of $\mathbf{x}(t+1)$. Thus,

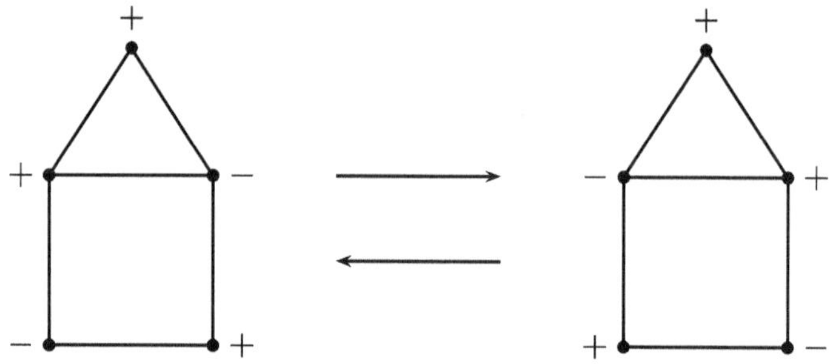


**Fig. 1** Periodicity 2

$$f(t) \equiv \mathbf{x}(t+1)^T A\mathbf{x}(t) = \max_{y_i \in \{\pm 1\}} \mathbf{y}^T A\mathbf{x}(t) = \mathbf{x}(t)^T A\mathbf{x}(t+1)$$

$$f(t+1) = \mathbf{x}(t+2)^T A\mathbf{x}(t) \geq \mathbf{x}(t)^T A\mathbf{x}(t+1) = f(t).$$

$f$ can attain only a finite number of values, so from some $t$, $f(t+1) = f(t)$. Thus, $\mathbf{x}(t+2) = \mathbf{x}(t)$.

**Remarks**

1. *The interplay between matrices and graphs is one of the important features of linear algebra and should be mentioned at an early stage of the course. Problem* 10 *was offered to the students after they had been asked to show, in an homework problem, that the number of walks of length k, from vertex i to vertex j, in a graph G is* $N_{ij}^k$, *where N is the adjacency matrix of G. The linear algebra requirement for the problem is to know that* $(AB)^T = B^T A^T$, *so the problem can be given early in the course. Similar to Problem* 3, *the strategy used in the solution is considering a Lyapunov function. Here the function is*

$$f(t) = \mathbf{x}(t+1)^T A\mathbf{x}(t) \ .$$

2. *An extension of the problem to infinite graphs is studied in Moran* (1995).

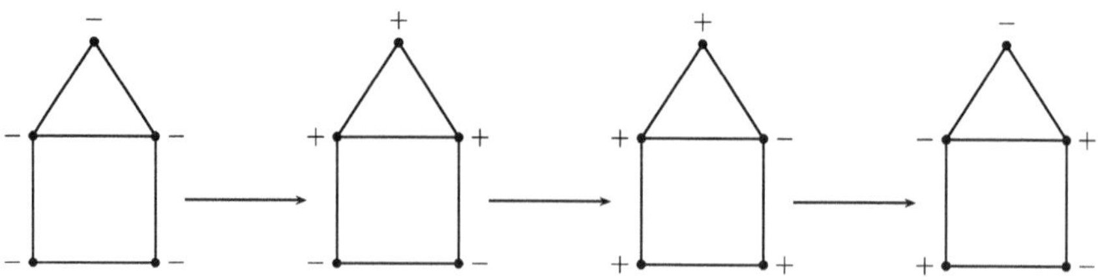

**Fig. 2** All lights "on"

**Problem 11** At each vertex of a graph there is a bulb and a knob. Pressing a knob activates it and its neighbors (Fig. 2).

Prove that for each graph one can start from all lights "off" and move to all lights "on".

**Solution** Let $A = N + I$. We have to prove that the system

$$A\mathbf{x} = \mathbf{e}$$

is solvable over $\mathbb{Z}_2$. Suppose it is not. Then $\big(A \mid \mathbf{e}\big)$ is equivalent to a matrix in which the last row is $\big(0 \cdots 0\ 1\big)$. Thus, there is an odd number of rows of $A$ such that in the corresponding submatrix, each column has an even number of 1's.
The same is true for the corresponding principal submatrix. This implies that the total number of 1's in the principal submatrix is even, but this principal submatrix is symmetric, its diagonal entries are 1 and its order is odd, so the total number of 1's is odd. Contradiction.

**Remark** *This problem is a known computer game. One can google FIVER to see a special case. The strategy used in the solution is parity checking. This strategy was used in the second solution to Problem 2.*

## 6 Suggested Research

I was lucky to have excellent students. Many became very successful in the high tech industry. Some of them credited learning through challenging problems for their high tech success. Of course, they wanted to please me. I was pleased but would like to know if the compliment is justified. **TtPS** in elementary and high school was studied, e.g. (Fi & Degner, 2012; Lester, 1994). In a recent doctoral thesis (Klinstern, 2016), Klinstern compares the way that teachers pose or choose problems for their teaching to the way this is done by TA's in the university. It is interesting to study the effect of using **TtPS** at the university level. How does it affect motivation, attitude, grades and knowledge? Is it time consuming? Is it good only for some students?
One of the ideas suggested in the discussion group on teaching linear algebra in ICME 13, was to identify research topics that could be studied jointly by researchers from different countries. I want to conclude the paper by suggesting "the effects of using challenging problems in teaching linear algebra" as such a topic.

## References


Berman, A., & Kon, B. (2002). *Linear Algebra* (in Hebrew) BAK, Haifa.

Carlson, D., Johnson, C., D. Lay, D., & Porter, A.D. (2002). *Linear Algebra Gems*. The Math Association of America, Notes Series No. 59.

Fi, C. & Degner, K. (2012). Teaching through problem solving, *Mathematics Teacher*, Vol. 105, No. 6, 455–459.

Halmos, P. R. (1995). *Linear Algebra Problem Book*, Dolciani Mathematical Expositions No 16, The Mathematical Association of America.

Klinstern, M. (2016). Problem Posing and Choosing Problems by High School and University Mathematics Teachers. *Ph.D dissertation*, Technion.

Lester, F. (1994). Musing about mathematical problem solving research: 1970–1994, *Journal for Research in Mathematics Education* 25, 660–675.

Matousek, J. (2010). *Thirty Three Miniatures Mathematical and Algorithmic Applications of Linear Algebra*, Student Mathematical Library, Vol. 53, American Mathematical Society.

Moran, G. (1995). *On the Period-Two-Property of the majority operators in infinite graphs*, Transaction of the American Mathematical Society. Vol. 347, No. 5, 1649–1667.

Polya, G. (2004). *How to solve it*, 1945; with foreword and added exercises by Y. Conway, Princeton University Press.

Prasolov, V. (1994). *Problems and Theorems in Linear Algebra*, Translations of Mathematical Monographs, Vol. 134, American Mathematical Society.

Schoenfeld, A. (1985). *Mathematical Problem Solving*, Academic Press.

Zhang, F. (1996). *Linear Algebra Challenging Problems for Students*, Johns Hopkins.

# Author Index





S. Stewart et al. (eds.), *Challenges and Strategies in Teaching Linear Algebra*,
ICME-13 Monographs, https://doi.org/10.1007/978-3-319-66811-6

# Subject Index





S. Stewart et al. (eds.), *Challenges and Strategies in Teaching Linear Algebra*, ICME-13 Monographs, https://doi.org/10.1007/978-3-319-66811-6

FSC
www.fsc.org
MIX
Papier | Fördert gute Waldnutzung
FSC® C083411

Kolibri 360 GmbH
Ferdinand-Jühlke-Straße 7
99095 Erfurt, Deutschland
produktsicherheit@kolibri360.de